History, Philosophy and Theory of the Life Sciences

Volume 29

W0080402

History, Philosophy and Theory of the Life Sciences is a space for dialogue between life scientists, philosophers and historians – welcoming both essays about the principles and domains of cutting-edge research in the life sciences, novel ways of tackling philosophical issues raised by the life sciences, as well as original research about the history of methods, ideas and tools, which constitute the genealogy of our current ways of understanding living phenomena.

The series is interested in receiving book proposals that • are aimed at academic audience of graduate level and up • combine historical and/or philosophical and/or theoretical studies with work from disciplines within the life sciences broadly conceived, including (but not limited to) the following areas: • Anatomy & Physiology • Behavioral Biology • Biochemistry • Bioscience and Society • Cell Biology • Conservation Biology • Developmental Biology • Ecology • Evolution & Diversity of Life • Genetics, Genomics & Disease • Genetics & Molecular Biology • Immunology & Medicine • Microbiology • Neuroscience • Plant Science • Psychiatry & Psychology • Structural Biology • Systems Biology • Systematic Biology, Phylogeny Reconstruction & Classification • Virology The series editors aim to make a first decision within 1 month of submission. In case of a positive first decision the work will be provisionally contracted: the final decision about publication will depend upon the result of the anonymous peer review of the complete manuscript. The series editors aim to have the work peer-reviewed within 3 months after submission of the complete manuscript. The series editors discourage the submission of manuscripts that contain reprints of previously published material and of manuscripts that are below 150 printed pages (75,000 words). For inquiries and submission of proposals prospective authors can contact one of the editors: Charles T. Wolfe: ctwolfe1@gmail.com Philippe Huneman: huneman@wanadoo.fr Thomas A.C. Reydon: reydon@ww.uni-hannover.de

Christopher Donohue • Charles T. Wolfe
Editors

Vitalism and Its Legacy in Twentieth Century Life Sciences and Philosophy

 Springer

Editors
Christopher Donohue
National Human Genome Research Institute
Bethesda, MD, USA

Charles T. Wolfe
Département de Philosophie & ERRAPHIS
Université de Toulouse Jean-Jaurès
Toulouse, France

ISSN 2211-1948   ISSN 2211-1956   (electronic)
History, Philosophy and Theory of the Life Sciences
ISBN 978-3-031-12606-2   ISBN 978-3-031-12604-8   (eBook)
https://doi.org/10.1007/978-3-031-12604-8

This Springer imprint is published by the registered company Springer Nature Switzerland AG
The registered company address is: Gewerbestrasse 11, 6330 Cham, Switzerland

Contents

Introduction: Vitalism and Its Legacies in Twentieth Century Life Sciences and Philosophy

Christopher Donohue and Charles T. Wolfe

Vitalism has spent most of the twentieth century, and part of the twenty-first, being perhaps the most misunderstood and reviled philosophy of life, with organicism being a close second (on the latter see (Martindale 2013), although some theorists seek to drive a wedge between the two in favor of a 'reasonable', less 'metaphysical' position often associated with organicism (Gilbert and Sarkar 2000). As a number of the essays in this collection point out (see especially the contributions by Donohue and Moir) vitalism has been conjoined to fascism and the Nazi horrors, and has been reduced to a series of ahistorical propositions. As both Moir and Donohue emphasize, such associations require more study, but at the same time are fundamentally misleading. Nonetheless, the traditional association of vitalism and fascism as well as vitalism and pseudoscience (or anti-science, as Shmidt underscores) has been remarkably pervasive, and still operates.

Part of the difficulty lies in the fact that vitalism has been accused of both everything and nothing at the same time (regarding its associations with fascism and irrationality.) Gilbert and Sarkar observe that vitalism's association with organicism has led to the latter's being associated with "fascism, communism, and New Age spirituality" which they observe "should be enough to bring down any philosophy" (Gilbert and Sarkar 2000, 5).

Similarly, philosophers since the Vienna Circle (particularly Schlick, Frank and later Nagel), have addressed (and oftentimes) castigated vitalism in various ways. Carnap himself was open to at least questioning the reducibility of biology to

C. Donohue
National Human Genome Research Institute, Bethesda, MD, USA
e-mail: donohuecr@mail.nih.gov

C. T. Wolfe (✉)
Département de Philosophie & ERRAPHIS, Université de Toulouse Jean-Jaurès,
Toulouse, France
e-mail: ctwolfe1@gmail.com

C. Donohue, C. T. Wolfe (eds.), *Vitalism and Its Legacy in Twentieth Century Life Sciences and Philosophy*, History, Philosophy and Theory of the Life Sciences 29, https://doi.org/10.1007/978-3-031-12604-8_1

physics and chemistry. He noted that "The philosophical problems of the *foundations of biology* (italics his) refer above all to the relation between biology and physics. "If yes" (he seemed undecided) all elements of biology can be reduced to physical occurrences." Carnap also described (in a relatively neutral way) that the core of the "vitalism-problem" was whether "the laws of biology be derived from the laws of the physics of the inorganic." He was much more dismissive of various ways to reduce psychology to physics (Carnap 1934, 18–19). Schlick, on the other hand, described the "psycho-vitalism" of Driesch as "wild speculations" (Schlick 1968, 79).

Phillip Frank is a bit more forgiving than Schlick (for more discussion of differing Vienna Circle attitudes towards Driesch and vitalism, see Chen 2019). Frank noted how both philosophers and scientists must be indebted to Driesch because he was the first to provide "a clear and unprejudiced formulation of the problem" of not only vitalism versus mechanism, but also whether "must we, in order to satisfy the law of causality in the domain of life, ascribe to the body besides the properties.... of physics and chemistry, also other, qualitatively different properties?" On Driesch's account of the *entelechy*, Frank noted that it was "not entirely convincing," as it was not entirely impossible that scientists had not discovered a combination of inorganic elements which could account for the particular characteristics of life (Frank 1949, 59–60) Nagel opined that "...vitalism has not proved to be a fruitful notion, and it no longer seems to present a live issue in the philosophy of biology" (Nagel 1979, 264). Nonetheless, Nagel underscored elsewhere that "the historically influential Cartesian conception of biology as simply a chapter of physics continues to meet resistance. Many outstanding biologists who find no merit in vitalism are equally dubious about the validity of the Cartesian program...." (Nagel 1961, 429).

Morris R. Cohen in his *Reason and Nature* took much of the same stance as Nagel (which is understandable considering that Cohen was Nagel's advisor and had a profound influence on him, see (Nagel 1957)). For Cohen, vitalists maintained that there was a "radical discontinuity between vital and non-vital phenomenon"; that biological laws were "different in character from the laws of the inorganic world" insofar as biological laws "add the fundamental category of purpose"; and that "the method of biology must differ essentially from that of physical science" (Cohen 1978, 248–9). While sympathetic to vitalism in some sense (Cohen observed that, "In general we may conclude that while the obvious and important differences between organic and non-organic....) he contended nonetheless that there did not exist a "logical proof that physics and chemistry will never be able to explain the phenomenon of life."

Cohen broke with the geneticist J.B.S Haldane when he further observed "that atoms, crystals" and other systems "do maintain their own structure" where "the maintenance of a certain form together with the splitting off of a part which forms the copy of an original is physically illustrated in the phenomena of vortex rings." It was also not clear for Cohen "where the progress of mechanical explanations will stop" (Cohen 1978, 269) Nor, according to Cohen, was it clear whether vitalists could decisively prove that "biological phenomena" resist all "physical analogies" (Cohen, ibid).

Likewise, Mikhail Bakhtin, appropriating the work of Nikolay Lossky, critiqued the 'neovitalism' of Driesch and Bergson as a too-strong ontological commitment to the existence of certain entities or 'forces' (Bakhtin 1992), over and above the system of causal relations studied and modeled by mechanistic science, which itself sought to express these entities or the relations between them in mathematical terms. Lossky himself observed that "The greatest difficulty for vitalism is to understand how this undoubtably immaterial factor, according to Driesch's research, interferes with the course of material-mechanical processes, *ordering* them" (Losskii 1923, 84) (emphasis Losskii's, my translation).

These comments concerning vitalism represent a common view of the subject, whether it is presented in positive terms, as a kind of commendable backlash against the de-humanizing, alienating trend inaugurated by the Scientific Revolution, which seeks to 'revitalize the world,' or in negative terms, as a kind of anti-scientific or 'para-scientific' trend which needs to be refuted (as in Francis Crick's rather confident pronouncement: "To those of you who may be vitalists, I would make this prophecy: what everyone believed yesterday, and you believe today, only cranks will believe tomorrow" (Crick 1966, 99)). And there is plenty of historical evidence that such a position existed (see Burwick and Douglass 1992). For sustained criticism of Crick's account of vitalism see (Waddington 1967 and Peterson's Chapter "A 'Fourth Wave' of Vitalism in the Mid-20th Century?" this volume).

But there has some significant scholarly 'pushback' against this orthodox attitude, notably pointing to the Montpellier vitalists of the eighteenth century (which is where the word 'vitalism' is first used), associated with figures like Diderot and the *Encyclopédie*, as has been shown in recent scholarship (Williams 2003; Reill 2005; Wolfe 2008; Normandin and Wolfe 2013; Wolfe 2019), but also, work on Driesch (Chen 2018, 2019 and Chapter "A Historico-Logical Re-assessment Of Hans Driesch's Vitalism" this volume, Bolduc, Chapter "On the Heuristic Value of Hans Driesch's Vitalism" this volume) has shown that even his 'neovitalism' is in need of reevaluation. So, there are different historical forms of vitalism, including in their relation to the mainstream practice of science (the topic of Wolfe and Normandin 2013, focusing however on the post-Enlightenment era and functioning in several respects as a predecessor volume to this one). Faced with this plurality, some adopt the strategy of presenting Enlightenment vitalism as somehow a different scientific paradigm – a holistic, non-reductionist project, directly opposed to mechanistic explanations which is sometimes enhanced philosophically as "mechanistic materialism" (Williams 2003, 177). This has echoes of Carolyn Merchant's 'Death of Nature' narrative (Merchant 1980). P.H. Reill's work presents a rather similar view: "Relation, *rapport*, *Verwandschaft* [affinity, kinship], and interconnection replaced mechanical aggregation as one of the defining principles of matter. By emphasizing the centrality of interconnection, Enlightenment vitalists modified the concept of cause and effect. In the world of living nature, each part of an "organized body" was both cause and effect of the other parts" (Reill 2010, 66; see also Reill 2005). A more nuanced picture, concerning the vitalist approach to reductionist scientific practices such as pathological anatomy, is presented in Wolfe (2013), as well as more broadly regarding the fruitful interplay between mechanistic and vitalistic models in early modern and Enlightenment science, and the emergence of biology (Wolfe, 2022, forthcoming).

But another option has so far been left out: what happens if we return to the challenge of the anti-vitalist arguments formulated by the Vienna Circle and its successors, and look at vitalism and its connections to biology, genetics, philosophy and medicine in the twentieth century and today, not just as a historical form but as a significant metaphysical and/or scientific model for working biology and medicine? Is it possible to grasp some of the conceptual originality of vitalism without either (a) reducing it to mainstream mathematicocentric models of science (in a kind of "victors' narrative") or (b) just presenting it as an alternate model of science? In other words, without either normalizing it or projecting a kind of 'weak messianic power' onto its supposed abnormality?

This volume seeks to promote dialogue and discussion about the historical and philosophical importance of vitalism and other non-mechanistic conceptions of life in the context of ancient philosophy to the late twentieth century. As such, these essays consistently move against the idea that the twentieth century biological sciences must be genetic and must be deterministic and reductionist. The idea that mechanism and materialism are essential to the development of the modern life-sciences has furthermore become a mantra, which frequently drives out understanding of holistic, organistic and more system-based approaches. As importantly, the contributions not only detail a broad engagement with a variety of nineteenth- and twentieth-century vitalisms and conceptions of life, but engage with important threads in the history of concepts in the United States and Europe, including charting new reception histories in eastern and south-eastern Europe. Most importantly, all of the contributions to this volume work against a reduced and monolithic account of vitalism. As several contributions make clear, even when vitalism is rejected or 'refuted,' such refutations and questionings often promote fruitful engagements which show the workings of vitalist ideas in a kind of negative reception (a kind of 'negative' vitalism and its influence). Last, this collection brings together the perspectives of scholars from across disciplines in the history and philosophy of biology, bridging critical *lacunae* in our understanding of key nineteenth and twentieth century figures, without reducing any of them to caricatures. This volume, drawing upon an already robust and transformative scholarship, not only illustrates the contemporary relevance of vitalism, organicism and holism to recent biology and medicine, but begins the characterization of these rich and complex perspectives on life.

The essays in the volume follow a roughly chronological order. We begin with Tano Posteraro's contribution. Posteraro, drawing on the significant progress of scholarship on Bergson, does much work on clarifying exactly what kind of vitalist Bergson is: his vitalism is certainly not that of Driesch and may even be argued as contra-Driesch. Next, Ghyslain Bolduc narrates how Driesch's vitalism was key to the progress of embryology in the twentieth century, often advancing concepts in order to refute Driesch's arguments. Bohang Chen argues that Driesch's vitalism should not be rejected due to its metaphysics, but rather because it produced no vital laws. Nonetheless, Driesch's account of the entelechy provides insights into both physics and theories of evolution.

Next Christopher R. Donohue argues that if there is indeed a receptivity towards vitalism in eastern and southeastern Europe, that this was the result of virulently anti-materialist polemic in Czech-speaking and Slovenian-speaking lands in the nineteenth and early twentieth centuries. Mazviita Chirimuuta contends that Cassirer's rejection of vitalism and his embrace of holism in his critique of Bergson tells us a great deal about many of his central philosophical and ethical commitments, placing him in conflict with logical empiricism and the Vienna Circle. Then Brooke Holmes demonstrates how Canguilhem appropriated an image of Greek philosophy as part of his discussion of vitalism. Holmes argues that Greek texts, rather than being static entities, function as entities that are consistently read and reread, becoming an essential structure in twentieth century accounts of the life sciences.

Arantza Etxeberria and Charles Wolfe examine Canguilhem's Kantian account of the living individual vis-à-vis vitalism. Regarding the naturalistic perspective of the logic of the living individual articulated by Maturana and Varela, both authors find instructive divergences with Canguilhem especially regarding the connection to vitalism. Sebastjan Vörös begins his contribution by noting that both Canguilhem and Merleau-Ponty independently developed a critique of the mechanical-behaviorist view of the life sciences, which generated for both an account of the organism that is not indifferent. Both authors filiations with vitalism (and in the case of Merleau-Ponty really in spite of himself) pave the way for what Vörös calls "ouroboric thought." This underscores that we as constituters of nature (or cognizers) live lives that are decisively changed through these acts of constitution.

Cécilia Bognon-Küss shows how what she calls the historical "crisis of the concept of metabolism," through a fresh interrogation of historical vitalism, allows for solutions to the corresponding "crisis" of biological identity and the limits of "autonomy" in the context of biological entities. Drawing from a deep discussion of the issues of "distinction" and "persistence" and matter and form as foundational problems of the organism, Bognon's contribution underscores the necessity of "ecologicizing" biology to account for life as a series of dynamic systems in flux and in openness with the world.

Moving into the realm of genetics, Erik Peterson underscores that Crick's (in) famous critique of vitalism targeted individuals who were often not vitalists at all. Rather Peterson underscores that many of the objects of Crick's ire were proponents of "bioexceptionalism" (or the idea that biology was irreducible to the laws of chemistry and physics). Although Crick was misinformed, his status cemented a reductionistic account of vitalism in the life sciences and beyond for decades. Victoria Shmidt argues that the critique of vitalism was key to the demarcation of health and disease. She contends that anti-vitalist polemic was a rhetorical resource in the post-war life sciences for ensuring the dominance of genetics, and its consequent reductionism and essentialism of the human person. Shmidt underscores further that epistemologies influenced by vitalism encourage the humanistic dissolution of these dichotomies.

Last, Cat Moir critically examines whether vitalism does indeed have politics, and whether that politics can be mapped to any kind of ideological framework. A closer look shows that vitalism is often confused with other frameworks (such as holism), and that any account of the political consequences of vitalism need to begin with a far narrower view of vitalism than has been usually admitted, which is nonetheless consistent with the historical and philosophical tenants of vitalism as it has developed.

Taken as a whole, whether or not these essays testify to the 'vitality of vitalism', we hope that they offer scholars from different domains and perspectives, some novel and challenging material – sources, arguments, connections between discourses and *problématiques* – that may contrast with, or simply complement, both earlier, somewhat flat emergence of 'science and its controversies' narratives or the distinctively (feverishly?) enthusiastic 'life-philosophy' ontologies and ontophanies.

References

Bakhtin, M. 1992. Contemporary Vitalism. Trans. Charles Byrd. In *The Crisis of modernism: Bergson and the Vitalist Controversy*, eds. Frederick Burwick and Paul Douglass, pp. 76–97. Cambridge: Cambridge University Press.

Burwick, F., and P. Douglass. 1992. *The Crisis in Modernism: Bergson and the Vitalist Controversy*. Cambridge: Cambridge University Press.

Carnap, R. 1934. On the Character of Philosophic Problems. *Philosophy of Science* 1 (1): 5–19.

Chen, B. 2018. A Non-Metaphysical Evaluation of Vitalism in the Early Twentieth Century. *History and Philosophy of the Life Sciences* 40 (3): 1–22.

———. 2019. Revisiting the Logical Empiricist Criticisms of Vitalism. *Transversal: International Journal for the Historiography of Science* 7: 1–17.

Cohen, M.R. 1978. *Reason and Nature: An Essay on the Meaning of Scientific Method*. New York: Dover Publications.

Crick, F. 1966. *Of Molecules and Men*. Tacoma: University of Washington Press.

Frank, P. 1949. *Modern Science and Its Philosophy*. Cambridge: MA, Harvard University Press.

Gilbert, S.F., and S. Sarkar. 2000. Embracing Complexity: Organicism for the 21st Century. *Developmental Dynamics* 219 (1): 1–9.

Losskii, N. O. 1923. *Материя и жизнь (Matter and Life)*, Обелиск (Obelisk).

Martindale, D. 2013. *The Nature and Types of Sociological Theory*. Taylor & Francis.

Merchant, C. 1980. *The Death of Nature: Women, Ecology, and the Scientific Revolution*. New York: Harper and Row.

Nagel, E. 1957. Morris R. Cohen in Retrospect. *Journal of the History of Ideas* 18 (4): 548–551.

———. 1961. *The Structure of Science: Problems in the Logic of Scientific Explanation*. New York: Harcourt, Brace & World.

———. 1979. *Teleology Revisited and Other Essays in the Philosophy and History of Science*. New York: Columbia University Press.

Normandin, S., and C.T. Wolfe, eds. 2013. *Vitalism and the Scientific Image in Post-Enlightenment Life Science, 1800–2010*. Cham: Springer.

Reill, P.H. 2005. *Vitalizing Nature in the Enlightenment*. Berkeley: University of California Press.

———. 2010. Eighteenth-Century Uses of Vitalism in Constructing the Human Sciences. In *Biology and ideology from Descartes to Dawkins*, ed. Denis R. Alexander and Ronald L. Numbers, 61–87. Chicago: University of Chicago Press.

Schlick, M. 1968. *Philosophy of Nature*. Greenwood Press.

Waddington, C.H. 1967. No Vitalism for Crick. *Nature* 216 (5111): 202–203.

Williams, E.A. 2003. *A Cultural History of Medical Vitalism in Enlightenment Montpellier.* New York: Ashgate.

Wolfe, C.T. 2008. Vitalism Without Metaphysics? Medical Vitalism in the Enlightenment. *Science in Context* 21 (4): 461–463.

———. 2013. Vitalism and the Resistance to Experimentation on Life in the Eighteenth Century. *Journal of the History of Biology* 46: 255–282.

———. 2019. *La philosophie de la biologie avant la biologie : une histoire du vitalisme.* Paris: Classiques Garnier.

———. 2022. Expanded Mechanism and/or Structural Vitalism: Further Thoughts on the Animal Economy. In *Mechanism, Life and Mind in Early Modern Natural Philosophy*, Archives internationales d'histoire des idées, ed. C.T. Wolfe, P. Pecere, and A. Clericuzio. Cham: Springer.

———. forthcoming. Vitalism and the Construction of Biology: a Historico-Epistemological Reflection. In *Philosophy, History, and Biology: Essays in Honor of Jean Gayon*, Boston Studies in the Philosophy of Science, ed. P.-O. Méthot. Cham: Springer.

Vitalism and the Problem of Individuation: Another Look at Bergson's *Élan Vital*

Tano S. Posteraro

Abstract Mikhail Bakhtin's 1926 essay, "Contemporary Vitalism," includes Bergson alongside Driesch in a short list of "the most published representatives of vitalism in Western Europe," and, indeed, Bakhtin's critique of Driesch is intended to undermine what he calls the "conceptual framework" of "contemporary vitalism" as a whole (The crisis of modernism: Bergson and the vitalist controversy. Eds. Frederick Burwick and Paul Douglass. Cambridge University Press, New York, 1992, p 81). The conceptual framework that Driesch and Bergson are supposed to have shared in common consists at bottom, for Bakhtin, in the ontological commitment to the autonomy of life, "its independence, its disconnectedness from physical-chemical phenomena" (81). This has long been understood as the defining mark of vitalism, at least in the mind of its critics: the contention that matter and the mechanical models that track it are insufficient to the reality of biological forms, and that the explanation of life therefore requires the postulation of a non-mechanical, possibly immaterial, uniquely vital principle, force, substance, or property. Recent scholarship has made considerable headway in complicating these pictures by attending to earlier and subtler forms of materialism, and by distinguishing between different types of vitalism and drawing out the heuristic or scientific utility of some of them (Wolfe, Eidos 14: 212–235, 2011, Antropol Exp 17(13): 215–224, 2017; cf. Wolfe and Normandin, Vitalism and the scientific image in post-enlightenment life science, 1800–2010. Springer, Dordrecht, 2013). The focus of some of this work has been on the critical revaluation of Driesch himself (Bognon et al., Kairos J Philos Sci 20(1): 113–140, 2018). Yet the status of Bergson's commitment to the existence of a vital principle remains underdeveloped. In the midst of what some are calling a "Bergson renaissance," I think that it calls for the same kind of critical reappraisal (Ansell-Pearson, Bergson: thinking beyond the human condition. Bloomsbury, New York, 2018: 1; cf. Lundy, Deleuze's Bergsonism. Edinburgh University Press, Edinburgh, p 5, 2018). The aim of this paper is to attempt the outline of an answer to that call. I begin with a brief summary of Driesch's vitalism, then I reconstruct Bergson's underappreciated critique of internal finality, or what

T. S. Posteraro (✉)
Department of Philosophy, Concordia University, Montréal, QC, Canada

© The Author(s) 2023
C. Donohue, C. T. Wolfe (eds.), *Vitalism and Its Legacy in Twentieth Century Life Sciences and Philosophy*, History, Philosophy and Theory of the Life Sciences 29, https://doi.org/10.1007/978-3-031-12604-8_2

Kant called inner purposiveness, and locate in it a subterranean criticism of vital principles of the Drieschian variety as well. Two consequences follow: first, if Bergson is to be considered a vitalist, it cannot be in the Drieschian sense and we are therefore wrong to associate the two; and second, if Bergson is to be considered a vitalist, then his vitalism has to be understood—somewhat counterintuitively, and certainly contra Driesch—on the basis of a principle external to the ostensible individuality of biological forms.

Hans Driesch is the primary target of Mikhail Bakhtin's 1926 essay "Contemporary Vitalism." But Bakhtin includes Bergson alongside Driesch in his short list of "the most published representatives of vitalism in Western Europe," and, indeed, Bakhtin's critique of Driesch is intended to undermine what he calls the "conceptual framework" of "contemporary vitalism" as a whole (1992: 81). The conceptual framework that Driesch and Bergson are supposed to have shared in common consists at bottom, for Bakhtin, in the ontological commitment to the autonomy of life, "its independence, its disconnectedness from physical-chemical phenomena" (81). Driesch locates the difference between this contemporary vitalism and its parent approaches not in the postulation of a special force whose action in the biological domain safeguards the irreducibility of life to mechanistic explanation, but in the fact that contemporary vitalists are concerned to justify that postulation with empirical data while the older vitalists "silently assumed [its] permissibility" (80). Contemporary vitalism is supposed to serve as a viable biological programme, capable not only of orienting research (as Bergson insisted [1977: 112, 115–116, 249]), but well-founded as a conclusion drawn from the evidence of existing biological research as well (as Driesch and Bergson both claimed). "For this reason," Bakhtin says, "we may term contemporary vitalism 'critical vitalism,' in contradistinction to the old vitalism" (1992: 80–81). Yet Bakhtin is clear that he thinks the critical aspirations of contemporary vitalism remain unrealizable, as vitalism is by its very nature a dogmatic position that cannot hope to justify itself with reference to empirical research (81). This is because in contending that biology is irreducible to physicochemical explanation, and that explanations of the specificity of life require something superadded to the material world, vitalism, "like any metaphysical theory," "uses subjective schemes beyond the scope of experimentation" (96).

This has long been understood as the defining mark of vitalism, at least in the mind of its critics: the contention that matter and the mechanical models that track it are insufficient to the reality of biological forms, and that the explanation of life therefore requires the postulation of a non-mechanical, possibly immaterial, uniquely vital principle, force, substance, or property. Every element of that contention is the artifact of a polemic: the vitalist is conceived as an anti-materialist by her critics, and the materialist is conceived as a necessary partisan of mechanism, usually reductionism as well, by her critics in turn. Recent scholarship has made considerable headway in complicating these pictures by attending to earlier and subtler forms of materialism, and by distinguishing between different types of vitalism and drawing out the heuristic or scientific utility of some of them (Wolfe 2011, 2017; cf. Wolfe and Normandin 2013). The focus of some of this work has been on the critical revaluation of Driesch himself (Bognon et al. 2018). Yet the status of Bergson's commitment to the existence of a vital principle remains underdeveloped. In the midst of what some are calling a "Bergson renaissance," I think that it calls for the same kind of critical reappraisal (Ansell-Pearson 2018: 1; cf. Lundy 2018: 5). The aim of this paper is to attempt the outline of an answer to that call.

I begin with a brief summary of Driesch's vitalism, though I intend it only as a preparation for Bergson's own response to Driesch and concede in advance that a more nuanced picture of the latter could and should be drawn. Then I reconstruct Bergson's underappreciated critique of internal finality, or what Kant called inner purposiveness, and locate in it a criticism of vital principles of the Drieschian variety as well. Two consequences follow: first, if Bergson is to be considered a vitalist, it cannot be in the Drieschian sense and we are therefore wrong to associate the two; and second, if Bergson is to be considered a vitalist, then his vitalism has to be understood—somewhat counterintuitively, and certainly *contra* Driesch—on the basis of a principle *external* to the ostensible individuality of biological forms.

The final sections of the paper deliver an account of what we might tentatively call Bergson's "external vitalism." The first step of this account consists in the *position* of Bergson's infamous *élan vital*, the vital impetus: that is, its externality to all constituted forms. The second step is a reconception of the *nature* of the *élan* on the basis of Bergson's own overlooked pragmatic minimalism regarding its ontological status. I insist upon two points: first, that the *élan vital* is not an actual force, property, substance, or principle, but a *tendency*—technically a "virtual" tendency—which means that when it is referred to as a principle, it is in abstraction from the concrete particularities in which it is embodied as a tendency to some degree of realization; and second, that understood in this way, the *élan* is an image drawn from the psychological register intended to best capture the nature of living systems. It is, in other words, *only an image for life*, though Bergson regards it as the best image we have available (1998: 257). I conclude by gesturing towards a discussion of some of the possible benefits of the Bergsonian account.

1   Driesch

Driesch's first major theoretical work in English, *The Science and Philosophy of Organism*, appeared in 1908 (but drew heavily from earlier works published in German). It offered a defense of vitalism on the novel basis of the experimental facts of regulation and regeneration. Driesch's idea was that only if undisturbed development was possible could everything about organisms be mechanistically explained (1929: 103). What he therefore set out to prove was that development could be interrupted without the individuality of the developing organism being compromised as a result. Cases of regulation and regeneration seemed to evince the point. Driesch argued that they could not be understood mechanistically, and that they testified as a result to the non-mechanical action of an immaterial force.

By compressing early sea urchin embryos between glass plates, Driesch was able to reconfigure the divisions in their eggs, reshuffling their nuclei so that some nuclei that would normally have produced dorsal structures were found in ventral cells instead. According to mechanistic (or preformationist) principles, the embryos should have developed in a disordered and unviable fashion. Yet Driesch famously obtained normal larvae from them, which meant for him that the early embryo was composed of pluripotent cells, and that the developmental processes through which they gave rise to differentiated organs must be self-regulating (cf. Sapp 2003: 100). Driesch linked the phenomenon of regulation to the already well-established facts of regeneration such as it occurred in his own experiments on salamanders, which are capable of regenerating the lenses of their eyes after they are removed.

Driesch argued that regulation and regeneration indicate the existence of an individualizing agency at work in the organism, distinguishing it from the mere mechanical assemblage of parts and securing its autonomy as an organized whole over and above changes in its constituent elements. Driesch's word for this agency was "entelechy," an Aristotelian term with its roots in the Greek *enteles* [complete], *telos* [end], and *echein* [to have]. Leibniz would later popularize the word with the definition of "something analogous to soul, whose nature consists in a certain eternal law of the same series of changes, a series which it traverses unhindered" (1989: 173). Driesch's entelechy is an immaterial force, acting to bring about the unified development of an organic individual from out of initially pluripotent cells. As a result of entelechy, "a sum (of possibilities of happening) is transformed into a unity (of real results of happening) without any spatial or material preformation of this unity" (Driesch 1929: 215).

Entelechy was what guided initially pluripotent cells to the specific structures in which their development culminated. Cellular pluripotency explained the fact that cells isolated at the two-cell stage of development in sea urchin eggs did not produce two half-embryos but two fully formed sea urchins. Entelechy explained the way those cells were guided towards their final forms, since their pluripotency seemed like evidence of the idea that no physical or chemical structures existed in order to determine development in advance. Driesch thought that cases of regeneration supported the existence of entelechy as well, since they demonstrate the way the individuality of the organism could be safeguarded against changes to its composition. Not only can living things self-regulate developmentally, but they can do so compositionally as well (153–154). The special force that brings about the individual whole from out of a pluripotent cellular field also secures the integrity of that whole once it is constituted. Regulation is regeneration for the adult organism. Driesch supposed both to be impossibilities for mechanical systems. He concluded that "embryological becoming is 'vitalistic' . . . it is impossible to comprehend it by the laws of physics and chemistry" (1914: 226). This is vitalism neatly stated: physics and chemistry are inadequate to the explanation of biological phenomena, and biology therefore requires the addition of a supplemental principle.

2 Bergson's Critique

As Driesch demonstrated, vitalisms typically consist in critical as well as constructive elements. Their criticisms target the scientific understanding of matter, usually mechanistically conceived, and argue that it is insufficient to the explanation of what is distinct about biological phenomena. Their constructive arguments advance varying positions regarding the new and irreducible principle, property, or force that has to be introduced in order to capture the specificity of life. Vitalism's critics purport to attack both, but it is only really a succession of variants of vitalism's constructive aspect that have been consistently discredited. Like many others, Bergson considers the critical moment worth taking seriously. Biology is for him irreducible to physicochemical explanation. In this respect Bergson is no doubt a vitalist, but this tells us little. The more interesting question is whether he advances his own constructive hypothesis as well, of the sort that would put him in line with Driesch.

Bergson does of course advance a positive theory of his own, but it is not one that puts him in line with vitalists of the traditional variety. Bergson is in fact an ardent critic of such theories. They share on his account an important deficiency with mechanism: both are human contributions and do not exist in nature independently. According to this criticism, Driesch's vital principle is an intellectual abstraction born of the projection of the manufacture model of organization onto the biological world. It understands organisms as if they were built artifacts and attempts to explain their composition on that basis. Here is Bergson (1998: 225):

> When we think of the infinity of infinitesimal elements and of infinitesimal causes that concur in the genesis of a living being . . . the first impulse of the mind is to consider this army of little workers as watched over by a skilled foreman, the 'vital principle,' which is ever repairing faults, correcting effects of neglect or absentmindedness, putting things back in place.

This first impulse is natural to the intellect, a product of our adaptation to acting on matter. It consists in treating the organism as if it were an object, its organization as if it were designed, and concluding that there must be a principle to account for that design, just as artifacts are constructed and repaired by external agents. The mistake is in thinking that organisms are complex in the same way that made things are complex. The appearance of that complexity is only "the work of the understanding" (250). It is not a fact, but a projection, and so does not require a superadded principle that would act as a designer in order to explain it.

That is the first problem with any vitalism that accounts for organization through the postulation of an organizing principle: its anthropocentric artifactualism. The second—and more important—is its supposition that determinate individuality is a biological reality. Bergson delivers a somewhat sophisticated critique of this position, though it is ultimately in service of his rejection of the theory of internal finality or inner purposiveness. I suggest nevertheless that the argument against biological individuality can be redeployed in the context of Bergson's engagement with Driesch.

Bergson understands internal finalism, or the theory of inner purposiveness, to have arisen as a consequence of the empirical difficulties faced by the external finalism that he attributes to Leibniz. The Leibnizian doctrine is supposed to consist—on Bergson's gloss—in the idea that "beings merely realize a programme previously arranged" (1998: 39).[1] On that account, the event of any actualization was preceded not only by its own specific possibility, but by a global set of possibilities together comprising the plan or programme on the basis of which it was realized. Bergson calls this externalist because it locates the goal, end, aim, or purpose orienting the actualization process external to any particular individual being actualized; it is attributed to world, not to the beings that populate it. There is more to say about this interpretation of Leibniz—as well as about the possibilism that it implies—but what is important for now is (1) the idea that doctrines of finalism have to locate ends, or purposes, *somewhere*; and (2) that external finalism consists in the location of purpose beyond or outside of the particular beings that realize it, in the whole instead of its parts.

Now, Bergson suggests that it is "the tendency of the doctrine of finality" to "thin out the Leibnizian finalism by breaking it into an infinite number of pieces" in response to the fact that "if the universe as a whole is the carrying out of a plan, this cannot be demonstrated empirically" (40). Indeed, "the facts would equally well testify to the contrary," for nature taken as a whole would seem to evidence as much disorder as order, chaos as harmony, as much retrogression as progress (40). Though it seems unlikely that finality might be reasonably affirmed of the whole of life as such, "might it not yet be true, says the finalist, of each organism taken separately?" (40). By further individuating its object, finalism locates the empirical reality of disorder in the clash of organisms with each other in order to preserve purposiveness at the level of each individual organism taken with respect to itself. Bergson calls this "internal finality" since it attributes purposiveness to the internal composition of the organic body as an explanation for the division of labour among its parts and their integration in service of the end of the individual whole (41).

When Bergson declares this "the notion of finality which has long been classic," I think he has Kant and Hegel in mind (41). It was Kant who first distinguished between external and internal purposiveness, the same terms that Bergson employs

---

[1] This is something of a caricature of Leibniz's defense of teleology (cf. Jorati 2017: 59–91). Leibniz serves to personify a certain position within Bergson's critical programme; that position is the elaboration to its conceptual conclusion of the idea that actual entities are the realizations of possibilities that pre-exist them. To the extent that the reality of time requires, for Bergson, a concomitant epistemic unforeseeability, it follows that any defense of preexistent possibilities at all is as mistaken as the extreme form of the position that he attributes to Leibniz. Consider also Bergson's methodology of tendency-analysis, according to which he extrapolates from a given tendency its fullest culmination and takes that to represent the core principle of the tendency itself (cf. 1998: 136).

in order to effectuate his own distinction in theories of biological finalism.[2] By external purposiveness, Kant meant the finality of artifactual manufacture, since the particular end served by the artifact lies outside of itself in its use; by internal purposiveness, on the other hand, he meant the particular kind of finality that qualifies living beings, for the ends served by the organization of their parts are *internal* to the wholes that they compose (2000: §82; 5: 425). The living being was to be conceived for Kant through itself, as self-organizing: it is, as a whole, the final cause of the efficient-causal relations among its parts, even as it is constituted by them recursively. And yet, the thought of purposive organization nevertheless required the thought of an external intention; this was the central antinomy of his "Critique of the Teleological Power of Judgment" (§70; 5: 387). Its resolution was to come by way of Kant's account of regulative judgments. Teleology was not, then, to be predicated of the organism constitutively, as if it really was a made thing; teleology was rather to be ascribed to it only regulatively, as a necessary constraint on the intelligibility of the organism as organized matter. Hegel later famously undertook to push the concept of internal purposiveness beyond its Kantian heritage by unbinding it completely from the yoke of externality (cf. Kreines 2004). The decisive Hegelian definition of life was therefore to consist in large part in the non-oppositional reciprocity between organization and purpose in the individual being (1969: §216, §219; cf. Kreines 2008). It is Hegel, then, that best represents the internalist culmination of the second tenet of Leibnizian finalism (cf. Michelini 2008). And so by 1907 it would have made sense for Bergson to consider "classic" the formulation of finalism that restricted the attribution of purposiveness to the organism qua self-organizing individual.

Bergson's contention that "finality is external or it is nothing at all" is therefore an audacious one; its foil might well be the entire history of the German Idealist philosophy of nature (1998: 41; cf. Wandschneider 2010: 71). I reconstruct the argument in four steps: discernment, criticism, and two inferences. Bergson begins by discerning in the theory of internal finality its dependence on a conception of the organism as a rigorously bounded individual. If its parts are to be subordinated to the organismal whole as the final cause of their organization, then there must be a determinate distinction between that whole and its outside; the whole must, in other words, have a definite shape. Purposiveness can only be internal with respect to a limit—a bounded individuality—that would differentiate that internality from what is external to it. The theory of inner purposiveness stands or falls with the individuality of the organism.

---

[2] Note that the organic body was, for Leibniz, to be understood as analogous to a human-made machine; the difference was that organisms were *infinitely* complex, while man-made machines eventually bottomed out into organized parts—which were, again, infinitely complex machines themselves (cf. Illetterati 2014: 89). Whereas the technical artifact was a product, for Leibniz, of human intelligence, the organic machine was to be understood as a product of the divine intellect instead (Smith 2011: 165–196). But both, on Kant's gloss, are equally made things; and therefore the purposiveness of the organization of both reside outside of them, in their respective makers (2000: §65; 5: 374). A properly *natural* purposiveness therefore required, for Kant, the overcoming of the utility-based teleology still prominent in Leibniz's externalism.

The second step to the argument is a criticism of this conception of the organism. It is motivated by two considerations: the relative autonomy of the parts, patterns, and processes that constitute the organism; and the continuity of the germ cells through their temporary instantiation in the soma. Bergson begins by noting that each of the elements of the organismal whole "may itself be an organism in certain cases," that "the cells of which the tissues are made," for instance, "have also a certain independence," and that, in sum, there is to be attributed to the organism's parts a relative autonomy from the whole (1998: 41–42). Thus, the same self-organizational powers that define the organism as a whole are characteristic too of the parts and processes that constitute it, as well as the subsystems whose interactions constitute them in turn. The unified individuality of the organism is perhaps better understood as the coordination of a set of self-organizing living systems that are each at the same time conceivable as unified biological individuals in their own right. It follows for Bergson that the inner purposiveness of the organism is in principle *external* to the inner purposiveness of each one of its parts when they are understood as self-organizing systems themselves. This is what Bergson means when he says that "the idea of a finality that is *always* internal is therefore a self-destructive notion" (1998: 41; cf. Ansell-Pearson 2002: 134).

Bergson prefers to put the point in terms of the impossibility for individuation, identity, or mereological closure to ever establish itself fully or finally in the organic domain. This formulation is represented in the second reason motivating Bergson's argument against the internalist's conception of the organism: given the facts of reproduction, organic individuation is always incomplete. Bergson makes passing reference here to the "Weismann barrier," the theoretically inviolable division between germinal and somatic cells (1998: 42). On Weismann's account, it is only germinal cells, or gametes, that have heritability functions: they pass information along their own line only (1893: 174). The somatic cells are an effect or product of the totipotent zygote, which is itself a product of the fusion of haploid gametes or germ cells. The germ cells are formed on the basis of a vital substance that Weismann called the germ-plasm, which remains continuous and unchanged through each iteration of this process (1893: 184). The causal line runs in one direction: the germ cells give rise both to themselves as well as to the somatic cells, while the somatic cells produce only cells that develop into the body of an organism (cf. Sabour and Schöler 2012: 716). This internal split rends the organic world in two, subtracting evolutionary significance from constituted organisms and relocating it in the pre-individual germ line that runs through them. This means that somatic mutations cannot be inherited; neither can habits, acquired characteristics, or associations. The germ line is deathless; individual organisms are its temporary excrescences, epiphenomenal byproducts deposited along the course of the germinal flow (cf. Bergson 1998: 26–27, 87). This is an extreme position, and has since been weakened and complicated by the epigenetic revolution in evolutionary theory (cf. Surani 2016: 136). Bergson thinks, in any case, that reproduction is on its own sufficient to furnish his conclusion that individuation is, in the organic domain, always necessarily unfinished (1998: 43, 27).

The implication is that if there is to be a vital principle, it cannot be indexed to individual organisms, because—if for no other reason—*organisms are never completely individual.* Germinally understood, reproduction undermines the closed individuality of the organism from behind by opening it onto its generating conditions. Since those conditions are developmentally continuous through it, the organism is less a thing of its own than a derivation from the material of its progenitors, a secondary effect. And the germ-plasm is, on Weismann's account, continuous not only through the processes of fertilization and development responsible for the formation of the adult organism, but through its entire phylogenetic lineage as well (184). This continuity must be what Bergson has in mind when he claims that in the attempt to locate and determine the principle of the beginning of an organism "gradually we shall be carried further and further back, up to the individual's remotest ancestors: we shall find him solidary with each of them, solidary with that little mass of protoplasmic jelly which is probably at the root of the genealogical tree of life" (1998: 43). If the organism is formed out of a combination of its parents' vital substances and determined by the unchanged germ-plasm continuous through them, then the same must go for each progenitor in turn. The parents on whose combined body the bud of the organism first sprouted are each themselves the flowers of budded parts of a combination of parental substances of their own. The process of their formation was determined and directed by germ-plasm that was continuous through them as well. This means that the principle of closure that would secure the determinacy of the organism as a distinct individual is deferred backwards through each of its generations, arriving finally at the last common ancestor shared by all extant forms of life, Bergson's "little mass of protoplasmic jelly" (43).

"Where, then," he asks, "does the vital principle of the individual begin or end?" (43). The question is unanswerable, for "each individual may be said to remain united with the totality of living beings by invisible bonds" (43). It follows as a consequence of this invisible (germinal) unity that "the individual is not sufficiently independent, not sufficiently cut off from other things, for us to allow it a 'vital principle' of its own" (42). Any attempt to individuate the vital principle is therefore going to collapse back onto the entire history of evolution and encompass "the whole of life in a single indivisible embrace" (43). Here is the anti-internalist logic running through this rebuke against vitalism: the vital principle is, when individuated, a variant of internal finality. When Bergson claims that the individual is not sufficiently independent for us to allow it a vital principle of its own, he means too that the individual is not sufficiently independent for us to allow it a *final cause*, or inner purpose, of its own either.

The closure required for the bounded individuality of the organism is frustrated by its reproduction and development just as it is by the relative mereological autonomy of its parts. Unifying and motivating these considerations is the same implicit premise: that internalism stands or falls with the complete reality of organic individuation. From this it follows, first, that in the absence of the ability to fully specify a limit between one living being and another—to individuate them—there can be no determinate distinction according to which purposiveness can be rendered internal. Second, and finally, we can conclude that if finalism—and any associated vital principle—is to be viably thought, it will have to be rendered *external* to any ostensibly individuated biological form. To say that finality is a necessarily external attribution is to say that it is to be attributed to all of life *indivisibly* (43). Internality requires that divisions be cut into the organic domain, and since they are always incompletely determinate, the locus of finality will always reside external to them, on the outer side of any individual so considered. At the limit, this externalism is a global phenomenon, qualifying the *whole* from out of which internalism attempts to dissociate its individual parts.

If by vitalism we understand the postulation of a specific x internal to the purposive organization of living things in order to explain their irreducible distinction from inorganic matter, then Bergson is no vitalist. There are no autonomous individuals in the organic domain. Yet Bergson does not conclude by rejecting the idea of a vital principle; he affirms instead that if it is to be attributed to the organic domain, then it has to be predicated of life as a whole.

3   External Finalism

This "external vitalism" is a variant of finalism because it still predicates purposiveness of life. It predicates it of the "whole" of life. It also inverts its position: purpose is not to be located at the end of a process of change, as the plan or program according to which the change is unfolded, but rather at its beginning, as the impetus or impulsion that sets off the process and provides it a kind of directional puissance without constraining its development in terms of a pre-existent end. There are three components to Bergson's external vitalism: (1) a particular conception of the "whole" to which externalism attributes purposiveness; (2) an inversion of the location of that purposiveness within the whole (that is, not as its end, but as its originating push); and (3) a reformulation of what it is that accounts for the existence of that purposiveness in the first place (that is, an impetus, not a plan, program, organizational unity, or otherwise).

3.1   The Whole

What is the whole, for Bergson, if it is neither the sum total of all currently present individuals (actualism), nor the pre-existent plan on the basis of which they are actualized over the course of time (Leibnizian externalism or "possibilism")? Here is what Bergson says (43):

> If there is finality in the world of life, it includes the whole of life in a single indivisible embrace. This life common to all the living undoubtedly presents many gaps and incoherences, and again it is not so mathematically *one* that it cannot allow each being to become individualized to a certain degree. But it forms a single whole, none the less; and we have

to choose between the out-and-out negation of finality and the hypothesis which co-ordinates not only the parts of an organism with the organism itself, but also each living being with the collective whole of all others.

To ask after the whole is to ask after what it is that is common to all the living, uniting every living being with every other at every level of organization. Bergson's explicit answer comes late in *Creative Evolution*: that which "links individuals with individuals, species with species, and makes of the whole series of the living one single immense wave flowing over matter" is, he says, a "unity" of "the *élan*" "passing through generations" (250 tm). The unity of the *élan* is a "movement," "a simple process" (250–251). Partially individuated organisms are to be understood as moments of a movement that traverses them, phases of a process that is realizing itself across them. It is that movement, that process, that unites them (128). The whole of life is therefore not comprised of all living beings past and present, taken together as one total collection. The whole of life is rather to be understood as the event of their progressive generation over evolutionary history. The whole of life is the movement of evolution considered as a single unfinished event, driven by a unified impulse (cf. Montebello 2012).

Life as a whole is external to any one of the ostensibly individuated biological forms that populate it because individual forms are only ever artificially stable perspectives on what is an event in the process of unfolding itself (128). It is in fact not exactly correct to speak of the event and its unfolding as separate things. The event of life is the very process of its unfolding, its movement, and nothing besides. Any determinate form is by definition an artificial stabilization of it. Consider the flight of Zeno's arrow, one of Bergson's favourite images. If the localization of the arrow at any of the spatial locations through which it passes is an artificial operation, this is because those locations are spatializations or freeze-frames of a qualitatively whole movement, and have no reality outside of it. The arrow's trajectory is external to any one of its possible locations in space as those locations represent possible stopping points and are therefore derivative on the movement as such. The movement, for this reason, is not only external to its possible stopping points; it is also immanent to them, coextensive with them. The same should be said of life. It is the qualitative whole of the event of evolution. Determinate organic forms are only its possible stopping points, and it is therefore external to any possible set of them. At the same time, and as a result, it is also immanent to all of them, for they are nothing outside of the movement through which they are formed (43–44).

3.2 Unity

If it is their shared history of development within the event of evolution that unites the extant plurality of life forms—what contemporary theorists call the "deep homology" beneath constituted organisms—then evolution unfolds in contradistinction to the way artifacts are constructed. This is to say that in the domain of life, unity resides prior to the differentiation of parts, or the proliferation of species. This model of life is at odds with what Bergson takes to be the Leibnizian finalism—or what we might better understand as preformationism—that understands the world to be harmonized in view of the ends that pre-exist and orient its trajectories of development, as if to integrate an initially disparate set of parts in view of a shared telos. Bergson's response is, again, not to deny purpose to the process, but rather to detach it from its telic location at the end, and relocate it at the beginning. Purpose—to put it this way—is a function not of final cause, but of efficient cause, of initial impulsion. For it is the latter that is shared in common across a plurality of different forms, it is the latter that unites them and provides a minimal directional constraint on their differentiation.

If there is unity in the domain of life—whether developmentally or evolutionarily determined—then it is to be located at the origin of a process of change, as change means differentiation and differentiation means the divergence of directions. It is never manifest in fact, but only in principle, as the state ever further away from which evolution is always in the process of developing (51). We can add now that if pre-existent ends unify *initially* disparate elements by *attracting* them from *ahead*, then common origins unify *eventually* disparate elements by *impelling* them from *behind* (103). That is the inversion.

Bergson's position is best described, I think, as an immanent finalism. Life "takes directions," he writes, "without arriving at ends" (102; cf. 16). Life is purposive because it is directional, which means that its shape is not entirely the result of a series of accidents pressed into form via the mechanical force of external circumstance. But the cause of life's directionality is not teleological; it does not pre-exist or reside outside of the contingent trajectories taken by life over the course of its own unfolding. The cause of evolutionary directionality is rather immanent to that directionality itself, interior to it. External causes shape, divert, and constrain it, but they do not explain it. This is the idea behind Bergson's claim that it is the "movement" through which novel forms are generated that "constitutes the unity of the organized world," and that the exterior force of "adaptation explains the sinuosities of the movement of evolution, but not its general directions, still less the movement itself" (105 and 102). Bergson offers the image of "the wind at the street-corner," dividing "into diverging currents which are all one and the same gust" (51; cf. Cunningham 1914: 649–650). The air owes a bifurcation in its current to the mechanical influence of its encounter with the corner, but the directionality of its movement, which both precedes and survives its division around the corner, has to be explained in another way. No matter how many times it is divided and diverted, each new current continues an original gust in a new direction. It is the unity of the evolutionary movement, by analogy, that each of its forms has in common. Just as their unity lies behind them, so too does the finalistic force that accounts for it, the originary impulsion that is prolonged through the movement that differentiates them. Bergson calls it an impetus, the *élan vital* (101).

3.3   Tendency

What is an impetus? Bergson is emphatic that it is first and foremost an image borrowed from psychology; that is, "it is only an image" for life, yet "no image borrowed from the physical world can give more nearly the idea of it" (257). This is because "the essence of the psychical is to enfold a confused plurality of interpenetrating terms," and in this sense it should be said that "life is of the psychological order" (257). There are five other mentions of the psychological nature of life in *Creative Evolution* (51, 54, 77, 86, 208).[3] One has to be careful about how to interpret them (cf. Ansell-Pearson 2002: 137, 2005b: 68). One instance in particular has attracted a lot of attention. This is the apparent definition of life as "consciousness launched into matter" (Bergson 1998: 181; cf. 261).[4] Bergson is clear, however, that life only *appears* "as if a broad current of consciousness had penetrated matter" (181). By consciousness here, Bergson means a "current"—elsewhere he says "wave" (250)—"loaded, as all consciousness is, with an enormous multiplicity of interwoven potentialities" (181). What makes the evolutionary movement look like a current of consciousness is this multiplicity of interpenetrating tendencies. In the equation of life with consciousness, it is the idea of interpenetrating tendencies that is at issue (257). Bergson does not contend that consciousness really did penetrate matter in the constitution of the first living cells, for example; he says rather that it is *as if* that is what happened. Life appears as such, it makes sense to consider it as such, but such is not in fact what it is.

Why the appearance? Life is essentially movement or mobility (128). The indivisibility of a movement means that its past swells with it as it endures. The elements of a continuous movement are tendencies, tending toward some limit, whose progressive materialization over the duration of the movement describe its evolving shape. When Bergson claims that "between mobility and consciousness there is an obvious relationship," or that "every [mobility] has a kinship, an analogy, in short a relation with consciousness," he has these attributes in mind (109 and 2004: 304–305). Understood as mobility, life can be seen therefore to evolve "exactly like consciousness, exactly like memory" (1998: 167). For consciousness and memory are defined by "continuity of change, preservation of the past in the present, real duration" (23). This is what it means to say that there is something consciousness-like in the evolution of life. The best way to grasp the analogy is via the materialization of tendencies over time. And the best model we have for how to understand tendency is that of our own consciousness (Bergson 1998: 54; cf. 201–202). A tendency is not yet an actuality; it is the potentiality to become some actuality if unthwarted by the other tendencies with which it conflicts (13). When Bergson

---

[3] Bergson makes similar claims, with similar qualifications, about matter as a whole and consciousness as well (cf. 2004: 292–293, 313, and 331).

[4] Alliez 2013: 69; Barr 1913: 646; Balz 1921: 637; Gunter 1999: 172; Kreps 2015: 171–172; Rignano 2014: 128. This is a small selection. The conception of life in terms of consciousness launched into matter plays a central role in most commentaries on *Creative Evolution*.

claims that tendency "cannot be resolved into physical and chemical facts" he is stating the obvious, for only actualities can be resolved into the physicochemical components that comprise them (1977: 114). This is one important reason why evolution cannot be exhaustively mechanistically explained. This is also why the idea of psychological impetus, the effort by which a psychological tendency is actualized in image or action, is the best analogue we have for life.

Tendencies are less like the component parts of an organism, less still like a group of objects placed side by side, and more like "psychic states, each of which, although it is itself to begin with, yet partakes of others, and so virtually includes in itself the whole personality to which it belongs" (Bergson 1998: 118). This is how we ought to understand the suggestion that life, as original tendency, is virtually instantiated in the developmental trajectory of each particular tendency that is dissociated from it. It is virtually immanent to them both as the global tendency towards temporalization *and* as a domain of interpenetrating tendencies from which every actual developmental trajectory can draw in evolving, even after they are externalized from out of that domain.

I have said that the unity of this movement is owed to its commonality of origin. We can now see that there are two senses to this claim. According to the first, the evolutionary movement is unified historically, as each lineage ultimately originates in a common ancestor, something like the way every marble sculpture human history has ever produced can be traced back to a developmental origin in the biogenic formation of minerals. In this sense, every distinct species and individual being is a moment in the ongoing elaboration of a single history, an unfinished event, and can be traced back to the one common origin of all evolvable terrestrial life. Its difference from the history of marble is that every present living being retains the whole evolutionary past in the form of organic memory. This is what it means to say that life endures (15). According to the second sense of the claim, the common origin of all life is in turn the manifestation of a densely heterogeneous tendency towards indetermination, implicating within itself an interpenetrating mass of virtual tendencies as well (258). This global whole is contracted and instantiated in the actuality of each extant evolutionary trajectory, unifying every living being with every other as dissociated parts from the whole that implicates them. The *élan vital* is both the originating condition of all living forms as well as the immanent unification of the divergent directions taken by the evolution of those forms via the register of virtual tendency that is instantiated across each (cf. Deleuze 2006: 94).

In sum, Bergson's form of finalism consists in (1) a reconception of externalism, (2) an inversion of its location of unity, and (3) a reconfiguration of what it is that accounts for the existence of that unity in the first place. (1) Bergson conceives the "whole" to which purposiveness is attributed not as the sum total of living entities in harmonious relation with each other, but as the evolutionary movement whose unfolding generates each of those entities from out of itself. As we will see, this means, for Bergson, that stable forms—whether of the organ, the individual, the species, or otherwise—are in fact best understood as the outlines of the directional movements that run through them. (2) Bergson detaches externalism's principle of unification from the teleological postulate of a pre-existent end, or plan, and relocates it in the originary impetus driving the evolutionary movement from which stable forms are to be derived. Finally, (3) he reconfigures the commonality of the origin of movement according to the psychological interpenetration of life by reformatting it through his modal mereology and positing it as both the register of interpenetrating tendencies as well as the essence of tendency as such, which is to extend itself to its limit, dividing in response to obstacles in order to extend itself ever further in divergent directions.

4 Conclusion

The problem with vitalism, for Bergson, is not that it insists on a difference between life and matter, but that it incorrectly individuates the difference-maker between the two. Bergson's difference-maker is something like Driesch's entelechy, though it is not an actual force—whether material or immaterial—but rather a virtual tendency. More importantly, its position is external to the ostensible individuality of any given biological form. Driesch's entelechy safeguards biological individuality; Bergson's *élan* explodes it. As a consequence, if Bergson is to be considered a vitalist, he should be positioned diagonally to the ideas of biological individuality, organization, and autonomy that are taken to characterize the "contemporary vitalism" at issue in Bakhtin's critique, for instance. Bergson is better understood as advocating for a tendency whose finality is immanent to it, and a directionality or orthogenicity that resides in the whole movement of evolution understood as a single event. For this reason I think that Bergson may be better situated in line with thinkers that problematize biological individuality than with the thinkers—the vitalists—that attempt to safeguard its reality and explain its irreducibility. Another way of putting the point is to say that what is at stake in Bergson's peculiar form of vitalism is the incomplete individuation of biological forms in the movement of evolution and their association with each other, both in organized quasi-wholes, and over time, in generationally striated processes of populational differentiation.

This, then, is Bergson's vitalist wager: contra Driesch, *there are no complete individuals in the biological domain*, and thus instead of having to account for their individuality, it is rather their trans-individuality, or their openness both to each other and to the evolutionary process through which their closure is effaced, that has to be accounted for—and Bergson does this through the idea of a tendency realizing a directional but open-ended progress across them. I think there are at least three possible benefits to this shift. One is that it avoids the problems that Bergson identifies in Driesch's brand of vitalism, though of course this only registers as a benefit if one accepts Bergson's criticism of Driesch. A second is that it seems nicely fitted to theories of deep homology and developmental constraint, as the *élan* is intended as an explanation of the deep continuity running beneath ostensible individuals, and it is an explanation that mediates the force of selection in accounting for the generation of heritable traits by positing directional constraints—in a process of what Bergson calls canalization—that trend and pattern the evolutionary movement.

Finally, a third possible benefit might be located in the fact that since the *élan* is in some ways an anti-individuating, or perhaps a trans-individuating tendency, then it might also be brought into a profitable alliance with recent work on the near-ubiquity of symbiosis, the microbiome, and holobiont or related revisions to biological notions of individuality. These remain for now only intimations of how Bergson's distinctive variety of vitalism might be thought today.

References

Alliez, Eric. 2013. Matisse, Bergson, Oiticica, etc. In *Bergson and the Art of Immanence: Painting, Photography, Film*, ed. John Mullarkey and Charlotte de Mille, 63–79. Edinburgh: Edinburgh University Press.

Ansell-Pearson, Keith. 2002. *Philosophy and the Adventure of the Virtual: Bergson and the Time of Life*. London: Routledge.

———. 2018. *Bergson: Thinking Beyond the Human Condition*. New York: Bloomsbury.

Bakhtin, Mikhail. 1992. Contemporary Vitalism. Trans. Charles Byrd. In *The Crisis of Modernism: Bergson and the Vitalist Controversy*, eds. Frederick Burwick and Paul Douglass, 76–97. New York: Cambridge University Press.

Balz, Albert G.A. 1921. Reviewed Work: *Mind-Energy* by Henri Bergson, Wildon Carr. *The Journal of Philosophy*. 18 (23): 634–643.

Barr, Nann Clark. 1913. The Dualism of Bergson. *The Philosophical Review* 22 (6): 639–652.

Bergson, Henri. 1977. *The Two Sources of Morality and Religion*. Trans. R. Ashley Audra and Cloudesley Brereton. Indiana: University of Notre Dame Press.

———. 1998. *Creative Evolution*. Trans. Arthur Mitchell. Mineola: Dover Publications.

———. 2004. *Matter and Memory*. Trans. Nancy Margaret Paul and W. Scott Palmer. Mineola: Dover Publications, Inc.

Bognon, Cécilia, Bohang Chen, and Charles Wolfe. 2018. Metaphysics, Function and the Engineering of Life: the Problem of Vitalism. *Kairos: Journal of Philosophy & Science*. 20 (1): 113–140.

Cunningham, Watts G. 1914. Bergson's Conception of Finality. *The Philosophical Review* 23 (6): 648–663.

Deleuze, Gilles. 2006. *Bergsonism*. Trans. Hugh Tomlinson and Barbara Habberjam. New York: Zone Books.

Driesch, Hans. 1914. *The History and Theory of Vitalism*. Trans. C. K. Ogden. London: MacMillan.

———. 1929. *The Science and Philosophy of the Organism: The Gifford Lectures delivered before the University of Abderdeen in the Year 2907 and 1908*. 2nd ed. London: Adam and Charles Black.

Gunter, P.A.Y. 1999. Bergson and the War Against Nature. In *The New Bergson*, ed. John Mullarkey, 168–183. New York: Manchester University Press.

Hegel, G.W.F. 1969. *Hegel's Science of Logic*. Trans. A.V. Miller. London: George Allen & Unwin.

Illetterati, Luca. 2014. Teleological Judgment: Between Technique and Nature. In *Kant's Theory of Biology*, ed. Ina Goy and Eric Watkins, 81–98. Boston: Walter de Gruyter GmbH.

Jorati, Julia. 2017. *Leibniz on Causation and Agency*. New York: Cambridge University Press.

Kant, Immanuel. 2000. *Critique of the Power of Judgment*. Trans. Paul Guyer and Ed. Allen Wood. Cambridge: Cambridge University Press.

Kreines, James. 2004. Hegel's Critique of Pure Mechanism and the Philosophical Appeal of the Logic Project. *Europoean Journal of Philosophy* 12 (1): 38–74.

———. 2008. The Logic of Life: Hegel's Philosophical Defense of Teleological Explanation of Living Beings. In *The Cambridge Companion to Hegel and Nineteenth-Century Philosophy*, ed. Frederick C. Beiser, 344–377. Cambridge: Cambridge University Press.

Kreps, David. 2015. *Bergson, Complexity, and Creative Emergence*. New York: Palgrave Macmillan.

Leibniz, G.W. 1989. *Philosophical Essays*. Trans. Roger Ariew and Daniel Garber. Indianapolis: Hackett.

Lundy, Craig. 2018. *Deleuze's Bergsonism*. Edinburgh: Edinburgh University Press.

Michelini, Francesca. 2008. Thinking Life: Hegel's Conceptualization of Living Being as an Autopoietic Theory of Organized Systems. In *Purposiveness: Teleology Between Mind and Nature*, ed. Luca Illeterati and Francesca Michelini, 75–96. Piscataway: Transaction Books.

Montebello, Pierre. 2012. La question du finalisme chez Bergson. In *Disséminations de l'Evolution créatrice*, ed. Shin Abiko, Hisahi Fujita, et al. New York: Olms Verlag.

Rignano, Eugenio. 2014. *The Nature of Life*. New York: Routledge.

Sabour, Davood, and Hans Schöler. 2012. Reprogramming and the Mammalian Germline: the Weismann Barrier Revisited. *Current Opinion in Cell Biology* 24 (6): 716–723.

Sapp. 2003, January. *Genesis: The Evolution of Biology*. New York: Oxford Unersity Press.

Smith, Justin E.H. 2011. *Divine Machines: Leibniz and the Sciences of Life*. Princeton: Princeton University Press.

Surani, Azim. 2016. Breaking the Germline-Soma Barrier. *Nature Reviews: Molecular Cell Biology* 17 (3): 136.

Wandschneider, Dieter. 2010. The Philosophy of Nature of Kant, Schelling, and Hegel. In *The Routledge Companion to Nineteenth Century Philosophy*, ed. Dean Moyar, 64–103. New York: Routledge.

Weismann, August. 1893. *The Germ-Plasm: A Theory of Heredity*. Trans. W. Newton Parker and Harriet Rönnfeldt. London: Walter Scott.

Wolfe, Charles. 2011. From Substantival to Functional Vitalism and Beyond: Animas, Organisms and Attitudes. *Eidos* 14: 212–235.

———. 2017. Materialism New and Old. *Antropologia Experimental* 17 (13): 215–224.

Wolfe, Charles, and Sebastian Nordmandin, eds. 2013. *Vitalism and the Scientific Image in Post-Enlightenment Life Science, 1800-2010*. Dordrecht: Springer.

On the Heuristic Value of Hans Driesch's Vitalism

Ghyslain Bolduc

Abstract In the first half of the twentieth century the harshest critics of Hans Driesch's vitalistic theory depicted it as an animistic view driven by metaphysical moods, while others merely saw it as a barren hypothesis. In the last decades the heuristic value of vitalistic principles was nevertheless suggested. In this chapter I examine the epistemic role of Driesch's critical vitalism in the progress of embryology. I first show that it did not contribute to falsify mechanical explanations of development such as Wilhelm Roux's mosaic theory and Driesch's own embryonic induction model. However, Driesch's argumentation for vitalism led to the final formulation of the most challenging developmental *explanandum* of the twentieth century: the harmonious-equipotential system (HES). I point out how major *explanans* like Charles M. Child's metabolic gradients, Hans Spemann's induction fields and Lewis Wolpert's positional information were conceived as promising answers to Driesch's problem.

Keywords Hans Driesch · Harmonious-equipotential system · Entelechy · Experimental embryology · Epigenesis · Gradient · Field · Wilhelm Roux · Positional information

1 Introduction

In the December 1913 issue of *Nature* the British morphologist Ernest MacBridge (1913: 400, 401) made critical remarks against Hans Driesch's main proof of vitalism; if the value of a biological theory "is its fruitfulness in connecting facts and in leading to the discovery of new facts" then the concept of *entelechy* – which represents a nonspatial organizing agent that drives the organism towards the realization

G. Bolduc (✉)
Centre interuniversitaire de recherche sur la science et la technologie (CIRST),
Edouard-Montpetit College, Longueuil, Canada
e-mail: ghyslain.bolduc@umontreal.ca

© The Author(s) 2023
C. Donohue, C. T. Wolfe (eds.), *Vitalism and Its Legacy in Twentieth Century Life Sciences and Philosophy*, History, Philosophy and Theory of the Life Sciences 29, https://doi.org/10.1007/978-3-031-12604-8_3

of its purpose – is "barren". Rudolph Carnap later rejected Driesch's vitalism for similar reasons. While the laws of science provide *"explanations* for observed facts" and "a means for *predicting* new facts not yet observed", Driesch's entelechy "does not give us new laws" and "did not lead to the discovery of more general biological laws" (Carnap's emphasis 1966: 16; Chen 2018). Philipp Frank (1932/1998: 85, 125), another logical positivist, even severely considered that vitalism comes from the surrender of scientific *rationale* driven by "metaphysical moods". If "the problem of the *method* of biology remains unaffected by the controversies between vitalism and mechanism" (Driesch's emphasis 1913), is the idea that living bodies are guided by nonphysical agents strictly metaphysical? It may not be after all a coincidence that Driesch became a philosophy professor in 1909 and gradually abandoned his empirical research.

However, metaphysical principles have historically guided the elaboration of fruitful hypotheses and have been key elements of successful research programmes (Lakatos 1976). As a "meta-theoretical commitment" (Normandin and Wolfe 2013: 5), vitalism may indeed have *epistemic virtues* such as a distrust of simple explanatory models and their expansive generalization (Hein 1972: 165). Dupré and O'Malley (2013: 312) have suggested that vitalism in general acts as a "heuristic that stimulates productive inquiries into the nature of living and non-living things". Does Driesch's vitalistic view had a significant heuristic value for the progress of biology? To answer this question, we must first distinguish negative from positive function of a potential vitalistic heuristic. Negative heuristic of vitalism must clearly contribute in revealing the falsehood or at least the insufficiency of existing or virtual mechanical models, i.e. their flaws and their basic inadequacy with known living phenomena. On the other hand, using vital principles as *explanans* does not alone constitute a positive heuristic. Vitalism must rather guide the elaboration of biological concepts, theories, hypotheses or methods that somehow are added to the toolkit of successful research programmes. These derived epistemic products must therefore be involved in fruitful *explanans* that may lead to the discovery of new facts.

In order to examine this hypothetical heuristic, I proceed in this chapter as follows; first I investigate the potential role of Driesch's vitalistic turn in the falsification of the Roux-Weismann thesis and Driesch's own embryonic induction model. In the next section I shall: (1) evaluate the epistemic function of Driesch's ultimate *explanandum*, the *harmonic-equipotential system* (HES), in the context of Charles M. Child's gradient theory, Hans Spemann's investigations on organizers and Lewis Wolpert's positional information theory and (2) determine the role of Driesch's vitalism in its formulation.

2 Driesch's Empirical Falsification of Mechanical Models

As an embryologist, Driesch deeply undermined ontogeny most successful mechanical explanation of the 1880s: the so-called "Roux-Weismann thesis" (Hertwig 1896) that a mosaic of self-differentiating parts is led by the unequal distribution of preformed chromatic factors during cell division. He also highlighted the insufficiency of the induction model that he had elaborated as an alternative to Roux's mosaicism. In both cases the *explanans* was highly challenged by experiments that occurred before Driesch's vitalistic turn in 1899. Here I then show that Driesch's vitalism did not act as a negative heuristic on existing explanatory models.

2.1 Entwicklungsmechanik *and the Roux–Weismann Thesis*

We must first look back at the early history of *Entwicklungsmechanik*, Driesch's research program as an embryologist. Development, claims Wilhelm Roux (1885: 414), the originator of the program, is "the production of visible manifoldness [*Mannigfaltigkeit*]" and the origin of this manifoldness remains ontogeny greatest mystery: Is development new formation "in the strongest sense, the real increase of an effective manifoldness" (*Epigenesis*) or "the mere expression of latent and pre-existing differences" (*Evolution*)[1] (414)? This is the problem this new program for embryology mainly tried to solve.

Roux (1883) suspected that most of this morphological and functional manifoldness was already latent in the newly discovered chromosomes – which were soon related to heredity and called *idioplasm*[2] by the Neo-Darwinian August Weismann and others. Typical development would then involve the transfer of complexity from the chromatic organization to the becoming somatic organization by means of indirect cell division. But first the *modus operandi* of differentiation had to be specified. The formation of the whole embryo or of one of its parts is *self-differentiating* if it depends only on its own inner factors, while its differentiation is *correlative* if it also depends on external determinations. Self-differentiating development led by the nucleus would mean the actualization of pre-existing manifoldness in the germ – while correlative differentiation may have "epigenetic" effects, the production of new manifoldness.

In 1885 Roux expected complete self-differentiation of the organism performed by partial self-differentiation of its components, and in order to test his hypothesis, he carried out in 1888 what became his most famous experiment: the production of half- and quarter-embryos after having punctured frog blastomeres of 2-cell and 4-cell stages with a hot needle. Spared 2-cell stage blastomeres became

---

[1] It has been shown since Karl von Baer that development consists in the metamorphosis of simpler forms into more complicated combination of parts, what Driesch (1908: I 25) calls epigenesis "in the descriptive sense". Roux's renewal of the old question of preformation (or "evolution") and epigenesis (Roger 1997; Duchesneau 2012) in the context of *Entwicklungsmechanik* conditioned the elaboration of "New preformationist" and "New epigenetist" theories of development (Maienschein 2005; Bolduc 2021).

[2] This concept originally comes from Karl von Nägeli's theory of heredity. Nägeli distinguished a highly organized molecular structure which he called "the idioplasm" from the cell nutritive plasma – the "trophoplasm".

semi-gastrula and semi-neurula while still being attached to the operated blastomere in disintegration. In Roux's eyes, the development of the healthy part could not depend on the remaining uncelled plasma because fixing and staining processes later confirmed that the targeted chromosomes were destroyed by the heat. Because self-differentiation seems simpler and less costly than correlation, Roux (1888/1895: 454, 455) concluded that *typical* development (without experimental disruption) of the frog embryo was also, at least until neurulation, a *mosaic* of at least four independently developing parts. He also assumed that these results would eventually be generalized to every embryonic cell and was convinced at this stage that cell division often involves what August Weismann (1893: 34) later called *heterokinesis* – the unequal distribution of the inherited idioplasm to somatic cells. Although empirical evidences showed that the distribution of the chromatic material to daughter cells is always quantitatively equal, Weismann postulated a not yet visible decreasing complexity of the somatic nuclear organization at each differentiating cell division. The formation of the somatic body would then rely on a stem tree of cells (Dröscher 2014) in which each branching division would be heterokinetic; the somatic idioplasm of each cell then only contains the *Anlagen* (morphogenic qualities) that its descendants need according to their preestablished fate. This was the best mechanical explanation for the inferred self-differentiation of blastomeres.

2.2 The Discovery of Part Formation

As soon as 1890 Driesch was convinced that *Entwicklungsmechanik* was the central discipline of biology but did not endorsed Roux's mosaic theory despite the most recent empirical results in its favor.[3]

Though Roux concluded from his experiments on frogs that development is the "metamorphosis of manifoldness" rather than its new formation (Driesch 1892a: 161), these had to be corroborated with similar experiments on other species. As Oscar and Richard Hertwig had already shown that vigorous shaking of unfertilized sea urchin eggs in water resulted in the separation of parts from each other (Maienschein 1994: 51), Driesch used this method to separate the first two blastomeres of cleaving sea urchin eggs. At the stage of sixteen cells he first witnessed half-embryos as expected, but the next morning he surprisingly found typical swimming larvae of half size (Driesch 1892a: 168). For Driesch this remarkable result showed that 2-cell stage sea urchin blastomere does not receive from the fertilized egg only half of the *Anlagen*. It is rather *totipotent* because "a normally formed whole larva can come from it; a part formation [*Theilbildung*], not a half-formation" (172).

---

[3] The French teratologist Laurent Chabry (1887) obtained even more conclusive results than Roux by puncturing blastomeres of ascidian eggs. For example, the formation of a "demi-individual" from a two-cell stage blastomere without "post-generation" of the missing part clearly supported the mosaic theory (Fischer 1991: 38).

Roux (1893/1895) nevertheless interpreted this result in line with mosaicism: before these half-blastulas started reacting to their missing halves with *post-generation* process, their initial formation clearly showed the self-differentiation of the first blastomeres. In fact, Roux himself had noticed in 1888 late regeneration of punctured, seemingly dead halves of frog embryos. He understood post-generation as functional adaptation to the disruption of typical conditions and thought it must be distinguished from "direct" development (Roux 1888/1895: 520): (1) by its modus operandi, correlative differentiation; (2) by the material involved – a *reserve* idioplasm that is located in the nucleus of each cell. In the case of his famous frog experiment, Roux assumed (without any empirical evidence) that this back-up nuclear material travelled from the developing half to the destroyed half during gastrulation before proceeding with nuclearization and cellulation of inert proto-plasmic substance. Weismann (1893) then suggested a mechanical model for regen-eration that was based on multiple "accessory" idioplasms; for example, because the same worm cells give rise to the tail-end or to the head-end of the worm accord-ing to whether they are situated on the anterior or posterior surface of the amputa-tion plane (126), he assumed that each worm cell had two accessory idioplasms (one for each end). According to this model each possible morphogenetic fate therefore relies on a separate idioplasm.

However, there is a crucial difference between these types of "post-generation" and sea urchin part formation: in the former the development of the healthy part is not affected by the regenerating part, while the latter involves the complete cellular *re-differentiation* of the isolated halves. Driesch did not witness the part formation itself but inferred that the opening of the half-blastula was closed by bringing together and merging the adjacent sides (M_o and M_u) (Fig. 1): the material which normally belongs to the median region would in this case form the right side, but at least no change of the embryo *polarity* would be required. It would involve never-theless a global redistribution of the cell morphogenetic fates.

Sea urchin blastomeres are indeed totipotent, admitted Roux, but only because they can rely on a reserve nuclear material that can take over "indirect" development

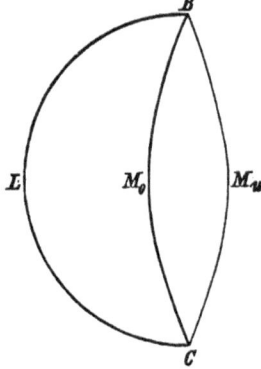

Fig. 1 Simplified illustration of a half-blastula hemisphere. **L** left side material; **B** and **C** poles of the median plane; M_o and M_u adjacent edges of the median region. (Reproduced from Driesch 1892a)

in case of "defect." Roux (1893/1895: 839) still acknowledged that how this material was triggered and at what scale cell correlation was taking place was totally unknown at this point. Driesch (1894: 11) however expressed his opinion on the current state of mosaicism with a vivid image: "Roux's theory is a pyramid stood on its point: below, the hypothetical basic notion, above auxiliary hypotheses pile up and finally as a heavy base on top – my simple experimental results". Even Weismann (1893: 137) sensed the threat of Driesch's crucial experiment for his whole Neo-Darwinian system, questioning the reliability of the experimental method (Churchill 2015: 418). The *ad absurdum* multiplication of accessory idioplams that was needed to match all possible morphogenetic fates of sea urchin blastomeres was definitely *ad hoc*. Idioplams then represented the unbearable epicycles of Weismann's system as the new regulative phenomena seemed impossible to save.

2.3   Driesch's Method, Axiom and Prospective Approach

When he published his result on part formation, Driesch (1892a: 161) thought that it "exceeded yet known physical (mechanical) phenomena" but expected that "it will probably be subordinated to the mechanistic view of the whole phenomenal world." This position, which he called "unmetaphysical vitalism," was in line with the mechanistic heuristic[4] that dominated research programmes in morphological and physiological sciences at the time. In his methodological essay for *Entwicklungsmechanik*, Driesch (1891) philosophically justified the biological quest for mechanical explanation appealing to Otto Liebmann's Neo-Kantism. If space is a pure form of perception and geometry is its science, then any natural phenomenon first ought to be geometrically – hence mathematically – represented (Waisse-Priven and Alfonso-Goldfarb 2009: 42). Natural events must secondly be expounded in physical terms. In this view, knowledge of organic forms first requires the mathematical formulation of the problem, while the latter is only solved *"when it is [...] reduced to the laws of mechanics and represented as a consequence of these laws"* (Driesch's emphasis, 1891: 9).

In light of part formation and other fruitful experiments such as the "compression effect" (see Posteraro, Chapter Vitalism and the Problem of Individuation: Another Look at Bergson's Élan Vital, in this volume), Driesch followed to some degree this methodological order by first formulating the problem using geometrical notions such as mathematical function – i.e. the correlation between two variables

---

[4]This mechanistic heuristic prohibits the insertion of a teleological cause into the *explanans* of phenomena and leads to a convincing or provisional explanation of high level regularities in terms of lower level regularities which are provided by physiology, chemistry, physics and their technical applications.

$(x=f\,[y])$ – and location in space. In this way he started elaborating a new *explanandum* – which he will later call HES – under the formulation of this axiom: "*The relative location of a blastomere in a whole will probably determine what will generally come out of it; if it is different, it will result in something else. In other words: its prospective relation* [morphogenetic fate][5] *is a function of* [is correlated to] *its position*" (Driesch's emphasis 1892b: 39). The formulation of this axiom entails the rejection of Roux's distinction between typical self-differentiation and adaptive correlation: if the fate of the blastomeres generally depends on their relative position within the embryonic whole, then the differentiation of their future cell lineage is *essentially* correlative. If the blastomeres of some species do in fact never re-differentiate and if some do it later than others, it may be due to unknown physical obstacles, such as the consistency of their egg protoplasm (Driesch 1908: I 73). The idea of unequal transmission of hereditary determinants through cell lineages was giving way to holistic determination.

This position led Driesch to rethink the problem of the origin of developmental manifoldness in a *prospective* way. Starting from a given visible manifoldness, Roux was asking *retrospectively* for its causal origin; a given form was either already implicit in the germ from the beginning (preformation) or a new formation (epigenesis). Driesch (1894: 75–78) now sees inherited manifoldness as the material conditions of morphogenetic *possibilities* – which he calls *prospective potency*. The increasing restriction of these potencies takes place during development until the morphological fate of the parts – or *prospective value* – is irrevocably defined. The question of preformation and epigenesis then takes this form: "Is the prospective potency of each embryonic part fully given by its prospective value in a certain definite case; is it, so to say, identical with it, or does the prospective potency contain more than what the prospective value of an element reveals in a certain case?" (Driesch 1908: I 77). For example, the regeneration of dissected sea urchin gastrulas shows that ectodermal and endodermal cells have reduced prospective potencies compared to the totipotent blastomeres; they also have different prospective potencies because an ectodermal cell cannot re-differentiate into an endodermal cell and vice versa. Furthermore, all (ectodermal or endodermal) cells that belong to the same germ layer have the same prospective potency – i.e. they are equipotential – because a whole dwarf layer can regenerate from its dissected parts. This new prospective view and its related concepts played a key role in determining the holistic properties and specificity of developmental systems. These systems were now asking for a convincing mechanical explanation. This second methodological step represented the embryologists' greatest challenge.

---

[5] Driesch (1899: 41) later specifies that the terms *prospective relation* (*Beziehung*) and *prospective value* (*Bedeutung*) both refer to the realized morphogenetic fate.

2.4   Driesch's Induction Model and Its Empirical Falsification

Because he acknowledged that eggs are purposeful arrangements of complex physico-chemical relations, Driesch never tried to *immediately*[6] account for part formation and other cases of re-differentiation in physical terms. In *Analytic Theory of Organic Development* (1894), he rather elaborates an explanatory model that reduces the functioning whole (the *explanandum*) to the inferred properties and activities of its parts. With his "machine theory of life" Driesch (1896) was the first to introduce the idea of cell induction as a differentiating mechanism and antici-pated the embryonic field theory.

Despite the precise ordering of mitotic and meiotic figures, Driesch considered that correlative differentiation cannot rely on the internal determination of a com-plex nuclear structure but rather on the increasing complexity of a three-cornered inductive network between (1) centers of formative stimuli that are specifically localized within the embryo, (2) the protoplasm of each cell and (3) its nucleus (Churchill 1969; Caianiello 2019). He postulated that all nuclei are heterogeneous mixtures of ferments that contain all necessary *Anlagen* for development. However, as catalytic-like materials these ferments do not directly induce cell differentiation but only give "direction" (*Leitung*) (Driesch 1894: 88) to the morphogenetic pro-cesses that take place in the protoplasm. The latter is then far from being passively formed by the organized expression of nuclear qualities, but rather acts as a "media-tor between the inductive cause and the nucleus" (81). In other words, it represents both a dynamic filter and a trigger that switches from a stimulus-specific responsive state to another depending on its current composition. Only under precise chemical conditions can certain stimuli from the extracellular environment modify these same conditions which can feed back on the nucleus by releasing specific nuclear ferments into the protoplasm; these would in turn reconfigure the protoplasm responsive state (90) allowing the reception of other external stimuli or a new dif-ferential release of nuclear ferments and so on.

This pattern explains how cells that differentiate into different tissues can have equipotential nuclei: the fate of each cell depends on the differential selection of extracellular stimuli and on the nuclear ferments that modified the protoplasm throughout the course of development. The protoplasm then becomes the core of a differentiation loop that connects the cell to the whole embryo. Driesch's theory also replaces the highly contested heterokinesis as the best explanation for the pro-gressive restriction of prospective potency; under determinate changes the proto-plasm would become gradually and irreversibly unresponsive to specific formative stimuli and types of ferments. And most importantly, it illustrates how the produc-tion of new qualitative manifoldness is possible:

---

[6]Causal analysis is still considered by Driesch (1908: I 119) as an *indirect* reduction to the laws of physics, because "the full analysis of morphogenesis into a series of single formative occurrences" may one day be completed by "the analysis into the elemental facts studied by the sciences of inorganic nature".

Development starts with a few ordered manifoldnesses, the ones that are given in the structure of the egg; but the manifoldnesses create, by interactions, new manifoldnesses, and these are able, by acting back upon the original ones, to create new differences and so on. For each effect there is immediately a new cause and the possibility of a new specific response, namely a new specific reactivity. We infer a complex form from a simple one given in the egg [...] consequently our theory is epigenetic *with respect to the origin of form as such* (Driesch's emphasis 1894: 86).

Driesch (1914: 197) will later believe that this kind of mechanical epigenesis violates the aprioristic ontological principle[7] that "the degree of manifoldness of a natural system can never increase of itself". This is why in the presumed absence of an equivalent pre-existing structural manifoldness, Driesch (1908: II 197) will appeal to "entelechy as an intensive manifoldness"; by purposefully suspending determinate physico-chemical reactions in the course of development, this vital agent achieves, like "the 'demons' of Maxwell" (198), what mere physical systems could never achieve by themselves.

In 1894 Driesch was still trying to mechanically account for the fact that the prospective value of a cell that belongs to an equipotential system is a function of its position. The first issue was how cells were localized. As experiments had shown that the egg polarity and the median plane set the direction of further cleavage, Driesch (1894: 14) inferred that they constitute a geometrical *coordinate system* (Driesch 1899: 49) that initially localizes cells within the embryo. As the latter grows, mechanical stimuli would come from cells under physical tension (*Zuginduktion*) (Driesch 1894: 83) and the emission of chemical stimuli would occur most likely from determinate points (such as poles) within the boundary region of the whole. Driesch assumed that, depending on its position and current reactivity, each cell receives various inductive stimuli *differentially*. Yet because each germ layer or tissue is typically delineated, the protoplasmic response cannot be gradual; cells would then only embark on the path of (ectodermal, epidermal, etc.) differentiation when the amount of received stimuli of a certain type exceeds a definite threshold.

But the greatest issue remained: when a part of the original whole is missing, how the new whole becomes the center of development (Driesch 1899: 20)? How do cells *harmoniously* re-differentiate, i.e. how do they "work together" (Driesch 1914: 209) to develop organs in proportion with the size of the new whole? As we previously saw (Sect. 2.2), the most likely scenario for sea urchin part formation is that while closing over, half-blastulas do not loose their polar axis; yet Driesch considers that their cells nonetheless loose their needed alignment with the poles. Because his analytic theory alone cannot explain how disoriented cells are still able to develop normally, he then postulated that the cell repolarization started after the disturbing event and was physically mediated by the magnet-like effect of the electrical charge of each individual blastomere (Driesch 1894: 22).

However, new experiments including Driesch's own dissection of a sea urchin gastrula in 1895 highlighted the insufficiency of this model as the weight of new

---

[7]Though akin to the second law of thermodynamics, this general principle is not limited to energetic "intensities" but rather encompasses "diversity of distribution" (spatial arrangements).

anomalies became unbearable. Driesch sliced a complete gastrula at the equator so that each half-gastrula contained the half of both ectoderm and endoderm (Driesch 1899: 9, 10); not only were the missing parts of both halves quickly restored but also the gut of each sub-product later showed a smaller but typical shape (the strict proportion between the fore-, mid- and hind-parts of the gut was strictly maintained). How this harmonious re-differentiation can happen even without one of the polar regions, which were viewed as essential parts of the coordinating system? Together with the hydra and the starfish embryo (20–24), the sea urchin embryo is a type of HES (45) that is somehow able to achieve what Driesch calls "secondary regulatory phenomena" (47), namely the prior reinstatement of a new coordinating system according to the new dimensions of the whole. And if we follow Driesch's induction model, the healed gastrula would then have to chemically reset all the protoplasmic filters and trigger the release of inductive stimuli from newly located emission points (Churchill 1969: 182). As a result, Driesch (1899: 37) not only concluded that there was an obvious "gap" in his analytical theory, but more significantly, that it was impossible to localize the causal processes that result in these regulatory responses. So instead of trying in vain to improve a model that was based on the "false dogma" of the machine theory of life (9), he abandoned it altogether. In 1899 he then spoke of "peculiar elementary lawfulness" (70) and "vitalistic causality" (71) but his vitalistic philosophy did not find its relatively complete form before the *Gifford Lectures* in 1907.

We now clearly see that Driesch's "metaphysical" vitalism was the consequence rather than the condition of the empirical falsification of past mechanical *explanans*. When Driesch discovered the secondary regulatory phenomena he still adhered to the "machine theory of life,"[8] but no longer saw how such a theory was possible.

3   The Challenge of HESs and the Positive Heuristic of Driesch's Vitalism

3.1   Critical Idealism and the Argument for Vitalism

Before starting his research program, Roux (1881: 229) said that the developing organism was like a music box that can learn new songs everyday while building itself. Within a decade experimental embryology however revealed the astonishing regulatory power of many invertebrates and amphibians; these could then rather have been compared with music boxes that still play the same old song even when

---

[8] Innes (1987) and Sander (1997) suggested that all along the 1890s Driesch was one step away from openly identify himself as vitalist because he was already acknowledging the "teleological" character of development. I think the idea that Driesch was a "closet vitalist" until 1899 is misleading. Driesch (1899: 36) clearly defines his past mechanical conception of life as "static teleology", which only refers to the purposeful arrangement of the egg starting structure. This view was also endorsed by Roux (1881: 2), who made the distinction between purposiveness and (nonmechanical) teleology.

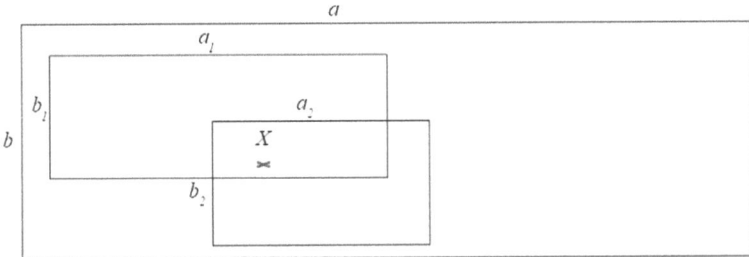

Fig. 2 One of Driesch's diagrams geometrically representing the HES formula. An element X can be part of the system $a\ b$ or $a_1\ b_1$ or $a_2\ b_2$. The prospective value of X would be different in each case. (Reproduced from Driesch 1908)

half of their parts have been removed! But after 1899 Driesch was convinced that HESs were not mere machines and he intended to prove it, not with further experiments, but with philosophical arguments.

It was then within the framework of his "Neo-"[9] vitalism that Driesch developed the more thorough analysis of HESs. He first updated the *explanandum* to demonstrate that it was really a problem which, if approached exclusively from a mechanistic point of view, was unsolvable. He first added a variable that was missing from the 1892 axiom (Sect. 2.3): the prospective value of a cell X is not only a function of X relative position (l), but also of the absolute size (s) of the system to which X belongs (Fig. 2). The new mathematical equation was at this point $(X) = f\,[l,\,s...]$ (Driesch 1908: I 124) but the "E factor" still had to account for the necessary realization of X according to l and s in every possible case. The final formula was $(X) = f\,[l,\,s,\,E]$ but the exact nature of E remained to be clarified.

This is precisely where Driesch subtly leaves the *explanandum* for the *explanans*, as he explains why the E factor cannot be a self-differentiating machine:

> Every volume [a b, a_1 b_1, a_2 b_2, etc.] which may perform morphogenesis completely must possess the machine in its totality. As now every element of one volume may play any possible elemental role in every other, it follows that each part of the whole harmonious system possesses any possible elemental part of the machine equally well, all parts of the system at the same time being constituents of different machines [...] you may ask yourselves if you could imagine any sort of a machine, which consists of many parts, but not even of an absolutely fixed number, all of which are equal in their faculties, but all of which in each single case, in spite of their potential equality, not only produce together a certain typical totality, but also arrange themselves typically in order to produce this totality (Driesch 1908: I 140, 153).

At this point Driesch argues that the E factor can be nothing but *entelechy* – a non-physical agent that purposefully uses the matter of the egg during development as a mean to achieve its typical form. Driesch's main proof of vitalism takes the logical form of the following *modus ponens*:

---

[9] Driesch differentiates his own vitalism from past "naïve" vitalism mostly by the fact that the former did not arise from the direct "contemplation of life's phenomena" (1914: 19), but rather *indirectly* by first making sure that mechanical causality was insufficient to explain these phenomena.

Premise 1: If HESs cannot be machines, then they are driven by an entelechy.
Premise 2: HESs cannot be machines.

Conclusion: HESs are driven by an entelechy.[10]

As we can see from the last quoted passage, Driesch also provides sub-arguments in support of the second premise: unlike HESs, no machine is contained in all its parts and no machine is fragmented without its functioning being impaired (Weber 1999). This of course applies to actual artefacts like "the phonograph" but also to "*any* sort of machine imaginable in physics and chemistry" (Driesch's emphasis, 1908: II 81). Does the concept of "possible machine" is defined by any machine "that Driesch has [subjectively] in mind" (Conklin 1929: 30)?

In Driesch's view, possible machines are virtually circumscribed by what he calls *singular or additive causality*, where the sum on the side of the cause corresponds to the sum on the side of the effect. The primary characteristic of mechanical reasoning is the localization in time and space of connected physico-chemical events, whereby the parts involved "are changed in themselves, irrespective of the others" (Driesch 1914: 199). Against Kant's transcendental analysis, Driesch also intended to show the apriority of another type of causality – the *individualizing causality* – by which "a distribution of the things in one system of the form of a mere *sum* would be transformed into a distribution that would be in some sense a *unity* or *totality*, without any spatial mechanical predetermination of this totality" (200). By first insisting on the insufficiency of all possible mechanical account of HESs, he ensures that, in order to be known, these empirical objects need to be subsumed under the pure concept of individuality, which proves that this type of non-spatial causality – and therefore entelechy – exists in nature (207).

"Neo-vitalism" therefore relies on a renewed critical idealism (Bolduc Forthcoming). Because entelechy is viewed as the product of critical philosophy, Driesch does not personally see it as a metaphysical dogma, but as a legitimate scientific fact (Bognon et al. 2018). However, if we acknowledge the possibility of *a priori* knowledge as a metaphysical assumption, then Driesch's vitalism is at least rationally grounded in metaphysics. In any case this does not make it scientifically irrelevant and its value as a positive heuristic remains to be determined.

3.2   HES: The Developmental Explanandum of the Twentieth Century

Although it was highly discussed among scientists in the first decades of the twentieth century, Driesch's concept of *entelechy* never positively integrated the theoretical framework of embryological research programmes. While it was mostly yet not

---

[10] My formal reconstruction of Driesch's argument was made in part from Driesch 1908: I 119, 187.

only criticized by the scientific community, its lack of heuristic value comes primarily from the impossibility of directly falsifying or corroborating Driesch's hypothesis. How could experience testify to something that is "not localisable at any point in space-time" and "being incapable of measurement, it cannot be a form of energy" (Needham 1936: 69)? Driesch himself conceded that is was "of quite limited application" as "vitalism has nothing to do with the progress of zoology as a pure science in the narrower sense of the word" (Driesch 1913).

3.2.1   Child's Metabolic Gradient Theory

At best, entelechy was used by biologists as a negative hypothesis to promote their own *explanans* for HES. Even the American zoologist Charles M. Child (1915: 24), who made relevant criticisms against the concept of HES (Sect. 3.3), thought that new vitalistic theories represented a real advance over Weismann's theory, for they "have at least the merit of recognizing and meeting squarely the real problem", namely "organic individuality" (Child 1916: 512). In Child's eyes, Driesch was right to stress on the need for a *dynamic* biological theory of the individual that "must deal primarily with processes, not structures, and with changes, not with static entities" (513). And Child explicitly meant that his metabolic gradient theory was the only plausible mechanistic alternative to Driesch's entelechy (519).

According to Child's theory, the process of organic individuation is based on stimuli coming from specific points in the protoplasmic mass (made of one or many cells) that increase the metabolic rate of the affected regions (these stimuli can be inherited or emerge *de novo* from external factors). Experiments showed that the metabolic activity resulting from points under stimulation is transmitted by axial gradients (Fig. 3) to other regions of the whole like the spreading of a wave. The intensity of a transmitted metabolic rate depends on the relative conductivity of the surrounded protoplasm and on the distance that separates a given region from the starting point (**a**); the energy of a gradient then relatively fades off as it is passed on

Fig. 3 Diagram illustrating a single axial gradient in a protoplasmic mass. a, the stimulation point. (Reproduced from Child 1915)

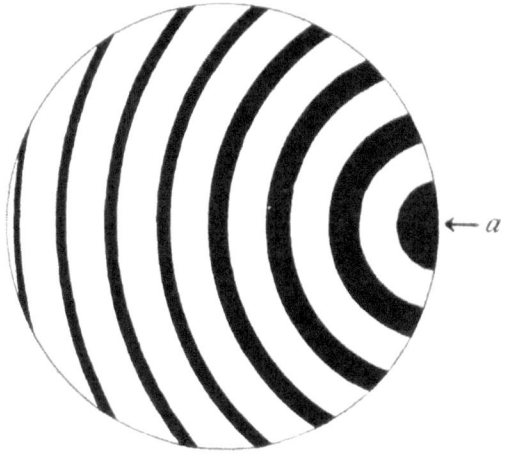

from one region to another. Many axial gradients coming from different stimulation points can simultaneously act upon the same region to a greater or lesser degree, so that the metabolic condition of the given region will result from its position in the overlapping gradients (Child 1915: 39). Most importantly, Child's major thesis is that *qualitative* differentiation comes primarily from *quantitative* differences of metabolic activity (Child 1916: 515). While irreversible differentiation is first caused by the persistence of gradients over time, changes in the stimulation pattern, mostly at the early stages of development, can transform the morphogenetic fate of the related parts. Child's gradients had a theoretical advantage over Driesch's first induction model: constellations of metabolic rates are more plastic and therefore more adaptable to new conditions than the chemical composition of cells. In sum, if a part of a lower organism that becomes isolated from the rest of the body does not loose the original gradient axes, the metabolic condition of its own parts will then proportionally tune with the size of the new whole; in the opposite case new gradients will emerge from the new external conditions. In both cases the piece can develop into a new individual (Child 1915: 46).

Because metabolic rate was in some way subject to empirical testing, Child's system had significant experimental and theoretical success during the 1930s, highlighting the role of metabolism in normal and abnormal development (Huxley and de Beer 1934/1963). Although its explanatory scope turned out to be more limited than first expected, recent studies on metabolic gradients in response to spatial stimuli in cytoplasm could still lead to a better understanding of the relationship between energetic and genetic factors in development (Blackstone 2008).

3.2.2   Hans Spemann, Organizers and Fields

Unlike Child, the German embryologist Hans Spemann openly adopted Driesch's HES concept as one of the main guiding problems of his research program (Allen 2004: 467, 468). By evolving the method of transplantation, Spemann's team was able to address the problem – first formulated by Driesch (Spemann 1938: 199) – of the prospective potency progressive restriction (epigenesis, Sect. 2.3). They, for example, exchanged a piece of presumptive neural tube from a newt young gastrula with a piece of presumptive external gills from another newt young gastrula. Each grafted piece developed not according to its origin but passively followed the development of its host (Spemann 1927: 179). These pieces were then equipotential at this stage as each one integrated perfectly into its new whole: "The development of the part is a function of its situation in the whole" (Spemann 1938: 348). Spemann also inferred that some factor was locally determining the fate of the transplanted pieces. This type of factor was famously discovered by Spemann's assistant, Hilde Mangold, when she took a piece from the upper lip of the newt blastopore and transplanted it into an indifferent region of another newt gastrula: not only did this grafted material not follow the development of its new host, but it also forced "the surrounding parts to follow its own direction" (Spemann 1927: 180), as a second embryo developed at the expense of the host material. Because it "induces a 'field of organization'" in its surrounding, Spemann named it the "organizer". Other orga-nizer phenomena were quickly found in echinoderms, insects, birds and mammals.

The organizer was behaving, stated Spemann, "like a harmonious equipotential system of Driesch" (183). This significant association meant more than the mere subsuming of Hilde Mangold's outstanding discovery under Driesch's concept: Spemann here suggests that the organizer represents the key clue to the riddle of HES. It is not "intensive manifoldness", Driesch's "idealistic" entelechy (Spemann 1938: 347), but rather a localized structure that induces a field of organization which is, like the field of physicists, "extensive manifoldness" (302) – a pattern that oper-ates spatially and hence materially. As transplantation experiments also showed the existence of second and third grade organizers – e.g. the *Triton* eyeball is itself induced by the mesoderm before it organizes a lens – the whole development appeared as "composed of single processes connected by organizers" (Spemann 1927: 186). Above all, the "field action" of these organizers was not a mere physical metaphor but a tool to describe and infer empirical facts (Spemann 1938: 305): in several cases the embryonic fields exceed the limits of the organ that they induce; they often overlap and when they do, one field must win the "rivalry" (311) for induction; some embryonic fields persist throughout development and during the adult stage (regeneration) even when there is no reacting material under their influ-ence; like Child's metabolic gradient, the power of an induction field appears to decrease from the source towards the borders; heteroplastic transplantations (graft-ing a piece from a donor of another species) suggest that fields release specific "genotypic potencies" (350) that were already latent in the reacting material.

Spemann admitted that biologists had "yet no real conception of what this means in the language of physiology" and that "the equipotential system capable of harmo-nious differentiation still remains as a real problem" before adding: "attempts to solve this problem, partly logical, partly experimental, induced several investigators to introduce into experimental embryology the conception of the 'embryonic field'" (347, 348, 366). This statement clearly highlights the epistemic value of Driesch's challenging *explanandum*.

3.2.3 Lewis Wolpert's Positional Information Theory

With the discovery of gene regulation in the 1960s, a convincing explanation of HESs looked within reach. By elaborating a hypothetical cybernetic model made of different regulatory and enzymatic genes, Jacques Monod and François Jacob (1961) had already addressed the problem of how embryonic cells with the same genetic code may differentiate. But according to the British developmental biologist Lewis Wolpert (1969: 4), almost no progress had been made in the area of pattern formation since the 1920s when concepts such as regulative development, gradient and field were elaborated. Wolpert, who considers Driesch as one of his precursors (Horder 2001: 121), reinterpreted these notions within the context of his positional information theory, which he inferred directly from HESs. He formalized the latter in terms of the "French Flag problem":

> This problem is concerned with the necessary properties and communications between units arranged in a line, each with three possibilities for molecular differentiation – blue, white and red – such that system always forms a French Flag irrespective of the number of units or which parts are removed [...] This abstraction of the problem corresponds quite well with experimental observations on the early development of sea urchin embryos, and regeneration of hydroids as well as a large variety of other systems (Wolpert 1969: 5).

This simple schematic illustration of morphogenetic fate specification whereby each band of colour represents a differentiated part (e.g. germ layer) displays the same property as Driesch's concept, namely "size invariant" regulation of typical patterns.

By referring to *The science of the organism*, Wolpert (1974: 674) pointed out that Driesch was the first to put forward the idea of position specification. In fact, Wolpert unprecedently inferred from the correlation between the position of a cell in the whole and its prospective value that: (1) there are *mechanisms* whose function is to specify the *positional value* of each cell – i.e. the position of each cell with respect to one or more points in the system; (2) these mechanisms are distinct from and operate prior to differentiation processes. Therefore, positional information is first "read-out" by a given cell and afterwards converted into molecular differentiation. The positional value of cells in HES is first specified by a set of reference points that form a coordinate system made of one or more bipolar[11] axes. In Wolpert's landmark paper of 1969, the physiological nature of these determinations remained (and in many cases still remains) to be clarified, but the author advanced that metabolic or molecular (morphogen) gradients may be involved because threshold effect, which explains reversal of polarity and field[12] dominance, can rely on both. He also made it clear that the interpretation of positional information depends "on the developmental history of the cell and its genome" (Wolpert 1969: 16). The role of the genome in positional signalling has been notably demonstrated with the discovery of the *bicoid* gene (Wolpert 1989: 5), which codes for the Bcd protein in *Drosophila* embryo. The patterning of this gradient along the anteroposterior axis exemplifies how positional information contained in a "morphogen gradient is transformed into discrete and precise patterns of target gene expression" (Crauk and Dostatni 2005: 1888).

At some point advances like this one led Wolpert (1985: 358) to believe that the *E* factor of Driesch's formula $(X) = f[p, t, E]$ was the genetic program contained in each cell. However, in the "postgenomic" era many theorists of biology (for example Robert 2004) consider this notion just as animistic as Driesch's entelechy. In fact, Wolpert (1989: 8) himself recognized that the original positional information theory "tried to do too much"; it is nevertheless still paradigmatic today as new versions of the French Flag model take into account the "patterning by several interacting, spatially coupled genes subject to intrinsic and extrinsic noise" (Hillenbrand et al. 2016).

---

[11] "Polarity is defined as the direction in which positional information is specified or measured" (Wolpert 1969: 1).

[12] A field is constituted "when cells have their positional information specified with respect to the same set of points" (Wolpert 1969: 7).

3.3 Is HES a Vitalist Concept?

The concept of HES may be "the most distinctive and novel thing" in Driesch's doctrine (Lovejoy 1911: 77) and clearly acted as a positive heuristic in the twentieth century embryology. However, one can still argue that, although Driesch's neologism coincides with his vitalistic turn in 1899, it is not essentially tied with his vitalistic view and the latter did not play a significant role in its development. After all, the essential characteristics of HESs had already been identified in 1892 (Sect. 2.3), when Driesch still embraced the "machine theory of life," like Roux and other embryologists. And according to Driesch (1899: 77, 78), the distinction between describing the *explanandum*, which results from answering, "What is actually happening here?," and finding the *explanans* – the solution to the problem "Why does this happen?" – is clearly made when the scientific method is properly followed, as Galileo and Newton have shown. Following this reasoning, one could then affirm that the concept of HES would still have been created if Driesch had kept his original mechanistic view.

This would nevertheless be a wrong conclusion mainly because Driesch did not only view HES as an *explanandum*, but above all as his crucial *proof for vitalism*: "in the theory of the harmonious-equipotential system", he stresses, entelechy "*must necessarily be applied*" (Driesch's emphasis 1913). As we previously saw (Sect. 3.1), in this peculiar case he blurred the demarcation between the problem and the solution as both were parts of the same argument. The formula with the "E" factor, the diagrams, the abstract regulative power: the more the behavior of HESs seemed far from pure mechanical capacity, the more the entelechy hypothesis looked convincing.

This is why HES was from the beginning a very controversial concept. As "Driesch himself remarks that it is only 'an approximate, as it were, figurative, method of speech,'" notices the American zoologist Herbert S. Jennings, "*no such thing as an equipotential system exists* among organisms" (Jennings' emphasis, 1918: 586). In fact, it was Child (1908: 580) who first made a highly critical analysis of Driesch's concept, which he called an "apparent problem" (*Scheinproblem*). Firstly, he raised that material likeness does not follow from prospective equipotentiality. Driesch's belief that the material basis of the system is a relatively formless *means* for achieving the end of entelechy makes him deny the existence of not yet visible material differences between elementary parts coming from their past histories or external conditions: the problem then falsely appears to be "the self-production of heterogeneity from homogeneity". Secondly, parts that are isolated from the whole are in fact never equivalent, as regional differences often remain from their previous differentiation and various internal and external changes result from their isolation. Finally, size invariance of regulative development or regeneration is only achieved *approximately* and with pieces of a certain size.

Spemann significantly pointed out what may be the paradox of HESs: although they "are perhaps never either harmonious or equipotential exactly in the sense of Driesch" (Spemann 1938: 347), they are still considered as a *real* problem. By abstracting and extrapolating essential properties from empirical events, by giving to these properties graphical and mathematical representations and by opposing them to existing and virtual mechanical realizations, Driesch failed to convince that entelechy was the answer but had nevertheless a lasting impact on how organic development was viewed and scientifically approached.

4   Conclusion

Driesch's vitalistic thesis, whereby HESs cannot be machines and hence are driven by non-spatial and purposeful "entelechies", did not falsify explanatory models nor reveal their flaws. The Roux-Weismann thesis was rather severely questioned by the discovery of the part formation in sea urchin embryo, which was achieved by Driesch under a pure mechanistic framework. Similarly, Driesch abandoned his induction model for his vitalistic thinking because it was not up to the newly discovered secondary regulatory phenomena. Nor did the concept of entelechy directly serve as a positive heuristic; because it was not empirically measurable nor testable, it could not guide insightful experiments towards the discovery of new facts. However, Driesch's notion of HES constituted an important positive heuristic for developmental biology throughout the twentieth century: though Child interpreted it as an unfaithful representation of empirical events, his metabolic gradient theory was nonetheless presented as an answer to the problem of organic individuation as conveyed by Driesch; among other biologists, Spemann openly considered the HES as the ultimate *explanandum* of embryology and saw the field theory as a promising step towards its explanation; Wolpert based his positional information theory on the French Flag problem, which is a formalized illustration of HESs. While it is true that the elaboration of the HES concept started with the axiom of part formation in 1892 and with the prospective approach to the problem of epigenesis in 1894, Driesch designed the reference and most impactful version of this *explanandum* as an essential part of his main proof for vitalism. In 1908, holistic or "harmonious" outcomes of development were: (1) systematized and abstracted from real material restrictions; (2) represented in persuasive diagrams and by the mathematical formula with the "E" factor; (3) defined as dynamically purposeful and irreducible to mechanical means.

I believe that these results provide further insights into the process of discovery at least in developmental biology. They show that, despite having often been labeled as unscientific, substantival vitalism (Wolfe 2011) can have a real scientific value; in this case it was able to influence the way the problem of development was understood by biologists and addressed within leading research programmes. As Child suggested, the specificity of vitalistic theories may consist in recognizing the *essence* of the problem. One reason for this may be that vitalists do not carry the burden of offering a convincing mechanical[13] solution to it. However, this study also highlights that the boundary between the *explanandum* and its *explanans* is often permeable; by aiming attention at the holistic and "purposeful" dimension of organic development, Driesch's concept of HES leaves in the dark the complexity of the chromatic structure and the material and functional manifoldness underlying (inter)cellular activity. But in this historical case at least, vitalistic and mechanistic thinking appear to have scientifically complemented each other, insofar as the former tended to establish what organisms can do and the latter, why or how they do it.

References

Allen, Garland E. 2004. A Pact with the Embryo: Viktor Hamburger, Holistic and Mechanistic Philosophy in the Development of Neuroembryology, 1927-1955. *Journal of the History of Biology* 37: 421–475.

Blackstone, Neil W. 2008. Metabolic Gradients: A New System for Old Questions. *Current Biology* 18 (8): R351–R353.

Bognon, Cécilia, Bohang Chen, and Charles Wolfe. 2018. Metaphysics, Function and the Engineering of Life: The Problem of Vitalism. *Kairos: Journal of Philosophy & Science* 20 (1): 113–140.

Bolduc, Ghyslain. 2021. *Préformation et épigenèse en développement. Naissance de l'embryologie expérimentale*. Montreal and Paris: Presses de l'Université de Montréal and Vrin.

Bolduc, Ghyslain. Forthcoming. Logique et développement. De la morphologie fonctionnelle au vitalisme critique d'Hans Driesch. In *Les Vitalismes. Histoire d'une équivoque ?*, ed. Bertrand Nouailles. Paris: Hermann.

Caianiello, Silvia. 2019. Mechanistic Philosophies of Development: Theodor Boveri and Eric H. Davidson. *Marine Genomics* 44: 32–51.

Carnap, Rudoph. 1966. *Philosophical Foundations of Physics*. New York: Basic Books.

Chen, Bohang. 2018. A Non-metaphysical Evaluation of Vitalism in the Early Twentieth Century. *History and the Philosophy of the Life Sciences* 40 (3): 50.

Child, Charles M. 1908. Driesch's Harmonic Equipotential Systems in Form-Regulation. *Biologisches Centralblatt* 28 (18): 577–588.

———. 1915. *Individuality in Organisms*. Chicago: The University of Chicago Press.

———. 1916. The Basis of Physiological Individuality in Organisms. *Science* 43 (1111): 511–523.

Churchill, Frederick B. 1969. From Machine-Theory to Entelechy: Two Studies in Developmental Teleology. *Journal of the History of Biology* 2 (1): 165–185.

———. 2015. *August Weismann. Development, Heredity, and Evolution*. London: Harvard University Press.

Conklin, Edwin G. 1929. Problems of Development. *The American Naturalist* 63 (684): 5–36.

Crauk, Olivier, and Nathalie Dostatni. 2005. Bicoid Determines Sharp and Precise Target Gene Expression in the *Drosophila* Embryo. *Current Biology* 15: 1888–1898.

---

[13] Here again, "mechanism" is taken in its *causal* sense. As Roux clearly stated, mechanical explanation of development mostly involves the analysis of *complex components* – which are not yet explicable in physico-chemical terms – into more elementary causal processes (Roux 1896; Needham 1936: 20, 21). Metabolic gradients, embryonic fields and patterns related to positional information are therefore complex components in Roux's sense and are subjected to causal analysis.

Driesch, Hans. 1891. *Die Mathematisch-mechanische Betrachtung Morphologischer Probleme der Biologie*. Jena: Gustav Fischer.

———. 1892a. Entwicklungsmechanische Studien. I. Der Werth der beiden ersten Furchungszellen in der Echinodermenentwicklung. Experimentelle Erzeugung von Theil- und Doppelbildungen. *Zeitschrift für Wissenschaftliche Zoologie* 53: 160–178.

———. 1892b. Entwicklungsmechanische Studien. III-VI. *Zeitschrift für Wissenschaftliche Zoologie* 55: 1–62.

———. 1894. *Analytische Theorie der Organischen Entwicklung*. Leipzig: Engelmann.

———. 1896. Die Maschinentheorie des Lebens. *Biologisches Centralblatt* 16 (9): 353–368.

———. 1899. *Die Lokalisation morphogenetischen Vorgänge. Ein Beweis vitalistischen Geschehens*. Leipzig: Engelmann.

———. 1908. *The Science and the Philosophy of the Organism*. 2 vols. London: Adam & Charles Black.

———. 1913. Philosophy of Vitalism. *Nature* 2301 (92): 400.

———. 1914. *The History & Theory of Vitalism*. Trans. C. K. Ogden. London: Macmillan.

Dröscher, Ariane. 2014. Images of Cell Trees, Cell Lines, and Cell Fates: The Legacy of Ernst Haeckel and August Weismann in Stem Cell Research. *History and the Philosophy of the Life Sciences* 36 (2): 157–186.

Duchesneau, François. 2012. *La Physiologie des Lumières: Empirisme, Modèles et Théories*. Paris: Classiques Garnier.

Dupré, John, and Maureen A. O'Malley. 2013. Varieties of Living Things: Life at the Intersection of Lineage and Metabolism. In *Vitalism and the Scientific Image in Post-enlightenment Life Science, 1800-2010*, ed. Sebastian Normandin and Charles Wolfe, 311–343. Dordrecht: Springer.

Fischer, Jean-Louis. 1991. Laurent Chabry and the Beginnings of Experimental Embryology in France. In *A Conceptual History of Modern Embryology*, ed. Scott Gilbert, 31–40. Baltimore: The Johns Hopkins University Press.

Frank, Philipp. 1932/1998. *The Law of Causality and its Limits*. Trans. Marie Neurath & Robert S. Cohen. Dordrecht: Springer.

Hein, Hilde. 1972. The Endurance of the Mechanism: Vitalism Controversy. *Journal of the History of Biology* 5 (1): 159–188.

Hertwig, Oscar. 1896. *The Biological Problem of To-day: Preformation or Epigenesis? The Basis of a Theory of Organic Development*. Trans. Chalmers Mitchell. New York: Macmillan.

Hillenbrand, Patrick, Ulrich Gerland, and Gasper Tkačik. 2016. Beyond the French Flag Model: Exploiting Spatial and Gene Regulatory Interactions for Positional Information. *Plos One* 11 (9): e0163628. https://doi.org/10.1371/journal.pone.0163628.

Horder, Tim J. 2001. The Organizer Concept and Modern Embryology: Anglo-American Perspectives. *International Journal of Developmental Biology* 45: 97–132.

Huxley, Julian, and Gavin de Beer. 1934/1963. *The Elements of Experimental Embryology*. New York: Hafner.

Innes, Shelley. 1987. *Hans Driesch and Vitalism: A Reinterpretation*. M.A. thesis. British Columbia: Simon Fraser University.

Jennings, Herbert S. 1918. Mechanism and Vitalism. *The Philosophical Review* 27 (6): 577–596.

Lakatos, Imre. 1976. Falsification and the Methodology of Scientific Research Programmes. In *Can Theories be Refuted?* ed. Sandra G. Harding, 205–259. Dordrecht: Springer.

Lovejoy, Arthur O. 1911. The Import of Vitalism. *Science* New Series 34(864): 75–80.

Macbridge, Ernest W. 1913. Philosophy of Vitalism [Reply]. *Nature* 2301(92): 400, 401.

Maienschein, Jane. 1994. The Origins of Entwicklungsmechanik. In *A Conceptual History of Modern Embryology*, ed. Scott Gilbert, 43–61. Baltimore: The Johns Hopkins University Press.

———. 2005. Epigenesis and Preformationism. In *Stanford Encyclopedia of Philosophy*. http://plato.stanford.edu/entries/epigenesis/. Accessed 10 June 2020.

Monod, Jacques, and François Jacob. 1961. General Conclusions: Teleonomic Mechanisms in Cellular Metabolism, Growth and Differentiation. *Cold Spring Harbour Symposia on Quantitative Biology* 26: 389–401.

Needham, Joseph. 1936. *Order and Life*. Cambridge: Cambridge University Press.

Normandin, Sebastian, and Charles Wolfe. 2013. Vitalism and the Scientific Image: An Introduction. In *Vitalism and the Scientific Image in Post-enlightenment Life Science, 1800-2010*, ed. Sebastian Normandin and Charles Wolfe, 1–15. Dordrecht: Springer.

Robert, Jason S. 2004. *Embryology, Epigenesis and Evolution*. Cambridge: Cambridge University Press.

Roger, Jacques. 1997. *The Life Sciences in Eighteenth-Century French Thought*. Stanford: Stanford University Press.

Roux, Wilhelm. 1881. *Der Kampf der Theile im Organismus. Ein Beitrag zur Vervollständigung der mechanischen Zweckmässigkeitslehre*. Leipzig: Engelmann.

———. 1883. *Über die Bedeutung der Kerntheilungsfiguren. Eine hypothetische Erörterung*. Leipzig: Engelmann.

———. 1885. Beiträge zur Entwicklungsmechanik des Embryo. *Zeitschrift für Biologie* 9 (3): 411–526.

———. 1888/1895. Ueber die künstliche Hervorbringung "halber" Embryonen durch Zerstörung einer der beiden ersten Furchungszellen, sowie über die Nachentwickelung (Postgeneration) der fehlenden Körperhälfte. In *Gesammelte Abhandlungen über Entwicklungsmechanik der Organismen*. 2, 419-521. Leipzig: Engelmann.

———. 1893/1895. Ueber Mosaikarbeit und neuere Entwickelungshypothesen. In *Gesammelte Abhandlungen über Entwicklungsmechanik der Organismen*. Volume 2, 818-871. Leipzig: Engelmann.

———. 1896. The Problems, Methods, and Scope of Developmental Mechanics. Trans. William M. Wheeler. In *Biological Lectures Delivered at the Biological Laboratory of Wood's Holl in the Summer Session of 1894*, ed. Charles O. Whitman, 149–190. Boston: Ginn.

Sander, Klaus. 1997. Hans Driesch's "Philosophy Really ab ovo", or Why to be a Vitalist. In *Landmarks in Developmental Biology 1883-1924*, ed. Klaus Sander, 35–37. Berlin: Springer-Verlag.

Spemann, Hans. 1927. Croonian Lecture: Organizers in Animal Development. *Proceedings of the Royal Society of London Series B, Containing Papers of a Biological Character* 102 (716): 177–187.

———. 1938. *Embryonic Induction and Development*. New York: Hafner.

Waisse-Priven, Silvia, and Ana Alfonso-Goldfarb. 2009. Mathematics ab ovo: Hans Driesch and "Entwicklungsmechanik". *History and the Philosophy of the Life Sciences* 31 (1): 35–54.

Weber, Marcel. 1999. Hans Drieschs Argumente für den Vitalismus. *Philosophia Naturalis* 36: 265–295.

Weismann, A. 1893. *The Germ-Plasm. A Theory of Heredity*. Trans. W. Newton Parker and Harriet Rönnfeldt. New York: Charles Scribner's.

Wolfe, Charles T. 2011. From Substantival to Functional Vitalism and Beyond: Animas, Organisms and Attitudes. *Eidos* 14: 212–235.

Wolpert, Lewis. 1969. Positional Information and the Spatial Pattern of Cellular Differentiation. *Journal of Theoretical Biology* 25 (1): 1–47.

———. 1974. Positional Information and Positional Signalling in Hydra. *American Zoologist* 14: 647–663.

———. 1985. Gradients, Position and Pattern: A History. In *A History of Embryology*, ed. Tim J. Horder, Jan A. Witkowski, and Christopher C. Wylie, 347–362. Cambridge: Cambridge University Press.

———. 1989. Positional Information Revisited. *Development* Supplement: 3–12.

A Historico-Logical Re-assessment of Hans Driesch's Vitalism

Bohang Chen

Abstract Today vitalism is widely dismissed as a metaphysical heresy. For instance, Brigandt and Love (Reductionism in biology. In: Zalta EN (ed) The stanford encyclopedia of philosophy, 2017) claimed that "the denial of physicalism by vitalism, the doctrine that biological systems are governed by forces that are not physico-chemical, is largely of historical interest" (p. 3). Perhaps the most "infamous" vitalist is the German biologist Hans Driesch. However, Driesch (In Rádl E (ed) Actes du Huitième Congrès International de Philosophie a Prague 2–7 septembre 1934. Comité d'Organisation du Congrès, Prague, pp 10–30, 1936) himself very explicitly stated that his vitalism is "neither 'mysticism'[…]nor 'metaphysics'" (p. 27). So, in order to address the mismatch between the present conception of vitalism and his own, I seek to offer a historico-logical re-assessment of Driesch's vitalism. From the historical point of view, I show that Driesch had provided long ignored theoretical reflections on the nature of entelechy (the central concept in his vitalism), especially those in relation to evolution and physics. From the logical point of view, following logical empiricists (Phillipp Frank and Rudolf Carnap), I indicate that Driesch's vitalism should be rejected due to its lack of vital laws, at least with respect to current biology; it is an unestablished theory rather than a metaphysical heresy. Ironically, some current theoretical biologists have proposed similar theories (or principles and laws) of life, even though they (incoherently) reject Driesch's vitalism. In the end, I briefly conclude that the failure of vitalism actually alludes to the fact that even today we understand very little about the nature of life (I mean, the pure concept/phenomenon of life!) (While I cannot elaborate here, it is of extremely importance not to conflate knowledge about the pure concept/phenom-

I thank Ghyslain Bolduc, Charles Wolfe and Christopher Donohue for their comments on earlier drafts of this chapter.

B. Chen (✉)
Department of Philosophy and Moral Sciences, Ghent University, Ghent, Belgium

School of Philosophy, Zhejiang University, Hangzhou, China

49

C. Donohue, C. T. Wolfe (eds.), *Vitalism and Its Legacy in Twentieth Century Life Sciences and Philosophy*, History, Philosophy and Theory of the Life Sciences 29, https://doi.org/10.1007/978-3-031-12604-8_4

enon of life and knowledge about objects predicable of life (Ben-Naim, manuscript, p. 281). For instance, it is common among philosophers of biology today to cite elementary knowledge in a particular biological discipline as offering a better understanding of life. Yet their promise fails to be delivered. At best, they are merely relying on knowledge about objects predicable of life (in most cases, merely knowledge about complex organizations of matter: about heredity, reproduction, development, metabolism, etc); but such knowledge has not been shown of any relevance to the pure concept/phenomenon of life).

1   Driesch's Vitalism: Introduction

The general attitude towards vitalism today is dismissive. While there is no need for the majority of working biologists to care about vitalism, it seems quite expedient for philosophers of biology to dismiss rival stances as vitalistic heresies. This phenomenon was summarized by Susan Oyama: "one need not be a scholar of vitalism to know that it can be a potent instrument of abuse, albeit one that often contaminates by association and innuendo rather than outright accusation" (2010, p. 401). Oyama offered a good example. As an anti-reductionist in philosophy of biology, Oyama has been classified as a vitalist by her reductionist opponents, especially for her enthusiasm for concepts like complexity and organization. But Oyama also shot back. For her, some reductionists are genuinely hard-core vitalists, because they reply on mysterious concepts such as information (Oyama 2010).

Vitalism is rejected by biologists and philosophers today because they treat physicalism (or materialism) as the default stance of modern science. In the entry "Reductionism in biology" of the *Stanford Encyclopedia of Philosophy*, Brigandt and Love gave an exhaustive summary of metaphysical or ontological issues relevant to the rejection of vitalism:

> Ontological reduction is the idea that each particular biological system (e.g., an organism) is constituted by nothing but molecules and their interactions. In metaphysics this idea is often called physicalism (or materialism), which assumes in a biological context that (a) biological properties supervene on physical properties (i.e., no difference in a biological property without a difference in some underlying physical property), and (b) each particular biological process (or token) is metaphysically identical to some particular physico-chemical process... Ontological reduction in this weaker sense is a default stance nowadays among philosophers and biologists...The denial of physicalism by vitalism, the doctrine that biological systems are governed by forces that are not physico-chemical, is largely of historical interest. (2017, p. 3)

This summary was written in 2017. In my view, it provides one of the most forceful illustrations of how vitalism is rejected today by presupposing metaphysical/ontological positions such as materialism or physicalism (also see Haraway 1976; Mayr 1982).

Despite the dismissive attitude towards vitalism, there has recently developed a new trend of reconsidering vitalism in its historical context, and the leading scholar is Charles Wolfe. Wolfe started with the French Montpellier school of medicine in the eighteenth century. Deepening the points made by Peter H. Reill (2005), Wolfe emphasized that at least some forms of vitalism in the eighteenth century went "without metaphysics" (Wolfe and Terada 2008; Wolfe 2008, 2017a). In addition, Wolfe identified what we might call the doctrine of Newtonian vitalism, inspired by Newton's conceptual innovations in classical mechanics. Importantly, Newtonian vitalism, according to Wolfe (2014), was also without metaphysical baggage. Moreover, Wolfe in another article cast doubts on the traditional dichotomy between vitalism and materialism. He identified a few vital materialists in the eighteenth century who attributed inherent vital forces to matter (Wolfe 2013). In the end, Wolfe also aimed to show that themes of vitalism were deeply embedded in the eighteenth-century Enlightenment context; vitalism was associated with issues like the continuation of the Scientific Revolution and anti-mathematicism (Wolfe 2011a, 2017b; Normandin and Wolfe 2013; for positive evaluations of vitalism, also see Lenoir 1982; Richards 2002; Zammito 2018; Steigerwald 2019).

Thanks to Wolfe's and others' efforts, now it has been confirmed that the term "vitalism" was used in many different meanings. So, it might not be incorrect to claim that there are a variety of vitalist doctrines in the history of science. Given all of these, a categorization of these different meanings and doctrines should be highly desirable. Based on his extensive works on the history of biology, Wolfe (2011b) proposed that there are three kinds of vitalism, substantival, functional and attitudinal. For the purpose of this article, only the first two will be discussed. According to Wolfe, substantival vitalism presupposes the existence of "a (substantive) vital force" (p. 212); functional vitalism indicates "an attempt to 'model' or 'describe' organic life without reducing it to fully mechanical models or processes" (p. 213). Regarding this distinction itself, my view is that it is valid, with respect to biologists' thoughts and attitudes in the history of science;[1] but here for our interest, it seems that Wolfe's attempt to distinguish between different meanings attributed to the term "vitalism" has also an implicit purpose of legislating vitalism. For Wolfe, the reputation of vitalism might be saved, if it can be shown that not every doctrine of vitalism in the history of biology was a metaphysical heresy.

The purpose of legislating the term "vitalism" becomes clearer when it comes to judging Driesch's vitalism. Since Driesch appealed apparently to entelechy as the living principle, Wolfe took Driesch's doctrine of entelechy to be a representative of (the bad) substantival vitalism, in contrast to (the good) functional vitalism. Yet, if Wolfe tried to save vitalism by disassociating this term from Driesch's doctrine of

---

[1] But to emphasize, this distinction might not be valid, with respect to biological knowledge itself. In that case, biologists (arguably most in the history of biology) who made the distinction would themselves be hopelessly confused. Such confusions are already detected in Wolfe (2011b). On the one hand, as Wolfe admitted, substantival vitalism was also "scientifically studied" (p. 212). Then, does this, even in the slightest sense, indicate that substantival vitalism can be used as a heuristic tool and therefore become functionalized? On the other hand, when functional vitalism claims to capture "the uniqueness of living bodies" (p. 218), does it succeed in doing that? Maybe in order to do that, it has to advance quasi-substantival claims as well?

entelechy, a few other scholars wanted to do more justice to Driesch and turn to his empirical achievements in biology (Freyhofer 1979; Innes 1987). For them, Driesch's achievements are underestimated. Indeed, Driesch's numerous experiments addressed relational and structural aspects of embryological processes (Driesch 1929; for an updated summary, see Bolduc's Chapter "On the Heuristic Value of Hans Driesch's Vitalism", in this volume), and in this respect there seems little difference between Driesch and Wolfe's functional vitalists. Further, as these scholars requested, Driesch's empirical achievements should be separate from his vitalism or his doctrine of entelechy. The historian Reinhard Mocek, for instance, even suggested that Driesch was almost an organicist or systems theorist, despite his unfortunate vitalist heresy (Mocek 1998; Nyhart 2000). In sum, even though the term "vitalism" can receive different interpretations compatible with modern biology, once it is presented as Driesch's doctrine of entelechy, it will be quickly dismissed as a metaphysical heresy. By terming Driesch's vitalism "substantival vitalism" or separating Driesch's empirical achievements from his doctrine of entelechy, Wolfe and other scholars have appeared to re-affirm the mainstream view. For all of them, it seems legitimate to refute Driesch's vitalism as a metaphysical heresy.

However, I would like to recall that this conception of Driesch's vitalism, unfortunately, goes sharply against his own evaluation of the doctrine of entelechy. In his lecture given in 1934 at the Eighth International Congress of Philosophy in Prague, Driesch emphasized that, in response to common misunderstandings, the choice of entelechy was "a matter of pure logic" and entelechy was "neither 'mysticism'[…]nor 'metaphysics'" (Driesch 1936, p. 27). Seen from this response, it seems that scholars often revoke Driesch's vitalism for their own purposes, despite Driesch's explicit denial. While it is of course inadequate to trust blindly everything he said, it seems more desirable to judge Driesch's doctrine of entelechy after checking its details. This clarifies the intent of this chapter, namely, a historico-logical re-assessment of Driesch's vitalism. In what follows, Sects. 2 and 3 elaborate Driesch's doctrine of entelechy from the historical perspective, with a focus on his theoretical speculations in relation to evolution and physics. Sects. 4 and 5 address the nature of entelechy from the logical point of view. Section 6 concludes by briefly comparing entelechy and life.

2   History I: Driesch on Entelechy and Evolution

It is well known that Driesch intended his doctrine of entelechy to explain some peculiar embryological phenomena about sea urchins. However, it is little known today that Driesch, in his doctrine of entelechy, also dealt with evolutionary phenomena. In Driesch's words, vitalism must concern "the relation of transformism in general to our concept of entelechy" (1908a, p. 293). For this, Driesch introduced a new phrase to study the relation between vitalism and evolution, that is, "systematics of entelechies" (p. 293). It is not difficult to appreciate Driesch's theoretical concern. As a biologist living more than half a century after Darwin popularized the thesis of descent, Driesch took facts of evolution for granted. Further, since Driesch also proposed the doctrine of entelechy, naturally he would have to face the following question: "we know that entelechy...uses material means in each individual morphogenesis...what then undergoes change in phylogeny, the means or entelechy?" (p. 295). Thus, according to Driesch, in a concrete case of organic transformation, either the same entelechy used a different mean to "construct" a new organism, or a different entelechy emerged in evolution to "give rise to" the new organism.

On this question Driesch opted for the latter, in favor of "entelechy, rather than "the means". Let us quote him in full:

> We know that entelechy, though not material in itself, uses material means in each individual morphogenesis, handed down by the material continuity in heritance. When then undergoes change in phylogeny, the means or entelechy? And what would be the logical aspect of systematics in either case?
>
> Of course there would be a law in systematics in any case; and therefore systematics in any case would be rational in principle. But if the transformistic factor were connected with the means of morphogenesis, one could hardly say that specific form as such was a primary essence. Entelechy would be that essence, but entelechy its generality and always remaining the same in its most intimate character, as the specific diversities would only be due to a something, which is not form, but simply means to form. But the harmony revealed to us in every typical morphogenesis, be it normal or be it regulatory, seems to forbid us to connect transformism with the means of morphogenesis. And therefore we shall close this discussion about the most problematic phenomena of biology with the declaration, that we regard it as more congruent to the general aspect of life to correlate the unknown principle concerned in descent with entelechy itself, and not with its means. Systematics of organisms therefore would be in fact systematics of entelechies, and therefore organic forms would be *formae essentiales*, entelechy being the very essence of form in its specificity. (p. 295)

For Driesch, in evolutionary history it is less probably the case that a single entelechy "gave rise to" the whole variety of organisms through different means; rather, it is more probable that in each concrete case of evolution, a new entelechy emerged to "construct" a new organism.[2]

However, as Driesch alluded, with facts of evolution he had a different concern with systematics. In particular, he asked for "a law in systematics" which would make a truly "rational" science. Then what did Driesch mean here? For this let us look at his discussion of vitalism and evolution in detail. To start, there is little doubt that Driesch accepted Darwin's "theory of descent" (p. 250) or "the idea of transformism" (p. 251), namely, "the hypothetical statement that the organisms are really allied by blood among each other, in spite of their diversities" (p. 251). Like us today, he agreed with Darwin that "there certainly is a great amount of probability" in "the idea of transformism" (p. 251). In particular, he classified "two different

---

[2] Yet later in 1927, Driesch frankly acknowledged the other possibility that there might be a "super-entelechy which possesses a fixed essence and does nothing but copy its essence once more into matter, and realize phylogeny in this way" (1927, p. 6). To emphasize, no matter which possibility Driesch choose, he strictly followed the logical rules. Indeed, regarding these two possibilities, Driesch concluded that with "empirical phylogeny as it is...we possess no means whatever to decide" (p. 7).

groups of facts" as supportive evidence, "the geographical distribution of animals and plants and to paleontology" (p. 251) and "similarities and diversities" observed "in the system of animals and plants" (p. 253).

While Driesch approved of the thesis of descent, he could not consent to the claim that the descent thesis explains "the diversities of the organism" (p. 255). On this point, Driesch raised strong criticisms of Darwinism and Lamarckism in his time. Driesch took Darwinism held by his contemporaries to be "dogmatism in one of purest forms" and very far removed from "the opinion of Charles Darwin" (p. 260). And he maintained that he was "speaking against Darwinism of the most dogmatic form only, not against Darwin himself" (p. 269). Dogmatic Darwinism, according to Driesch, consisted of two theoretical parts, natural selection and "fluc-tuating variation" (p. 264). First on natural selection. For Driesch, it was self-evident that natural selection "always acts negatively only, never positively" (p. 262). So, Driesch continued, it could be correct to state that natural selection explained "why certain types of organic specifications, imaginable a priori, do not actually exist" (p. 262). But Driesch stressed that it was misleading to assert that natural selection explained "the existence of the specifications of animal and vegetable forms that are actually found" (p. 262). For this, Driesch turned to "fluctuating variation the alleged cause of organic diversity" (p. 264). But he remained deeply unsatisfied. In response to contemporary Darwinians in the biometric tradition, Driesch agreed with them that the concept of fluctuating variation might be useful to explain "new quantitative differences" (p. 266); yet he insisted that it was illegitimate to merely use fluctuating variation to explain "organic diversities" (p. 269, for reasons see below).

Driesch also criticized the doctrine of Lamarckism or Neo-Lamarckism. Again, Driesch emphasized that he was not against the opinion of Lamarck himself; instead he only attacked Lamarckism "in its dogmatic modern form" (p. 271). For Driesch, Neo-Lamarckism had two basic parts, the idea of the "inheritance of acquired char-acters" (p. 275), and "the hypothesis of storing and handing down contingent varia-tions" (p. 282). On the first Driesch said little as he viewed it as basically an empirical question. He concluded that "not one case is known which really proves the inheritance of acquired characters", yet it is not advisable to "deny the possibil-ity...in an absolute and dogmatic manner" (p. 278). Driesch focused on the second part, which was open to similar objections made against the Darwinian concept of fluctuating variation. For Driesch, "the whole anti-Darwinistic criticism...may also be applied to Lamarckism with only a few changes of words" (p. 288).

Indeed, Driesch took issues with the Darwinian concept of "fluctuating varia-tion" and the Lamarckian assumption of "storing and handing down contingent variations" *for exactly the same reason*. According to Driesch, Darwinism and Lamarckism in his time were similarly problematic in that they "shake hands on the common ground of the contingency of organic forms" (p. 288): Darwinism tended to explain the diversity of organic forms through fluctuating variation and natural selection, Lamarckism tried to do it by referring to contingent needs of organisms in their living environment. But for Driesch, the appeal to contingency was unac-ceptable in science, and "other principles (were) wanted" (p. 281). Here Driesch requested "law(s) of systematics" (p. 295), even though he knew well that these laws might only be available "at some future date" (p. 295).

According to Driesch, laws of systematics were concerned with "the totality of possible forms" (p. 296). Ideally, with such laws all possible organic forms and species could be placed into a unified system, and knowledge about then could accordingly be deduced from laws of systematics. Then, what would such laws of systematics be like? On this Driesch wrote, with an illuminating reference to physics and chemistry (and Kant!):

> In physics and chemistry no perfect rational systems have been established hitherto, but there are many systems approaching the ideal type in different departments of these sciences. The chemical type of the monohydric saturated alcohols, for instance, is given by the formula $CnH2n+1OH$, and in this formula we not only have an expression of the law of composition which all possible alcohols are to follow, but, since we know empirically the law of quantitative relation between n and various physical properties, we also possess in our formula a general statement with respect to the totality of the properties of any primary alcohol that may be discovered or prepared in the future. But chemistry has still higher aims with regard to its systematics : all of you know that the so-called "periodic law of the elements" was the first step towards a principle that may some day give account of the relation of all the physical and chemical properties of any so-called element with its most important constant, the atomic weight, and it seems to be reserved for the present time to form a real fundamental system of the "elements" on the basis of the periodic law by the aid of the theory of electrons. Such a fundamental system of the elements would teach us that there can only be so many elements and no more, and only of such a kind...we are dealing here with some of the most remarkable properties of the so-called synthetic judgments a priori in the sense of Kant...(p. 244)[3]

Further, Driesch compared phylogeny (as the study of the evolutionary history) and rational systematics. For Driesch, in contrast to the pursuance of rational laws, the study of the evolutionary history only attended to "actual diversities" (p. 264) of organic forms, both existent and extinct, found in the history of nature. But importantly, according to Driesch, the study of the history did not contradict rational systematics. Driesch wrote:

> Is there no contradiction between historical development and a true and rational system...by no means. A totality of diversities is regarded from quite different points of view if taken as the material of a system, and if considered as realized in time. We have said that chemistry has come very near to proper rational systematics, at least in some of its special fields; but the compounds it deals with at the same time maybe said to have originated historically also, though not, of course, by a process of propagation. It is evident at once that the geological conditions of very early times prohibited the existence of certain chemical compounds, both organic and inorganic, which are known at present. Non the less these compounds occupy their proper place in the system. And there may be many substances theoretically known to chemical systematics which have never yet been produced, on

---

[3] It might be helpful to show what Driesch (and Kant) meant here by "a priori" here. Important theories in science, like Newton's thee laws in classical mechanics and the periodic table of elements in physico-chemistry, are of course a posteriori, as empirical achievements in science; yet they are also a priori, in relation to more concrete works in classical mechanics and physico-chemistry. In sum, these are both a posteriori and a priori, but only at different levels. I thank my friend Ghyslain Bolduc for raising this issue.

account of the impossibility of arranging for their proper conditions of appearance, and nevertheless they must be said to "exist". "Existence", as understood in systematics, is independent of special space and of special time, as is the existence of laws of nature: we may speak of a Platonic kind of existence here. Of course it does not contradict this sort of ideal existence if reality is added to it. (pp. 257–8)

Driesch's reference to the laws of systematics in chemistry (e.g., the periodic table of elements) is illuminating. For Driesch, similar to those in chemistry, laws of biological systematics in relation to evolution, if they were available, would only determine "possible forms", from which the evolutionary history would pick up "actual forms". In other words, the evolutionary history should proceed partly in accordance with a pre-determined system of possible species governed by laws of systematics.

To connect evolution with his vitalism, recall that Driesch came up with the concept of "systematics of entelechies" (p. 293). Indeed, it is clear that, in Driesch's doctrine of entelechy, the system of entelechies must be correlated with the system of all possible organic forms. To put it more explicitly, it is not incorrect to say that evolution is partly governed by the system of entelechies. Hence in Driesch's doctrine of entelechy the relation between vitalism and evolution becomes clear. He summarized this as follows:

Thus the problem of systematics remains, no matter whether the thesis of descent be right or wrong. There always remains the question about the totality of diversities in life: whether it may be understood by a general principle, and of what kind that principle would be. As, in fact, it is most probably by history, by descent, that organic systematics is brought about, it of course most probably will happen some day that the analysis of the causal factors concerned in the history will serve to discover the principle of systematics also. (p. 258)

For Driesch, if Darwin's doctrine or even Darwinism was understood merely as the thesis of descent, then it was the most plausible answer to the question of actuality: how "organic systematics is brought about". However, the question of possibility, that is, the question of "the totality of diversities", remained entirely untouched, and here laws yielded by the systematics of entelechies must be requested.

Finally, to emphasize, it should be kept in mind that the systematics of entelechies or the systematics of all possible species or organic forms, even till today, have not been given in any rational form, i.e., as a system of rational laws. Indeed, Driesch was not the first, nor the last to envisage such rational systematics for evolution. Rather, Driesch's request of laws of systematics represents the central concern of theorists of evolution in more than two hundred years, both before and after Darwin. Before Darwin we have the great name of Kant and other German philosophers of nature such as Schelling (for more, see Chen 2019). After Darwin, as we will see later (Sect. 5), the theoretical biologist Brian Goodwin has also come up with similar research projects. Therefore, it should be clear now that the relation between vitalism and evolution envisaged by Driesch is not the product of some boundless speculations offered by a discredited biologist.

3 History II: Driesch on Entelechy and Physics

As in Sect. 2, I begin with Driesch's *The Science and Philosophy of the Organism* (1908a, b). We have seen that in the first volume (1908a) Driesch with much rigor discussed the relation between vitalism and evolution in detail. In the second volume, in like manner, he went on to dealing with issues in relation to vitalism and physics. In his own words, he was going to show "how our concept of entelechy as an elemental natural factor is related to those concepts of general ontology which play any part in the science of inorganic nature" (1908b, p. 153). Here Driesch demonstrated his non-trivial learning in physics, admittedly rare among biologists at any time.

Driesch first touched on the principle of the "conservation of energy" (p. 158), which is now termed the first law of thermodynamics. In relation to his doctrine of entelechy, Driesch raised the following question, "how stands entelechy to the concept of energy itself?" (p. 164). According to Driesch, some contemporary physicists such as the distinguished Wilhelm Ostwald conjectured that "in cases of morphogenesis...some unknown potential forms of energy may be at work" (p. 167) and entelechy was just vital energy peculiar to life. But Driesch endorsed an unequivocal rejection of this supposition. Two reasons were advanced by Driesch to support his rejection. First, "at least in all [functional and developmental] cases where the economic equation [the principle of the conservation of energy] is fulfilled there would seem to be no place for a 'new' energy" (p. 168). Second, "all energies, actually known to exist or invented to complete the general energetical scheme, are quantities" (p. 168), but "entelechy lacks all the characteristics of quantity" and it is "order of relation and absolutely nothing else" (p. 168). Clearly, Driesch, in his own words, "decline(d) any kind of 'energetical' vitalism" (p. 170).

Driesch next dealt with the principle of the "augmentation of entropy" (p. 158) or the principle of "dissipation" (p. 174), which is now termed the second law of thermodynamics. Unlike many physicists and philosophers who have always been producing mysterious interpretations of the second law, Driesch took it to be "a mere fact that is encountered in almost all fields of physics" (p. 174). Since this principle was merely empirical, according to Driesch, "it of course offers no special ontological problem with regard to entelechy" (p. 176). Then regarding possible functions of entelechy in relation to physico-chemical factors, Driesch came up with some rather imaginative and entertaining suggestions. According to Driesch, entelechy might be able to "suspend possible becoming" (p. 179) in physico-chemical systems. More concretely speaking, suppose a chemical reaction, with all its necessary chemical compounds ready, was about to take place; yet at this time entelechy was able to suspend the chemical reaction. On this Driesch wrote:

> ...Entelechy is able, so far as we know from the facts concerned in restitution and adaptation, to suspend for as long a period as it wants any one of all the reactions which are possible with such compounds as are present, and which would happen without entelechy. And entelechy may regulate this suspending of reactions now in one direction and now in the other, suspending and permitting possible becoming whenever required for its purposes. (p. 180)

Driesch took "this faculty of a temporary suspension of inorganic becoming" (p. 180) to be "the most essential ontological characteristic of entelechy" (p. 180). In other words, entelechy was able to interfere the principle of the augmentation of entropy in physico-chemical systems, as "the non-physico-chemical agent" (p. 180). After this, Driesch moved to speculating about the function of entelechy in organic systems. Driesch started with his famous concept of the "harmonious-equipotential system", which he considered to be a "system of equally distributed potentialities" (p. 192). Then in the embryological process, according to Driesch, "entelechy transforms a 'homogeneous' distribution of given different elements and given possible reactions into a 'heterogeneous' distribution of effect" (p. 193). In other words, "entelechy...is capable of augmenting its diversity of distribution in a regulatory manner" (p. 192), as "an intensive manifoldness, embracing a real system of pre-existing diversities in itself" (p. 197).

According to the principle of the augmentation of entropy, the increase of entropy signifies a decrease of diversity. Given this principle, did Driesch ever consider the role of entelechy to be contradicting the entropy principle? Not at all. For Driesch, the entropy principle only held in physico-chemical systems, empirically. Then regarding a biological system, Driesch suggested, even though the entropy principle still tends to eliminate its inner diversity, entelechy, on the contrary, elevates the diversity of the system as a biological agent. As a result, being biological, "organic systems may acquire a higher degree of diversity of distribution" (p. 197), due to the presence of entelechy. In a positive formulation, Driesch could say, entelechy and the entropy principle operate simultaneously, although in opposite directions, to give rise to relevant effects in relation to the biological system. So, for Driesch, "there is no opposition between inorganic and vital phenomena" (p. 196). To summarize, in Driesch's doctrine of entelechy, it is perfectly fine to conclude that a biological system, thanks to the presence of entelechy, might be able to overcome the principle of the augmentation of entropy only ascertained in physico-chemical systems.

More generally speaking, for Driesch, in biology entelechy gave rise to diversities not found in physico-chemistry, just like gravity did with respect to Cartesian mechanics (p. 193) and human mind did in creating artifacts from chaos in nature (p. 194). Entelechy is just a new concept, which introduces a new type of causal factors. Since the presence of entelechy was assumed by Driesch to be unique in biology, it could then justify the autonomy of biology and revive the organic-inorganic distinction "forgotten by physics and chemistry" (p. 196). With entelechy, Driesch reminded us that:

> There also was a great contrast between vital phenomena and the complete "science of inorganic or spatial becoming" that is to be written in the future. Entelechy, as endowed with the faculty of enlarging the amount of diversity in the distribution of given elements, was in opposition to that future science. (p. 225)

By the way, given Driesch's clear formulation, no one will fail to recognize the similarity between entelechy and the famous Maxwell's Demon. Indeed, Driesch commented twice on Maxwell's Demon. First, regarding mere physico-chemical systems, "of course, the empirical law of the dissipation of energy would be contradicted by Maxwell' fiction" (pp. 199-200). Second, In the case of entelechy, "the work of the 'demons' of Maxwell is here regarded as actually accomplished" (p. 225).

Driesch's discussion on entelechy and thermodynamics, or more generally, "vital and physical principles" (p. 198), alludes to many interesting, but long-ignored historical details about vitalism. Although these details cannot be presented here due to the lack of space, there is an important lesson to learn from the logic of science. As I have shown, according to Driesch, even though the effects of entelechy run against the second law of thermodynamics, it does not necessarily follow that the existence of entelechy should be forbidden in science. Further, in no way it follows that vitalism should be banned as a metaphysical heresy. Then, why is vitalism still not a scientifically valid theory? To answer this question, we must now execute a logical analysis of Driesch's doctrine of entelechy.

4   Logic I: The Logical (in Contrast to the Metaphysical) Refutation

As I have shown in the Introduction, vitalism is now almost universally refuted as a metaphysical heresy by presupposing materialism and physicalism. Sometimes it is even used as "a potent instrument of abuse, albeit one that often contaminates by association and innuendo rather than outright accusation" (Oyama 2010, p. 401). In this section I first make the structure of the metaphysical refutation clear. To do this, let us first quickly review a few negative remarks on vitalism, from contemporary biologists and philosophers:

> To those of you who may be vitalists I would make this prophecy: what everyone believed yesterday, and you believe today, only cranks will believe tomorrow. (Crick 1966, p. 99)
>
> Both scientists and philosophers take ontological reductionism for granted. Vitalism is dead. organisms are nothing but atoms, and that is that. (Hull 1981, p. 282)
>
> [Vitalism] virtually leaves the realm of science by falling back on an unknown and presumably unknowable factor... (Mayr 1982, p. 52)
>
> No "vital forces" exist, and all living phenomena consist only of chemical and physical processes. Such an ontologic position (i.e., a stance as to what exists in the universe) is called materialism, and it provides the basis for contemporary natural science. (Gilbert and Sarkar 2000, p.1)
>
> [Vitalism] faded as the mechanistic side of biology advanced. (Godfrey-Smith 2014, p. 10)
>
> The denial of physicalism by vitalism, the view that biological systems are governed by forces that are not physico-chemical...is largely of historical interest... (Brigandt and Love 2017, p. 3).

These passages contain both "innuendos" and "outright accusations". In sum, especially in Gilbert and Sarkar (2000), vitalism is rejected with forceful statements asserting that vitalism violates an overarching worldview legitimized by modern science. Hull termed this worldview "ontological reduction", Gilbert and Sarkar termed it "materialism", and most recently Brigandt and Love went for "physicalism". In spite of this terminological diversity, it is clear that vitalism today is refuted by presupposing materialism and physicalism, and it is a metaphysical refutation.

A few more remarks on the metaphysical refutation. In the first place, my impression is that scientists are more used to the term "materialism", and they are very inclined to accept Hull's ontological statement, "organisms are nothing but atoms, and that is that" (I have encountered no exception!). In contrast, philosophers prefer "physicalism" with the intent of being more precise. From "materialism" to "physicalism", on the one hand, philosophers take apparently non-materialistic concepts (gravity, electro-magnetism, field, energy, etc.) in modern physics into account; on the other hand, they can claim that physicalism relies on ongoing physical sciences, rather than knowledge in physics in a particular time.

In the second place, I anticipate that some will raise objections to my characterization of materialism and physicalism, since today there are diverse "philosophical" doctrines of materialism and physicalism (Stoljar 2021). I cannot clarify all relevant issues here, but I do think that such a diversity is more or less a result of terminological confusions. Further, the essential point I want to make is that materialism and physicalism, whatever they mean, are used in this context to issue dismissals of vitalism. In sum, the existence of life forces, entelechies and other similar vital entities are judged as violating the metaphysical tendency of current scientists and philosophers, and they all receive a metaphysical refutation.

However, there is a different way of rejecting vitalism, and I call it a logical refutation. This refutation was endorsed in the early twentieth century by a few biologists and philosophers (for details see Chen 2018). In this logical refutation, first of all, vitalism is not treated as a metaphysical heresy; rather, it is received as a promising (or at least not-too-bad) scientific theory. According to the Russian biologist Alexander Gurwitsch, "practical vitalism claims the right to be restricted in formulating hypotheses only by postulates of logic and of the general theory of knowledge, and by nothing else" (1915, p. 765). It is also important to notice that logical empiricists essentially agreed with Gurwitsch. For instance, Philipp Frank wrote in his early years:

> To be sure, Driesch shows that we can assume for the living processes a specific state variable, not that we must. For it is not possible to foresee every trick that one might invent in the fiction of hidden combinations of inorganic state variables. In favor of vitalism I should like to remark that, just as I cannot force someone who regards heat as a specific state variable to consider it as a motion of particles, so I cannot force the adherents of entelechy to replace it by fictitious state variables. (Frank 1941/1908, pp. 26–7)

Frank did not refute vitalism by presupposing any metaphysical materialism or physicalism, as present biologists and philosophers do.

According to the logical refutation, second, vitalism is defective for a different reason. On this Carnap wrote:

> Driesch did not give laws. He did not specify how entelechy of an oak tree differs from entelechy of a goat or giraffe. He did classify his entelechies. He merely classified organisms and said that each organism had its own entelechy. He did not formulate laws that state under what conditions an entelechy is strengthened or weakened… the notion of an entel-

echy does not give us new laws, it does not explain more than the general laws already available. It does not help us in the least in making new predictions. (Carnap 1966, pp. 15–16)

According to Carnap, the central defect of vitalism was that Driesch failed to give new laws after he advanced the concept of entelechy.[4] Without such laws, even today, we simply have no idea about what the nature of entelechy is.

In this section I have contrasted the metaphysical refutation of vitalism with its logical refutation. I also want to endorse the view that the logical refutation is to be preferred. While there is no space to elaborate here (but see Chen 2019), the main defect of the metaphysical refutation, in my opinion, is that it relies on a rather problematic conception of matter, in which matter is treated as a Kantian thing in itself, rather than ever-changing physico-chemical principles. As a result, it has not, even in the slightest, captured the historical development of science. Indeed, in the history of science metaphysical versions of materialism and physicalism were often obstacles to scientific progress.

5   Logic II: Theoretical Biologists and Their Envisaged Vital Laws

In this section, I show one, arguably the most important, implication of the logical refutation of vitalism. In current biology as well as the history of biology, there have been theoretical biologists who take the metaphysical refutation of vitalism for granted and propose speculative research projects to search for principles of life. For instance, Ludwig von Bertalanffy, once influential for his general systems theory, took biological organisms to be far-from-equilibrium open systems; based on his knowledge in thermodynamics, Bertalanffy was eager to have "a thermodynamic criterion that would define the steady state in open systems in a similar way as maximum entropy defines equilibrium in closed systems" (1968, p. 151). Bertalanffy himself was not so hostile to vitalism (1933, p. 30), but he still complained that vitalism "refers…to a metaphysical or psychical factor and consequently renounces the possibility of a natural scientific explanation" (1933, p. 46). More recently, similar to Bertalanffy, the well-known theoretical biologist Stuart Kauffman envisaged "the fourth law of thermodynamics" (2000, p. xi) for life. Kauffman did not think highly of vitalism, either. Obsessed with holism and emergentism, he further claimed that "no vital force or extra substance is present in the emergent, self-reproducing whole" (1995, p. 24).

Yet is Driesch's vitalism really different from Bertalanffy's and Kauffman's research projects to find either a new thermodynamic criterion for open systems or even the fourth law of thermodynamics? My view is that, there is little difference,

---

[4]This refutation can also be found in other contemporary authors, such as Morris Cohen (1931, pp. 253–4). I thank Christopher Donohue for this reference.

from the logical point of view. Recall Driesch's discussion on how entelechy might function in the physico-chemical world; in its very least, the search for either a new thermodynamic criterion or the fourth law of thermodynamics can be read as an attempt to give laws to capture the effects of entelechy. In other words, these are just vital laws envisaged by Driesch. Therefore, despite a shared but actually unfounded metaphysical superiority, their research projects, from the logical point of view, can be received as disguised forms of vitalism. Meanwhile, needless to say, like Driesch's search for vital laws, so far Bertalanffy's and Kauffman's ambitious proposals have not found anything even close to a new thermodynamic criterion or the fourth law of thermodynamics.[5]

In addition to Bertalanffy and Kauffman, another theoretical biologist with interest in a general or universal theory of life is Brian Goodwin. Goodwin stood out among theoretical biologists as a rare admirer of Driesch (Webster and Goodwin 1981, pp. 5–6). In his research project, Goodwin not only proposed "biological field theory" (1982), he also made essential use of the analogy between the system of possible biological forms and the periodic table of elements: "(we need) a theory of organismic form analogous to the theory of atomic structure" that is, "the universal laws or constraints which define what is possible" (1982, p. 45). Like Driesch, Goodwin wrote:

> The rational taxonomy which could emerge from a logical classification of these forms would be quite independent of the actual historical sequence of appearance of species, genera and phyla, just as the periodic table of the elements is independent of their historical appearance, and is compatible with a great variety of possible sequences. (1982, p. 51).

In Goodwin's rational morphology, every possible biological organization is a dynamic whole governed by its own unique (morphogenetic) field equations. Then, all these unique field equations are further unified into a common scheme. For all these, Goodwin's project cannot fail to remind us of Driesch's "systematics of entelechies". In addition, it is clear that Goodwin's proposed "biological field theory" correspond to Driesch's envisaged vital laws.

Finally, to emphasize, vital laws envisaged by Driesch, like Goodwin's biological field theory and Bertalanffy's and Kauffman's new thermodynamic principles, have never been obtained till today. Indeed, Driesch's claim that his concept of entelechy must be invoked to explain embryological phenomena is untenable. On this point Carnap's criticism of vitalism is temporarily adequate. However, Driesch would not be wrong, if he merely proposed to look for vital laws. As a matter of fact, if vitalists could offer such vital laws, Carnap's criticism of vitalism would fail (and he would acknowledge this, for sure) and vitalism would consequently be as valid as any respectable theory in science. While it might not be very promising to attain such vital laws even in future biology, but the possibility of such vital laws can never

---

[5]To emphasize, this is not to dismiss other important but more concrete contributions made by theoretical biologists. Yet, it is also important to understand that those more concrete contributions so far have not given any genuine theory of life. Driesch also had genuine contributions in theoretical biology; however, those did not help his doctrine of entelechy, which in its essence tends to be a theory of life (see below).

be denied decisively. The validity of vitalism relies on vital laws, and this is a logical point. As we have seen, a lack of clarity on this point misleads some theoretical biologists (e.g., Bertalanffy and Kauffman) to envisage theories and principles with fancy names, which, from the logical point of view, turn out to be scarcely different from Driesch's vital laws.

6   Conclusion: Entelechy and Life

The logical refutation shows that at least in current biology vitalism is invalid. In other words, the statement that "the organism is governed by its entelechy" should be viewed with suspicion. Due to the lack of vital laws, this statement cannot be treated as a serious piece of biological knowledge. But now consider a different statement, "the organism is alive". Indeed, even though this statement is treated as trivially true, its logical status is similar to that of the previous statement about entelechy. Unfortunately, due to a lack of principles or laws of life (in Bertalanffy's and Kauffman's words), the statement that "the organism is alive", although presupposed, is of little real use in biology. It stands very isolated from all other well established biological statements. As a matter of fact, it seems possible to treat the two previous statements as synonymous (but note that vitalism is still invalid, since to explain life by appealing to entelechy amounts to nothing but a tautology), and the invention of entelechy can be read as a desperate and failed attempt to explain life (I mean, the pure concept/phenomenon of life). Nevertheless, such a failure should also remind us of the gloomy fact: even in current biology our knowledge about life does not go beyond our knowledge about entelechy, because both amount to almost nothing (for an elaboration of this point, see Chen 2019).

References

Ben-Naim, A. manuscript. Can Entropy be Defined for, and the Second Law Applied to Living Systems? https://arxiv.org/abs/1705.02461
Bertalanffy, L.V. 1933. *Modern Theories of Development: An Introduction to Theoretical biology.* London: Oxford University Press.
———. 1968. *General Systems Theory: Foundations, Development, Applications.* New York: George Braziller.
Brigandt, I., and A. Love. 2017. Reductionism in Biology. https://plato.stanford.edu/entries/reduction-biology/
Carnap, R. 1966. *Philosophical Foundations of Physics* Ed. M. Gardner. New York: Basic Books.
Chen, B. 2018. A Non-metaphysical Evaluation of Vitalism in the Early Twentieth Century. *History and Philosophy of the Life Sciences* 40 (3): 1–22.
———. 2019. *A Historico-Logical Study of Vitalism: Life and Matter.* PhD thesis. Universiteit Gent.
Cohen, M. 1931. *Reason and Nature: An Essay on the Meaning of Scientific Method.* London: Kegan Paul.

Crick, F. 1966. *Of Molecules and Men*. Seattle: University of Washington Press.

Driesch, H. 1908a. *The Science and Philosophy of the Organism*. Aberdeen: Printed for the University.

———. 1908b. *The Science and Philosophy of the Organism*. London: Adam and Charles Black.

———. 1927. Emergent Evolution. In *Proceedings of the Sixth International Congress of Philosophy*, ed. E.S. Brightman, 1–9. New York: Longmans, Green.

———. 1929. *The Science & Philosophy of the Organism*. London: A. & C. Black.

———. 1936. Naturwissenschaft und Philosophie. In *Actes du Huitième Congrès International de Philosophie a Prague 2–7 septembre 1934*, ed. E. Rádl, 10–30. Prague: Comité d'Organisation du Congrès.

Frank, P. 1941/1908. The Law of Causality and Experience. In *Between Physics and Philosophy*, 17–28. Cambridge: Harvard University Press.

Freyhofer, H.H. 1979. *The Vitalism of Hans Driesch*. PhD thesis. Los Angeles: University of California.

Gilbert, S.F., and S. Sarkar. 2000. Embracing Complexity: Organicism for the 21st Century. *Developmental Dynamics* 219: 1–9.

Godfrey-Smith, P. 2014. *Philosophy of Biology*. Princeton: Princeton University Press.

Goodwin, B.C. 1982. Development and Evolution. *Journal of Theoretical Biology* 97 (1): 43–55.

Gurwitsch, A. 1915. On Practical Vitalism. *American Naturalist* 49: 763–770.

Haraway, D.J. 1976. *Crystals, Fabrics, and Fields: Metaphors of Organicism in 20th-Century Developmental Biology*. New Haven: Yale University Press.

Hull, D. 1981. Philosophy and Biology. In *Contemporary Philosophy: A New Survey*, ed. G. Fløistad, vol. 2, 281–316. The Hague: Martinus Nijhoff.

Innes, S. 1987. *Hans Driesch and Vitalism: A Reinterpretation*. Master thesis. British Columbia: Simon Fraser University.

Kauffman, S. 1995. *At Home in the Universe: The Search for Laws of Self-Organisation and Complexity*. London: Penguin.

Kauffman, S.A. 2000. *Investigations*. Oxford: Oxford University Press.

Lenoir, T. 1982. *The Strategy of Life: Teleology and Mechanics in Nineteenth Century German Biology*. Chicago: University of Chicago Press.

Mayr, E. 1982. *The Growth of Biological Thought: Diversity, Evolution, and Inheritance*. Cambridge: Harvard University Press.

Mocek, R. 1998. *Die werdende Form: eine Geschichte der kausalen Morphologie*. Marburg: BasiliskenPresse.

Normandin, S., and C. Wolfe, eds. 2013. *Vitalism and the Scientific Image in post-Enlightenment Life Science, 1800-2010*. Dordrecht: Springer.

Nyhart, L. 2000. Book Reviews. *Journal of the History of Biology* 33: 194–197.

Oyama, S. 2010. Biologists Behaving Badly: Vitalism and the Language of Language. *History and Philosophy of the Life Sciences 32*: 401–423.

Reill, P.H. 2005. *Vitalizing Nature in the Enlightenment*. Berkeley: University of California Press.

Richards, R.J. 2002. *The Romantic Conception of Life: Science and Philosophy in the Age of Goethe*. Chicago: University of Chicago Press.

Steigerwald, J. 2019. *Experimenting at the Boundaries of Life: Organic Vitality in Germany Around 1800*. Pittsburgh: University of Pittsburgh Press.

Stoljar, D. 2021. Physicalism. https://plato.stanford.edu/entries/physicalism/

Webster, G., and B.C. Goodwin. 1981. History and Structure in Biology. *Perspective in Biology and Medicine* 25 (1): 39–62.

Wolfe, C.T. 2008. Introduction: Vitalism without Metaphysics? Medical Vitalism in the Enlightenment. *Science in Context 21*: 461–463.

———. 2011a. Why was There No Controversy Over Life in the Scientific Revolution? In *Controversies within the Scientific Revolution*, ed. M. Dascal and V.D. Boantza, 187–219. Amsterdam: John Benjamins Publishing.

———. 2011b. From Substantival to Functional Vitalism and Beyond: Animas, Organisms and Attitudes. *Eidos* 14: 212–235.

———. 2013. Sensibility as Vital Force or as Property of Matter in Mid-Eighteenth-Century Debates. In *The Discourse of Sensibility*, ed. H.M. Lloyd, 147–170. Cham: Springer.

———. 2014. On the Role of Newtonian Analogies in Eighteenth-Century Life Science: Vitalism and Provisionally Inexplicable Explicative Devices. In *Newton and Empiricism*, ed. Z. Biener and E. Schliesser, 223-261. Oxford/New York: Oxford University Press.

———. 2017a. Varieties of Vital Materialism. In *The New Politics of Materialism. History, Philosophy, Science*, ed. S. Ellenzweig and J.H. Zammito, 44–65. London/New York: Routledge.

———. 2017b. Vital Anti-mathematicism and the Ontology of the Emerging Life Sciences: From Mandeville to Diderot. *Synthese*: 1–22. https://doi.org/10.1007/s11229-017-1350-y.

Wolfe, C.T., and M. Terada. 2008. The Animal Economy as Object and Program in Montpellier Vitalism. *Science in Context 21*: 537–579.

Zammito, J.H. 2018. *The Gestation of German Biology: Philosophy and Physiology from Stahl to Schelling*. Chicago: University of Chicago Press.

"A Mountain of Nonsense"? Czech and Slovenian Receptions of Materialism and Vitalism from c. 1860s to the First World War

Christopher Donohue

Abstract In general, historians of science and historians of ideas do not focus on critical appraisals of scientific ideas such as vitalism and materialism from Catholic intellectuals in eastern and southeastern Europe, nor is there much comparative work available on how significant European ideas in the life sciences such as materialism and vitalism were understood and received outside of France, Germany, Italy and the UK. Insofar as such treatments are available, they focus on the contributions of nineteenth century vitalism and materialism to later twentieth ideologies, as well as trace the interactions of vitalism and various intersections with the development of genetics and evolutionary biology see Mosse (The culture of Western Europe: the nineteenth and twentieth centuries. Westview Press, Boulder, 1988, Toward the final solution: a history of European racism. Howard Fertig Publisher, New York, 1978; Turda et al., Crafting humans: from genesis to eugenics and beyond. V&R Unipress, Goettingen, 2013). English and American eugenicists (such as William Caleb Saleeby), and scores of others underscored the importance of vitalism to the future science of "eugenics" (Saleeby, The progress of eugenics. Cassell, New York, 1914). Little has been written on *materialism qua materialism* or vitalism *qua vitalism* in eastern Europe.

The Czech and Slovene cases are interesting for comparison insofar as both had national awakenings in the middle of the nineteenth century which were linguistic and scientific, while also being religious in nature (on the Czech case see David, Realism, tolerance, and liberalism in the Czech National awakening: legacies of the Bohemian reformation. Johns Hopkins University Press, Baltimore, 2010; on the Slovene case see Kann and David, Peoples of the Eastern Habsburg Lands, 1526-1918. University of Washington Press, Washington, 2010). In the case of many Catholic writers writing in Moravia, there are not only slight noticeable differences in word-choice and construction but a greater influence of scholastic Latin, all the more so in the works of nineteenth century Czech priests and bishops.

C. Donohue (✉)
National Human Genome Research Institute, Bethesda, MD, USA
e-mail: donohuecr@mail.nih.gov

© The Author(s) 2023
C. Donohue, C. T. Wolfe (eds.), *Vitalism and Its Legacy in Twentieth Century Life Sciences and Philosophy*, History, Philosophy and Theory of the Life Sciences 29, https://doi.org/10.1007/978-3-031-12604-8_5

In this case, German, Latin and literary Czech coexisted in the same texts. Thus, the presence of these three languages throws caution on the work on the work of Michael Gordin, who argues that scientific language went from Latin to German to vernacular. In Czech, Slovenian and Croatian cases, all three coexisted quite happily until the First World War, with the decades from the 1840s to the 1880s being particularly suited to linguistic flexibility, where oftentimes writers would put in parentheses a Latin or German word to make the meaning clear to the audience. Note however that these multiple paraphrases were often polemical in the case of discussions of materialism and vitalism.

In Slovenia Čas (Time or The Times) ran from 1907 to 1942, running under the muscular editorship of Fr. Aleš Ušeničnik (1868–1952) devoted hundreds of pages often penned by Ušeničnik himself or his close collaborators to wide-ranging discussions of vitalism, materialism and its implied social and societal consequences. Like their Czech counterparts Fr. Matěj Procházka (1811–1889) and Fr. Antonín Lenz (1829–1901), materialism was often conjoined with "pantheism" and immorality. In both the Czech and the Slovene cases, materialism was viewed as a deep theological problem, as it made the Catholic account of the transformation of the Eucharistic sacrifice into the real presence untenable. In the Czech case, materialism was often conjoined with "bestiality" (*bestialnost*) and radical politics, especially agrarianism, while in the case of Ušeničnik and Slovene writers, materialism was conjoined with "parliamentarianism" and "democracy." There is too an unexamined dialogue on vitalism, materialism and pan-Slavism which needs to be explored.

Writing in 1914 in a review of O bistvu življenja (Concerning the essence of life) by the controversial Croatian biologist Boris Zarnik) Ušeničnik underscored that vitalism was an speculative outlook because it left the field of positive science and entered the speculative realm of philosophy. Ušeničnik writes that it was "Too bad" that Zarnik "tackles" the question of vitalism, as his zoological opinions are interesting but his philosophy was not "successful". Ušeničnik concluded that vitalism was a rather old idea, which belonged more to the realm of philosophy and Thomistic theology then biology. It nonetheless seemed to provide a solution for the particular characteristics of life, especially its individuality. It was certainly preferable to all the dangers that materialism presented. Likewise in the Czech case, Emmanuel Radl (1873–1942) spent much of his life extolling the virtues of vitalism, up until his death in home confinement during the Nazi Protectorate. Vitalism too became bound up in the late nineteenth century rediscovery of early modern philosophy, which became an essential part of the development of new scientific consciousness and linguistic awareness right before the First World War in the Czech lands. Thus, by comparing the reception of these ideas together in two countries separated by 'nationality' but bounded by religion and active engagement with French and German ideas (especially Driesch), we can reconstruct not only receptions of vitalism and materialism, but articulate their political and theological valances.

Keywords Vitalism · Materialism · Pantheism · Neovitalism · Eastern Europe · Slovenia · Czechia

Although a rich historiography attends the histories of vitalism and materialism in Western Europe, to my knowledge there have been no systematic attempts to discuss vitalism and anti-materialism in eastern Europe (or more properly east central Europe, according to recent scholarly conventions). The reasons for this are puzzling and can only be partially explained in the context of this paper. A significant reason, I suspect, is presentism. Charles Wolfe has been instrumental in providing a number of contexts and contextualization for vitalism and materialism in Western Europe (Normandin and Wolfe 2013; Wolfe 2016, "Canguilhem and the Logic of Life", see also this volume) and arguing that the histories of vitalism and materialism add to our "philosophical technology" (my term) for understanding contemporary science and more importantly, the implications of those life sciences now undergoing (one may argue). Vitalism in "Eastern Europe" or "East Central Europe" has been in English language literature more or less ignored or exoticized (Chirot 1991). Marius Turda has likewise referenced "Balkanism" and "Orientialism" (2011).[1]

This is due in part to what I have recently called "occidentalism" (Donohue, 2020) in which Eastern Europe is placed outside of the development of Western science and culture and onto the path of an alternative or "backward" modernity. Even for very recent scholarship somewhat more responsible in the anglophone world, "science" in Czech speaking lands is science in the "German" language and, according to Michael Gordin, reducible to physics, where writing in the Czech language was the consequence of "linguistic purism" and a "dilemma" (Gordin 2020: 219). Those who identified as "Czechs" were conflicted over the use of German or Czech in their scientific publications. Although Gordin's work is here singled out, such almost ethnographic accounts of Czech-language intellectual output are widespread in the field.

Although the processes of transformation of the scientific languages of the Czech and Slovenian speaking lands are not the central focus of this paper, the subsequent material discussed will illustrate that many Czech especially and Slovenian writers were quite comfortable moving between their native languages, French, German and even occasionally English. It is also worth noting that Gordin as a historian of physics, is focused on physics journals in Central Eastern Europe, and thus would not necessarily be aware of journals in the biological, theological and sociological sciences such as *Ceska mysl* (Czech mind) begun in 1900 as an international pedagogical review in the social and the biological sciences, with articles in the physical sciences as well.

---

[1] A key issue in the literature on the national awakening was to what degree it was particularistic versus universal. Based on a Cold War framework, nationalist movements in the nineteenth century were divided into "ethnic" and "civic" varieties, with civic nationalists being those of the "West" (the classic case being France, the failed exemplar being German state unification) with states and peoples in "Eastern" Europe being subject to a vague kind of exclusionary belong which made universalism impossible. In the Soviet period and afterwards, the strength of ethnicities and "tribal loyalties" has supported a consistent narrative of "backwardness." This helped to explain the embrace of Communism as well as the rather uneven pace of modernization throughout most of the twentieth century. On this see among others (Chirot 1991; Gerschenkron 1965).

Magazines such as *Czech Mind* as well as a number of theological periodicals discussed below used a number of languages which showed Czechs critically engaged with a number of European intellectual currents. As importantly, while Czech physics was attempting to migrate into the German language, Czech biology, social sciences and theology remained intermixed with languages and a reception to pan-European and American intellectual currents. Incidentally, the Habsburg influence on Catholic theology in Czech speaking lands lead to some interesting vocabulary differences between the predominately Catholic areas of Moravia and the "standard" Protestant version of literary Czech used in the universities in and around Prague see (Bartoš 1905).

Vitalism itself has only recently been subject to sustained critical analysis. For a generation of scholars, particularly after the Second World War, vitalism was synonymous with irrationalism and political messianism and among the progenitors of the Nazi racial state and of the Holocaust or Shoah. Many years ago, Emile Brunner wrote bluntly that "National Socialism springs from a view of man which may be described as vitalism" (Brunner 1947, 56.) "Vitalism" has also not fared particularly well in the work of the past generation of intellectual historians (and intellectuals) who sought to understand the ideas driving the totalitarian regimes of the twentieth century and the eras of fascism. Georg Lukács, commenting on Ludwig Klages in his *Destruction of Reason*, noted that the philosopher and physiologist Klages "transformed vitalism into an open combat against reason and culture" (Lukács 1981, 248). Roger Griffiths locates Mussolini's "ethical revolution" in "Fascist vitalism" which would "lift the apathetic, cynical individual…into a new spiritual orbit" (Feldman and Griffin 2008, 8). Various efforts have been made more recently to connect various kinds of vitalism with French fascism for example in the work of Maurice Barrès (Soucy 1972), and the varieties of Spanish fascism (Priorelli 2020) and the elitist and eugenic political theory of vitalism and 'life-force' in George Bernard Shaw (Linehan 2000).

Moir ("What Is Living and What Is Dead in Political Vitalism?", this volume) underscores that the links between fascism and vitalism in the twentieth century are often tenuous and vitalism itself is often unfairly conjoined with totalitarian and other forms of political and social reaction. Oftentimes, Moir argues, it was not vitalism itself, but conceptions often viewed as closely related to vitalism, such as holism in the case of National Socialism, monism or animism. Today, vitalist understandings of the distinction between life and non-life (properly defined) often lead to precisely the kinds of pluralistic politics and anti-reductionistic accounts of social life and of the person which work against totalitarianism and its brethren. Vörös (in the chapter "Is There Not a Truth of Vitalism? Vital Normativity in Canguilhem and Merleau-Ponty", also this volume) underscores that vitalism- even in the work of its skeptics or opponents- amounts to an openness to complexity and to spontaneity where vitalism is capable of both establishing as well as transcending norms.

Materialism from the second half of the nineteenth century to the First World War in the Czech-speaking lands especially, enjoyed the same dubious reputation of being connected with unsavory forms of politics and social life. Vitalism in both Czech and Slovene contexts was viewed, as I will show in the following pages, was a complex, perhaps a bit frustrating alternative to materialism by both secular and religious intellectuals in the Czech case (discussions of the Slovene case will be more limited). The relative openness to vitalism (of all kinds) and to understanding vitalism in their late nineteenth century present as a historical development of scientific inquiry in the Czech case, especially, allows for a discussion of the richness of international conceptual inquiry. As importantly, the virulent reaction of Czech writers to materialism, from the very mainstream of the literary and cultural 'national awakening' and national literature, as well as the unity of Catholic religious opinion from the 1840s and 1850s to the First World War presents a reasonable explanation for the openness of Czech writers from the 1840s to the First World War to sophisticated accounts of vitalism. As importantly, there is a never articulated sense that materialism was connected (at least in the latter portion of the nineteenth century) with "Hapsburg" or "Germanness," and this cultural eversion, while not sufficient, perhaps contributed.

A full study of anti-materialism in Catholic thought in so-called "Eastern Europe" remains to be written. If anything, the rather colorful language used to describe materialism and its social, political and spiritual consequences would provide resources for rhetorical analysis and its function in scientific polemic. To take one example, perhaps the most important Polish Catholic theologian of the nineteenth century, Piotr Semenenko (1814–1886) described philosophical materialism as an "enemy" (*nieprzyjaciel*) "who is in our houses" who uses his "poisoned sorcery" to make the faithful believe it was not God who created the world out of nothing. Materialism, he continued, which Semenenko often conflates with pantheism, was "a bad taste, a rotten consolation" (*zgnila pociecha*) "which does not dare come to an honest tongue, but lives there in a despondent heart" (Semenenko 1885: 116–118). On 'materialism' Semenenko is somewhat vague. Materialism and pantheism were "that unhealthy science" "mainly from Germany" "smuggled to us mostly through France" (Semenenko 1885 :119). Regarding materialism (most probably Büchner's), Semenenko underscored that modern materialism (as opposed to the materialism of the Greeks) is much more "audacious" or "brash" (*zuchwały*) as it "combines atoms and motion, matter and force into one…thereby abolishing the core difference between what is passive and what is active in itself, the most intrinsic and aboriginal difference of all" (ibid, 124.)

For Semenenko,"Materialism wanted to derive the whole world from atoms" and in order to do so, according to Semenenko, it "begins with a core intellectual contradiction" which "endows atoms with motion, and makes passive things active without any reason or law" (*żadnej racyi i prawa*) only because the materialist wishes it. According to Semenenko, a body cannot have the properties of matter and force at the same time; it is either matter or force, not both. He declared: "If the materialist choses the active atom, [it is] a motion endowed by itself….it is motion (*ruch*), it is force, it is a center (*srodek*) of force, but it is not atoms!" (ibid, 129.)

Czech writers were no less sparing and colorful regardless of religious observance. Josef Durdík (1837–1902) was among the most important philosophers of the natural and the biological sciences in the nineteenth century. He produced, among other texts, one of the first history of philosophy in the Czech language in

1870, the "Historical Outline of Modern Philosophy" (*Dějepisný nástin filosofie novověké*). He observed in his "On the progress of the natural sciences" that Darwin was "as little a materialist as all men who have faced the truth…matter is a mystery, therefore materialism does not explain anything" (Durdík 1874: 232). Durdík understands modern materialism as emerging in part from the "naturalism" of Ludwig Feuerbach (or more accurately Feuerbach exemplifies for Durdík the naturalism of the day). This "war over the soul" with Rudolf Wagner struggling against Karl Vogt, and Moleschott against Liebig. But for Durdík, materialism was exemplified by its "Bible" Buchner's *"Kraft und Stoff"* (*Síla a hmota*) (Durdík 1876: 25).

Durdík responded to the work of the materialist philosopher George Stiebeling. Stiebeling wished to refute Hartmann's "philosophy of the unconscious" and to a certain extent, to promote "realism" in philosophy. Durdík observed that Stiebeling's work *Natural Science Against Philosophy* made the intentions of materialism clear, to conflate natural science and the scientific method with materialism. Durdík contended that this was not true (as many branches of science did not depend upon materialism) and many scientists were not themselves materialists (Durdík 1876). As importantly Durdík also observed that Stiebeling when discussing philosophy had a very narrow account of philosophizing in mind, namely, the philosophy of Hartmann and the dialectic of Hegel. Durdík underscored, contra Stiebeling, that "Philosophy is broader," containing not only the metaphysics of Hegel but others (albeit equally problematic). Stieberling, moreover, constructs his own system, in many ways equally obtuse. Durdík observed that Stieberling "mechanizes a view of the world, and zealously upholds it, glorifying it as a panacea for future prosperity" as have many philosophers of a "lesser order" (Durdík 1876: 12).

Matěj Procházka (1811–1889) was a Moravian church historian and social philosopher. In his rather polemical work for high school students (approved by his colleague and close collaborator Antonín Lenz (1829–1902) whom I will say more about below) he opened: "In our times, materialism makes brazen attacks on the fundamental truths of religion, an odd enlightenment forgetting of spirit for matter, an unwillingness to believe anything that they can not see, hear, or taste; (materialism) wages war (*brojí*) against the teachings of the Church about God, about creation, about man, about redemption…." Of materialism itself Procházka complains that the doctrine is (literally) "full of resistors" (*plno odporů*). He continued that "Materialists" (such as Vogt) claim that "atoms are eternal (without transformation), that they do not have certain properties, but that they are different from each other, that they are unconditional in themselves, and yet that they co-condition, that is, they are conditioned" (Procházka 1876: 9).

For Procházka, as for Lenz, materialism was the doctrine not only attacking the Catholic faith, but the ideology confronting both the political order as well as the order of nature. For Procházka, materialism was synonymous with Darwinism and liberalism. He declared that "odd liberalism" (*lichý liberalismus*) "is holding the scepter of government" and where materialism has "reached its zenith" it has begun "terrorizing in an imperial tone all naturalists and other writers in other fields of science, proclaiming all scientists as idiots and fools who do not yet proclaim:

Omnia in majorem Darvini gloriam!" and whom do not " wish to venerate the golden calf of matter" (*se klaněti zlatému teleti hmoty*) but to recognize the existence of a higher world (Procházka 1876: 17).

Materialism too according to Procházka could not explain the basic phenomena of life and consciousness. He observed "by what right can materialists just style mere movements or brain tremors (*záchvěje mozku*) thoughts?" What was the connection, Procházka asked, between thought and "brain matter"? How was it possible that a "mere physical process at once and immediately be transformed into a spiritual process, so that mere movements of matter can be transformed immediately into movements of spirit, i.e. into thoughts, feelings, and desires?" What, Procházka asked, was occurring in this transformation, in which for him, the cause of this transformation from matter into spirit was also its effect? (Procházka 1876: 328–9).

But what was as concerning as the political, Darwinian, and reductionistic dimensions was materialism's theological connotations. Against the "murderous attacks" of materialism against the Catholic faith, Procházka and others such as Lenz had written a number of works. Just how disturbing Procházka believed materialism to be was revealed by his catechismal discussion of the Book of Genesis where he described in a long footnote how materialism led to pantheism and bestiality. Footnoting a discussion of how God himself created the world "by the freest will" (*nejsvobodnější vůlí*), Procházka argued that materialism and its close corollary pantheism were both equally pernicious. Procházka wrote that while "ancient teaching" sought to "elevate man", the school of pantheism and materialism "drives him directly into bestiality (*zvířeckosti*)" (Procházka 1876: 3, 35).

The association between "bestiality" and materialism was a serious association for Catholic theologians in Czech speaking lands. Their furor came from two sources: theological and political. On the first, according to Lenz, a materialist could not philosophically believe in the miracle of transubstantiation or in the divinity of the person, of in life after death. Materialism then threatened for Czech Catholic theologians (and others) both sacramentology and philosophical anthropology. According to Lenz, "Man is a compound *ex materia et forma*", he is not simply matter but whose "essential form" was an "intellectual or immortal soul." And it was this immortal soul which is the principle of a "a living, psychic, and spiritual human life." The doctrine of materialism and the materialist does not make a real, substantial difference between things and beings (*mezi věcmi a bytostmi*) or between a plant, an animal and a human being" (Lenz 1889: 28).

Because of the lack of distinction between these elements, materialists could not believe in the doctrine of the transubstantiation of the eucharist, whereby the eucharistic wafer and wine becomes the actual body and blood of Christ. As Lenz explained, it was through transubstantiation that the bread and wine are changed "on the altar" (*na oltáři*) into the body and blood of Jesus Christ. On the altar, the bread and wine underwent a change in substance (where the accidental features of the bread and wine itself remained unchanged) where "a new essence arises after the change" while the matter, the material of the bread and wine stay the same. This was very different, according to Lenz, than a *transformation* of the bread and of the wine as it enters our blood, and there was a great distance between a natural change and a miraculous one (Lenz 1889: 28–9). According to Lenz, it was this kind of fundamental change which the materialist could not understand nor could his philosophy address. Drawing on the trope which connected materialism with bestiality too when turning to the political, Lenz thundered that any adoption of the materialist doctrine and acceptance of its "lies" "would lead to…atheistic and rationalistic stupidity [and] barbarism, which seeks to drive humanity to the troughs (*žlebům, old Czech koryto,*) of beasts and cattle" (Lenz 1878: 11).

Lenz, again discussing the political implications of materialism, commented on the syllabus of Pius IX "on errors" promulgated in 1864, underscored that materialism, pantheism, liberalism, atheism will all "unite…" against "the dogmatic, social, moral and political direction of the Church." Lenz continued that there was a "war of Christian civilization against materialism" in which Darwinism and atheism, along with materialism "have no truths in themselves." All of these doctrines, along with various political and philosophical systems (including the systems of Fichte, Schelling and Hegel, which had nothing to do with materialism but were viewed, by Lenz and others as introducing a naturalistic and then a pantheistic view of the world) according to Lenz were to be rejected because they either "rejected the Divine Being" or imposed on God Himself qualities which he cannot possess (Lenz 1889: 37–38). For writers such as Lenz and Procházka any stance towards nature which was either materialist, atomist, "dynamist"[2] was a stance towards politics and immortality.

For Lenz, as for Procházka, there was also an intimate connection between materialism, atheism, and political liberalism. For Lenz, both liberalism and nihilism were "modern heresies" and intimately connected to the reformation of Martin Luther (who according to Lenz paved the way for all other modern revolutions). For Lenz materialism, Lutheranism, and nihilism were all political-theological heresies, such as those which bedeviled the early Church, presenting views of nature and of politics which were untenable to Catholic dogma, insofar as it reduced soul and consciousness to base matter and motion, and could not admit any change of substance without fundamental change in accident (Lenz 1889, 5). Discussing in detail Pope Pius IX's "Syllabus of Errors", published in 1864, now infamous for its condemnation of liberalism, modernism, secularization and other modernizing forces, Lenz underscored that materialism and liberalism were "incompatible with the Christian" (Lenz 1891: 3).

For both authors, if one did not believe in the "miracle" of the Eucharist, or in the immortality of the soul, then one was radical and subversive politically. In the case of both authors, theology, politics and morality were intertwined. For both authors, whose rhetoric underscored their convictions, materialism was a philosophy (or

---

[2] "Dynamism" was, for Czech writers, a kind of materialism, or at least a doctrine "that stood entirely on materialistic grounds" Dynamism referred to the tendency of materialist authors to use "mere matter and forces" to explain mental life and consciousness spiritual or mental "being" through recourse only to atoms. For a rather general and polemical definition see (Pospíšil 1885, 344).

even ideology) which not only exhibited an untrue view of the natural world, but which also would lead to moral depravity and bestiality.

The association between materialism, pantheism and "bestiality" was not merely a connection made by Catholic theologians. The Czech politician and writer, one of the founders of the Czech "national awakening," František Palacký, whose work and writings have influenced generations of Czech writers and statesmen (including Thomas Masaryk), also conjoined, as it was then understood, "bestiality" (*zvířeckosti*), materialism in order to denigrate his opponent Alfons Šťastný (Baár 2010; Trencsenyi et al. 2018; Baran-Szołtys et al. 2020).

Šťastný was a writer and political theorist who supported the agrarian movement in the Czech lands, meaning that he was a political and economic liberal, e.g. demanding a reform to agricultural laws and customary duties, education for farmers, and the relaxation of property regulations (Miller 1999: 19). Šťastný's "Concerning Salvation After Death" (*O spasení po smrti*) was part of a larger program which combined elements of materialism and a philosophical anthropology which attempted to ground man as a reasoned and thinking being without any reference outside of the human self (as he explained in his "Concerning Intellectual Morality" (*O mravnosti rozumove- Über die Verstandesmoral*)) Drawing upon Spinoza and others, Šťastný attempted to articulate an account of morality in the nineteenth century which did not depend in any way upon the external notion of salvation.

Paraphrasing Spinoza, Šťastný declared that "Man can only want his own good and he cannot wish for his own doom, assuming a man of common sense and sound reason." He argued that morality and reason could both exist without the immortality of the soul or without life after death (Šťastný 1874: 5, 6). Šťastný wanted to construct a morality based on entirely common-sense and empirical foundations, based upon his supposition that any reform of government and of society would have to begin with an account of man that did not depend on any theological structures.

Šťastný too believed that the God of Christianity was too anthropomorphized (echoing in many ways Feuerbach). Šťastný reasoned that if God created man and the world out of his own beatitude, rapture or blessedness (*blaženost*), then there was a fundamental contradiction between the attributes of God or a deity and his works, as a thing with needs could only be something which is mutable, changeable, and alive. As God had no needs this fundamentally pointed to a higher being, if he existed, which possessed the characteristics of non-life rather than life (*není života nýbrž neživot*) (Šťastný 1874, 90)

For this František Palacký accused Šťastný of not only political and social radicalism but amoralism and "bestiality." Palacký accused Šťastný of "denying everything that mere reason does not understand" not wanting to know "God as a personal being" nor giving humanity a "moral vocation." Šťastný 's teachings and the school of materialism "drives man directly into bestiality (*zvířeckosti*)".

Šťastný immediately objected that it was not at all true that materialism and pantheism lead to bestiality, rather pantheism and materialism lead not to "bestiality" nor to atheism, but to "humanity", in the sense of humane feeling of togetherness, brotherhood and humanity (*lidskost*). Pantheism as such was for Šťastný "a safe harbor or haven (*útulek*) for those who can no longer defend God as a being, but who wish to have God in name at least" (Šťastný 1874: 91, 92).

Very different was the reception of vitalism in the Czech context. In a 1906 issue of "Czech Mind" a review of Driesch's *Der Vitalismus als Geschichte und als Lehre* was favorable, although it complained of Driesch's obtuse style. The reviewer approvingly described Driesch's "entelechy" (which incidentally had normal declension rules and was thus part of the accepted Czech scientific vocabulary, there was addition of a Latin or German word in order to clarify the scientific meaning, and the word appears to have a normal declension pattern, which suggests wide usage.) The "entelechy" was a self-regulating, autonomous (*svézákonnost*) body, meaning the "autonomy of living bodies" where the entelechy "in the broadest sense" was a "real elementary natural agent that appears in living bodies." The entelechy was moreover a "constant," "an elementary constant" and "as such is (fundamental) element of natural extended reality." As such "vitalism has a place of honor in the history of the natural sciences." The reviewer continued that entelechy must be subject to the "general principle of causality" and was in the broadest terms "a real, natural, elemental agent which appears in living bodies" (Čáda et al., 1906: 455).

For the reviewer, however, "Nothing definite could be said about the relation of the entelechy to the theory of descent." Nonetheless, the reviewer observed that "vitalism has a place of honor in the history of the sciences" having both roots in antiquity and a new phase called "neo-vitalism." The older vitalistic tradition, of which its very first representative was Aristotle, "culminated in Blumenbach" and ended with the work of Johannes Müller. For the reviewer, the vitalistic tradition found new life in the critique of Darwin (Čáda et al., 1906: 456).

"Neo-vitalism", while having some roots in "older vitalism" was nonetheless a new movement. Able representatives of this new inquiry were scientists like Bunge, Hartmann and Otto Liebmann. Edmund Montgomery deserved particular mention for his solving "of the problem of the self and of individual organization, where Montgomery, approvingly turns against "any theory of mechanism" (ibid, 558).

The reviewer continued that so called "neo-vitalism" had "overcome the era of materialism" and could serve as a critique of Darwinism. This was among the most favorable attributes of vitalism in the years before the First World War. The reviewer continued that of the major problems which attracted the attention of neo-vitalists (such as Driesch, Fritz Noll, Eugen Albrecht) was the "psychological and organization problem of integration." The reviewer concluded that Driesch's work was "the best work on the nature of vitalism and its historical development" and although it was not appropriate for a general audience, for naturalists this was "an essential text" (Čáda and Krejčí 1908: 457–8).

For the physiologist and philosopher František Mareš, vitalism was a specific antidote to both materialism and the various kinds of mysticism (spiritualism) which had propagated in the years before the First World War. For Mareš (1857–1942) there was a great "mystery" to life which was not appreciable by our external senses. As thinking, feeling beings we were only able to understand the phenomenon of life by appealing to our "inner sense" (Mareš et al. 1897: 165). He declared ,"Naturalists cry out: old mysticism, old vitalism is returning to physiology. No gentlemen. The nature of these phenomena cannot be determined empirically. It is only possible to describe the phenomena and determine their lawfulness, empirical exactness is no longer possible (Mareš et al. 1897, 167).

Mareš noted elsewhere, referring to the controversy that his writings on vitalism inspired, that the magazine "Life or Liveliness (Živa) wanted him to conclude that life was nothing more than a "very complex chemical process." Now Mareš was more than happy to admit that certain aspects of metabolism could be reduced to chemical analysis, and such analysis was essential for the science of physiology. However, Mareš quipped "even the world's physiologists are also living bodies, and I don't think they would recognize themselves as mere chemical processes." Mareš continued that "The living entity (*zivá bytost*)" while approachable and cognizable to physio-chemical analysis, "is also as yet something more." The analysis of the "living being" while of course belonging to physio-chemical research, cannot proceed only through physio-chemical analysis (Mareš 1904: 11).

Mareš also observed that "Mechanical-causal beliefs have dominated biology for the past half-century, and only in the last decade has there been strong opposition to it in the form of vitalism." Moreover, the dispute (*spor*) between mechanism and vitalism "was coming to an end" with the numbers of supporters of vitalism increasing, where, in the last few years supporters of vitalism outnumbered proponents of materialism and mechanism (Mareš 1904: 12).

Mareš specifically singled out the work of Wilhelm Roux. Roux was a celebrated anatomist but who according to Mareš "also places great emphasis on inorganic experiments as explanatory analogies of life-forming processes." Roux "rejects vitalism" and like all other mechanists and materialists, "based [his] beliefs not on scientific facts" since according to Mareš, "not the slightest process of life has been explained physio-chemically, or mechanically," not even the "circulation of blood" for example.

Mechanists, according to him, based this rejection of vitalism on the simple logical error that there is no action which is not "physio-chemical" and no action which is not "mechanical." However, in the natural world "concepts such as: power, action, cause, and so on, are vitalist in nature." The neo-vitalism of the present, according to Mareš, understood that physio-chemical explanations were indeed needed, but that mechanism and materialism did not capture all of the happenings (*dění*) of life such that "...life demonstrated its own specific laws, which are not in inanimate nature" (Mareš 1904: 12).

The better-known (than the other Czech figures discussed here) philosopher and historian of biology Emanuel Rádl (1873–1942) was discussed by Georges Canguilhem. Canguilhem underscored in his discussion of vitalism, with reference to Rádl, "that vitalism was an imperative rather than a method and more of an ethical system, perhaps, than a theory" (Delaporte 1994: 292). This is certainly true, but Canguilhem's discussion reduced the depths of Rádl's engagement with vitalism. Rádl, in his discussion of the history of vitalism from Leibniz to Stahl, observed that there was a great difference between "organism" and "mechanism" where movements occur upon a mechanism, while upon an organism, those movements happen within an organism and are due to its own cause (*způsobovány*). This corresponded for Rádl to the distinction between *facere* (something else doing or making) and *efficere* (agent-activity working out, causing to occur). The latter is willed while the former is not as the latter supposes an agent which is doing, making, producing, or carrying out (Čáda and Krejčí 1905: 264).

For Rádl, this conforms to the Kantian account of the organism whereby anything which is "caused" in the domain of the organism "does not make sense in and of itself, but only makes sense for the purpose for which it takes place" (Čáda and Krejčí 1905: 265). Because of this presence of purposefulness which is intrinsic to the organism, the sciences of physics and chemistry "obscure (*zastírá*) the true essence of biology." Physics as well only addresses "the most general descriptions of matter, and these abstractions only cover only a few real phenomena."

Physics, unlike biology does not address "the true and active causes (*pravé a činné příčiny*) of these phenomena." Rádl wrote that "today's physics places its emphasis on matter...but it ignores movements, forces and actions....the intensity of these movements and their changes" (Čáda et al. 1906: 265). He observed that the old adage of "where the physicist ends, the doctor beings" should be supplemented with the observation that today the physicist "does not care" where at the "very least" they would do well to tell us how the body "could be damaged (*poškozeno*) at all" (Čáda and Krejčí 1905, ibid).

In a similar way, Rádl noted that "it was not possible to build a general theory of life on chemistry, as there existed foundational differences between chemical processes and life processes." Living bodies, he continued, "are never sustained for an unlimited amount of time (*po neomezenou dobu*)" (ibid.) Living bodies moreover sustained themselves much longer than predicted by the chemical elements which compose them and while chemical processes in inorganic nature only depend on the chemicals themselves, in living organisms there was a process of regulation to keep the organism alive and "in its entirety" (*a ve své celosti*) (ibid, 266).

Writing a bit later in 1909, Radl observed that after 1880, the most current form of vitalism underscored that "the purposefulness of life is a phenomenon *sui generis*" which "cannot be explained as the result of a summation of chance effects." The organism, according to Radl, "is quite different in kind from any inorganic substance." The organism is "not an aggregate...but it is a unit which exerts definite control over the actions and reactions between itself and its environment." Radl wrote that Driesch, reacting against the "decline of naturalism" and the rise of materialism and the mechanistic account of life, came into the view that "life 'is a law unto itself.'" Rádl, mirroring Mareš' language, noted approving that Driesch rejected that view that life "is only a very complicated chemico-physical process" and that he underscored the "autonomy of living processes" (Rádl and Hatfield 1930: 356-7, 382).

Continuing with Driesch and contemporary vitalism specifically, Rádl underscored that Driesch went further and was more successful than other vitalists. Driesch was ignored, according to Rádl, because of his opposition to Darwinism. Rádl is quick to observe later that Driesch adhered to the same critique of Darwinism as that of Otto Liebmann. For Liebmann, fundamental error of Darwin's theory was that although it served as a theory of development, such a theory of development, the concepts of reproduction, heredity and variability, upon which that theory is based "are completely unclear" (*jsou úplně nejasny*) (Rádl 1908: 35).

Driesch commanded Rádl's respect because he conducted his own experiments and also examined the philosophical consequences of his work (Rádl 1908: 36). Rádl underscored that in his break with Roux, Driesch first attempted to explain living phenomena only through reference to physics and chemistry, but eventually had to abandon that scheme and then, in his articulation of vitalism opposed "the whole of modern biology" (*proti celé moderní biologii*).

In opposition to other vitalists (Bunge and Crossman) whose doctrines he calls "static teleology," Rádl underscored that for Driesch biological phenomena needed both causal and teleological explanations, and in this sense Rádl believed Driesch to be greatly indebted to Kant, although this indebtedness changed throughout his career (Rádl, 1908: 37). For Driesch, although some of life's processes could be explained through recourse to physics and chemistry, many organic phenomena are subject to "a special, pure vital lawfulness and demonstrate the autonomy of living actions" (ibid, 38.) Empirical evidence from among other subjects, sea urchins, underscored for Rádl the correctness of Driesch's "dynamic vitalism" which among other characteristics, yielded the specific qualities by which living organisms differ from non-living entities.

Driesch's dynamic vitalism was also defined by his attempt to describe those laws which only pertained to living processes (ibid, 39.) For example, for Driesch (and Rádl) the organism was itself a constant, and as such, the organism had differing properties than the inorganic. Rádl noted that as a developing entity, the chicken egg (and later chick, and then adult chicken) had not only a capacity to respond specifically to external stimuli, but also as a developing organism, could change its reactions over time. This was unlike inorganic matter, which exhibited the same reactions to environmental changes, i.e. an iron always becomes hot in the presence of heat (ibid, 40.)

Among the most important ideas that Driesch articulated was the principle of holistic organization. Driesch compared an organism to an army (*armade*) and understood that both the organism and the body have a hierarchy which followed a strict chain of command for organization, movement, and other activities. Driesch viewed the soul as a kind of "commander-in-chief" of the organism, in the same way that a commander is at the top of the chain of a particular army. The organism in its functions was controlled by a number of entelechies, some higher and some lower, some very specialized (though some are governed by merely mechanical processes, i.e. the action of the lens in the eye) (ibid, 41.)

Methodologically, Rádl noted that Driesch was against induction. For Driesch, conceptual analysis and science made progress by being against induction. Like Kant, Driesch wished to exclude all 'metaphysics' and to place science on a true foundation. However, such a stance against metaphysics and in support of Kant made the move back to Aristotle more surprising.

This was the case since, for Rádl, Driesch's "Naturphilosophie" (*naturfilosofie*), a kind of metaphysics,[3] represented the progress of a philosophy of development. For Rádl, Hegel understood development in too metaphysical a sense, divorced from experiment and from empirical data. Darwin understood life in too mechanical a way, and too materialistically (Rádl here was at odds with Durdík). Because of this, Driesch took the "radical step back to Aristotle" passing through *Naturphilosophie*. Because Driesch represented the culmination of the doctrine of development, it would "have great consequences for the future" (ibid, 123–124). This was the case since Driesch was able to develop an account of development which was neither dialectical nor metaphysical, rooted in empirical data, but against induction. In this way, Rádl believed that Driesch's vitalism had been purified through its overcoming of the prior limitations of numerous systems of thought and inquiry, made possible through a return to the Greeks. This is a challenging account of Driesch's work which almost presumed a Heideggerian turn.

Rádl's vitalism, and his appreciative, detailed glosses on Driesch make more sense if one considers Mareš and Rádl as working within a nineteenth and early twentieth century space (which was appreciative of vitalism as an inquiry, and for the autonomy of biology as subject to its own specific laws). Both men's vitalism underscored, some rhetoric aside, that while mechanism and physics and chemistry were able to explain some organismal properties, there was an additional modality to life and to the organism which could not be explained through reference to physics and chemistry alone.

Both Rádl and Mareš' hostility to mechanism and materialism and their insistence on the autonomy of organism, biology and physiology is rendered more explicable when it is understood that there exists a tradition against materialism in Czech philosophy and theology. As discussed above there was a continual emphasis placed on the supposed connections between materialism, liberalism, barbarism and bestiality. This was fertile soil for vitalism in Czech and Slovene speaking lands.

Josef Durdík and Rádl (though writing a bit later) differed in their reception but not in their engagement with Darwin, both were united in their understanding of life as not subject to materialist explanations. For Durdík, the inquiry into man and organic nature was moving further and further into a knowledge of the whole. He declared that as these sciences become more refined "the more they realize the detail does not destroy the whole." He continued that "life, organism, development" have been met with similarly spectacular social and political developments (Durdík 1876: 21).

For Durdík, Darwin's theory and the theory of evolution, was unlike the rest of the natural sciences, for he understood Darwin's theory as an aesthetic and artistic one as well, as opposed to other explanations of the modern world. For Darwin not only depicted an animal form as a naturalist, but also much like a painter or a poet, ensured that the specific features of the horse are well-represented in their

---

[3] The connections between Driesch and *Naturphilosophie* have been outlined in more detail by (Rosenstock 2017).

distinctiveness. The horse, or any creature, will be considered more beautiful the more it was distinctive (Durdík 1876: 38). It was much the same with man, where man, especially in the realm of Darwinian theory, was both an aesthetic as well as a biological subject. It is worth noting here that the union of aesthetics and nature, art and biology was one of the main premises of Kant's *Critique of Judgement*.

Durdík's statement that "materialism explains nothing" was quite like Matěj Procházka's famous observation that materialist theories were a "mountain of non-sense" (Procházka 1876). Perhaps most importantly, Czechs believed themselves to not be Germans (or more correctly Hapsburgs) and rejecting "German material-ism"; this is clear in the writings of Jaroslav Goll (1897) and in the writings of Palacký.

The writings of Slovenian theologians and intellectuals were not as colorful as Piotr Semenenko's, and their discussion of vitalism was more subdued than their Czech counterparts, but with a discussion of both there appears to be a pattern of acceptance of vitalism in "eastern Europe" for reasons that remain a bit mysterious. Aleš Ušeničnik (1868–1952) writing in "Čas" (The Times) observed in his review of Boris Zarnik's "Concerning the Essence of Life" (*O bistvu življenja*) that the choice in modern times really was between mechanism and vitalism. Ušeničnik, after surveying the discussions of vitalism and materialism since Aristotle, observed that vitalism must have some power to it, otherwise it would not keep returning to the surface (*površje*). Scientists like Driesch are to be commended, he continued, are to be commended because they are not simply "rechristening" older scientific worldviews.

Life cannot be understood, except according to a "super-mechanical principle," that is, according to other principles than the "mechanical." Vitalism was then in many ways a negative concept based on its embrace of non-mechanical explanation. Vitalists were correct, Ušeničnik observed, that living beings do indeed have some-thing quite unique "that is nowhere to be found in inanimate matter." Life, Ušeničnik continued, and living organisms had individuality, while "inanimate substance is indifferent." Life itself, according to modern science, is highly organized at the level of even the most simple cell, "it has different parts with different functions, all of which are in the service of the whole." (Ušeničnik 1914: 140–144). But for all of his agreements with vitalism Ušeničnik was clear that when vitalism morphed into a kind of pantheism, "where all nature is one", as it did in the writings of Zarnik, then the philosophy was "charlatanism" (Ušeničnik 1914: 152).

One of the consistent frustrations in the history of science is the limitation of science to one kind of practice (physics or chemistry) and to limit locality. Existing narratives which depend on a truncated account of "Europe" or the "West" or sci-ence itself tend to obscure the complex realities of those intellectual movements outside of self-imposed scholarly boundaries. The Czech and Slovene literature on vitalism is rich and especially illustrative, not least because it mirrors those debates going on elsewhere in Europe and the United States at the same time.

In the latter part of the nineteenth century Czech theologians especially were concerned with "materialism from Germany" which must be understood not only with the European-wide debate for and against modernizing forces but meshed with developing Czech national consciousness which was anti-German (and anti-Habsburg).

This intersected, in Czech speaking intellectual circles, with a deep understanding of German philosophy and a reliance on the German language itself as a medium of international philosophical communication. Slovenian theologies were certainly engaged in the same work against modernism, liberalism and secularism, as well as Darwinism and atheism, but this anti-German element (because of the particularities of language, culture, and history) was less strong than in the Czech case.

In the case of Czech intellectuals such as Durdík, Rádl and Mareš there was undoubtably a reception of eighteenth and nineteenth century philosophy, but as in the Slovenian case, vitalism and the critique of materialism, the rejection of modernism and liberalism, was in a clear sense an effort to develop a specifically Slovene or Czech intellectual life and scientific program. Although the reception of Darwinism was mixed in the Czech context, Rádl's suspicion of Darwinism melded well with Driesch's own anti-Darwinian stance. The combination of the critique of mechanism and Darwinism provided fertile soil for vitalism in Czech intellectual and scientific circles.

This stance towards vitalism melded with tradition not in the reductive sense, but with the very individual opinions of Czech and Slovene intellectuals at the time. Vitalism and the critique of materialism not only had deep roots in the Czech and Slovene cases, but also allowed for a specifically Czech and Slovene philosophical contribution, which melded well with social and political commitments. In an era were politics and philosophy and theology were indistinguishable, the position against a doctrine or ideology was also productive (i.e. anti-Darwinism) insofar that a position against Darwinism or against materialism, was also for vitalism or for a particular view of life. Thus, intellectuals in "Eastern Europe" were not merely reactive but developed those reactions into a positive program which developed alongside full expressions of scientific and philosophical systems, which had their roots in the many enlightenments of Europe, the many romanticisms and the many industrial revolutions.

References

Baár, M. 2010. *Historians and Nationalism: East-Central Europe in the Nineteenth Century.* Oxford: Oxford University Press.

Baran-Szołtys, M., J. Wierzejska, K. Kotyńska, A. Woldan, L. Cybenko, F. Solomon, N. Weck, D. Sosnowska, I. Voloshchuk, and H. Witoszynska. 2020. *Continuities and Discontinuities of the Habsburg Legacy in East-Central European Discourses since 1918.* Vienna: Vienna University Press.

Bartoš, F.Š. 1905. *Dialekticky slovnik moravsky (A Dialect Grammar of Moravian)* Prague: The Czech Academy of Emperor Franz Joseph.

Brunner, Emile. 1947. *Man in Revolt: A Christian Anthropology.* Philadelphia: The Westminster Press.

Čáda, F., and F.V. Krejčí. 1905. *Ceska mysl: casopis filosoficky.* Prague: Jan Laichter.

———. 1908. *Česká mysl (Czech Mind).* Prague: Jan Laichter.

Čáda, F., et al. 1906. Review of "Der Vitalismus als Geschichte und Lehre" *Česká mysl: Svazek 7* (The Czech Mind, volume 7), ed. František Čáda et al., 453–458. Prague: Jan Laichter

Chirot, D. 1991. *The Origins of Backwardness in Eastern Europe: Economics and Politics from the Middle Ages Until the Early Twentieth Century*. Berkeley: University of California Press.

David, Z.V. 2010. *Realism, Tolerance, and Liberalism in the Czech National Awakening: Legacies of the Bohemian Reformation*. Baltimore: Johns Hopkins University Press.

Delaporte, Francois, ed. 1994. *A Vital Rationalist: Selected Writings from Georges Canguilhem*. Boston: Zone Books.

Donohue, Christopher. 2020. Social borrowings and biological appropriations: Special issue introduction. *Studies in History and Philosophy of Science Part C: Studies in History and Philosophy of Biological and Biomedical Sciences* 83: 101309. https://doi.org/10.1016/j.shpsc.2020.101309.

Durdík, J. 1874. *O pokroku přírodních věd (On the Progress of the Natural Sciences)* Prague: Kolar.

———. 1876. *Rozpravy filosofické (Philosophical Transations)* Prague: I.L. Kober.

Feldman, M., and R. Griffin. 2008. *A Fascist Century: Essays by Roger Griffin*. London: Palgrave Macmillan.

Gerschenkron, A. 1965. *Economic Backwardness in Historical Perspective: A Book of Essays*. New York: Frederick A. Praeger Publishers.

Goll, J. 1897. *Čechy a Prusy ve středověku (Bohemia and Prussia in the Middle Ages)*. Prague: Bursík & Kohout.

Gordin, M.D. 2020. *Einstein in Bohemia*. Princeton: Princeton University Press.

Kann, R.A., and Z. David. 2017. *Peoples of the Eastern Habsburg Lands, 1526-1918*. Washington: University of Washington Press.

Lenz, A. 1878. *Syllabus jeho svatosti Pia IX.: jejz vyklada s povinnym zretelem ku syllabu (Syllabus of His Holiness Pius IX: in which he interprets with obligatory attention to the syllabus)*. Prague: Cyrillo-Method'sche Buchdr.

Lenz, Antonín. 1889. *Petra Chelčického Učení o sedmeře svátosti a poměr učení tohoto k Janu Viklifovi (Peter Chelčicky's Teachings on the Seven Sacraments and the Relation of this teaching to Jan Wycliff)*. Prague: Cyrillo-Methodius Press.

———. 1891. *Syllabus jeho svátosti Pia IX (Syllabus of His Holiness Pious IX)*. Prague: Cyrillo-Methodius Press.

Linehan, T. 2000. *British Fascism, 1918-39: Parties, Ideology and Culture*. Manchester: Manchester University Press.

Lukács, G. 1981. *The Destruction of Reason*. Atlantic Highlands: Humanities Press.

Mareš, František. 1897. "Mechanism a mysticism" (Mechanism and Mysticism) Živa, Časopis přírodniský edited by František Mareš and Bohuslav Rayman *(Life: A Magazine of the Natural Sciences) V Praze TISKEM A NÁKLADEM J. Otty* (Prague: Printed and Edited by Jan Otto)

Mareš, František. 1904. *Naturalism a svoboda vůle (Naturalism and Freedom of the Will)*. Prague.

Marius Turda, Christian Promitzer and Sevasti Trubeta 2011. *Health, Hygiene, and Eugenics in Southeastern Europe to 1945*. Budapest: Central European University Press.

Miller, D. 1999. *Forging Political Compromise: Antonín Svehla and the Czechoslovak Republican Party, 1918–1933*. Pittsburg: University of Pittsburgh Press.

Mosse, G.L. 1978. *Toward the Final Solution: A History of European Racism*. New York: Howard Fertig Publisher.

———. 1988. *The Culture of Western Europe: The Nineteenth and Twentieth Centuries*. Boulder: Westview Press.

Normandin, S., and C.T. Wolfe. 2013. *Vitalism and the Scientific Image in Post-Enlightenment Life Science, 1800-2010*. Cham: Springer.

Pospíšil, Josef. 1885. Kritika moderniho atomismu (A Criticism of Modern Atomism). *Časopis katolického duchovenstva (Journal of the Catholic Clergy)* 26

Priorelli, G. 2020. *Italian Fascism and Spanish Falangism in Comparison: Constructing the Nation*. New York: Springer.

Prochazka, M. 1876. *Katolicka verouka pro vyssi skoly stredni a vzdelanejsi obecenstvo vubec. (A Catholic Catechism for highschools and for more educated audiences in general)*. Budejovice: Stropek.

Rádl, Emmanuel. 1908. Filosofické názory Drieschovy (The Philosophical Opinions of Driesch). *Česká mysl (Czech Mind)* 7.

Rádl, E., and E.J. Hatfield. 1930. *The History of Biological Theories*. Oxford: H. Milford, Oxford University Press.

Rosenstock, B. 2017. *Transfinite Life: Oskar Goldberg and the Vitalist Imagination*. Urbana: Indiana University Press.

Saleeby, C.W. 1914. *The Progress of Eugenics*. New York: Cassell.

Semenenko, P. 1885. *Credo: chrzescijanskie prawdy wiary (Christian Truths of Faith)*. Lvov.

Soucy, R. 1972. *Fascism in France: The Case of Maurice Barrès*. Berkeley: University of California Press.

Stastny, A. 1874. *O mravnosti rozumove (Concerning the Morality of Intellect)*. Vol. Prague: Grager.

Trencsenyi, B., M. Kopeček, L.L. Gabrijelčič, M. Falina, and M. Baár. 2018. *A History of Modern Political Thought in East Central Europe: Volume II: Negotiating Modernity in the 'Short Twentieth Century' and Beyond, Part II: 1968-2018*. Oxford: Oxford University Press.

Turda, M., S. Antohi, C. Huang, and J. Rüsen. 2013. *Crafting Humans: From Genesis to Eugenics and Beyond*. Goettingen: V&R Unipress.

Ušeničnik, Aleš. 1914. *"O bistvu življenja" ("Concerning the Essence of Life")* Čas *(Time)*. Ljubljana: Katoliške Tiskarne (The Catholic Printing House).

Wolfe, C.T. 2016. *Materialism: A Historico-Philosophical Introduction*. Cham: Springer.

The Critical Difference Between Holism and Vitalism in Cassirer's Philosophy of Science

M. Chirimuuta

Spinoza would not have spent so much time considering a drowning fly if this behavior had not offered to the eye something other than a fragment of extension; the theory of animal machines is a "resistance" to the phenomenon of behavior. Therefore this phenomenon must still be conceptualized.
Merleau-Ponty The Structure of Behaviour (1942/1967, 126–7).

But we cannot deny that the equation between these two Weltanschauungen has not yet been found, even though it would provide us with all that we are striving for in the relationship between thought and reality. Perhaps it is erroneous to search for a stable balance between them; maybe it characterises the actual rhythm and formula of modern life that the boundary between the mechanistic conception of the world and the Goethean – whether we term that metaphysical, artistic or vitalistic – remains in constant flux. Simmel Kant and Goethe: on the History of the Modern Worldview (1906/1916).

Abstract This chapter surveys Ernst Cassirer's responses to the vitalist and holist/organicist movements in biology during the early decades of the twentieth century. I argue that examination of the combination of Cassirer's enthusiasm for holism, and rejection of vitalism, puts into relief many themes and preoccupations that are consistent across Cassirer's philosophical career, and aids the interpretation of his philosophy of symbolic forms. I propose that it is useful to read the third volume of the *Philosophy of Symbolic Forms* as a critical response to anti-rationalistic tendencies in the philosophy of Henri Bergson, and other proponents of *Lebensphilosophie*. Hence the availability of holism, as a purportedly less obscure alternative to vitalism, suits this broader agenda. At the same time, Cassirer's acceptance of holism depends on a commitment to the autonomy of biology which is at odds with the

M. Chirimuuta (✉)
Department of Philosophy, University of Edinburgh, Edinburgh, UK
e-mail: m.chirimuuta@ed.ac.uk

© The Author(s) 2023   85
C. Donohue, C. T. Wolfe (eds.), *Vitalism and Its Legacy in Twentieth Century Life Sciences and Philosophy*, History, Philosophy and Theory of the Life Sciences 29, https://doi.org/10.1007/978-3-031-12604-8_6

physicalism of the Vienna Circle, but consistent with Heidegger's favourable response to holism in comparison with vitalism. Yet, in the end we are left with an interpretative puzzle about how Cassirer proposes to avoid the encroachment of physicalism into theorising in the biological and human sciences while maintaining his view that progress in science is the result of increasing quantification.

1 Introduction

In the recent outpouring of works examining Ernst Cassirer's contributions to philosophy of science, far less attention has been paid to his writings on biology than to his numerous publications on mathematics and physics.[1] However, as I aim to show in this chapter, Cassirer's reflections on the science of biology and the phenomenon of life offer an interpretative perspective which affords a unified view of many of the features of Cassirer's philosophy that would appear otherwise disconnected. Another lesson of my study is that the influential view of Michael Friedman's (2000), of Cassirer as a philosopher holding an intermediate position between the logical empiricism of Carnap and the phenomenology of Heidegger, is questionable when one takes into account Cassirer's philosophy of biology – a topic not reviewed in Friedman's otherwise valuable book.

The quick statement to be made on the topic of Cassirer and vitalism, is that the philosopher rejected vitalism and endorsed holism or organicism.[2] The quick explanation for the choice of this position is that Cassirer, staying true to his neo-Kantian heritage, reacted negatively to the ontological inflationism of vitalism in biology – the positing of *entelechies*, etc.[3] – and to the anti-intellectual tone of vitalism in philosophy, exemplified in the writings of Henri Bergson. Yet, as I propose here, Cassirer's position deserves a more nuanced and involved account than this, and it is worth the trouble constructing it, because of the way that vitalism and the issues around to it – holism, the autonomy of biology, the very notion of vital phenomena – link together some of the disparate, but consistently held motivations for the philosophy of symbolic forms.

---

[1] But see Paci (1973), Krois (2004), and Stjernfelt (2011).

[2] For the purposes of accounting for Cassirer's opinion, it is not necessary to distinguish between these labels, except to say that 'holist' is a term applicable to a range of scientific disciplines (e.g. Gestalt psychology and field theory in physics), whereas organicism is more specific to biology. I will say more about them in Sect. 3. I classify as vitalist those theories that posit some kind of non-physical vital force, and hence the irreducibility of biological to physical laws, and as organicist/holist those theories that assert the autonomy of biology through the positing of irreducible biological concepts of the whole, structure, or organism.

[3] For a more positive assessment of Driesch see the chapters in "On the Heuristic Value of Hans Driesch's Vitalism" and "A Historico-Logical Re-assessment of Hans Driesch's Vitalism" this volume by Bolduc and Chen.

In Sect. 2 I discuss Cassirer's engagement with *Lebensphilosophie* – the modish (according to Heinrich Rickert)[4] philosophy of life of the first decades of the twentieth century. Bergson is often (inaccurately) classified as a vitalist in the same manner as Driesch, as well as a philosopher of life.[5] Given that the culminating third volume of the *Philosophy of Symbolic Forms (Phenomenology of Knowledge)* begins with a rejection of Bergson's guiding notion of the immediate intuition of life, one might deduce that Cassirer's rejection of vitalism is conditioned by his setting this major philosophical project against Bergson, and *Lebensphilosophie* more generally. However, that would be to ignore the many ways that the agenda of the *Philosophy of Symbolic Forms* is continuous with the currents of *Lebensphilosophie,* especially the work of Georg Simmel who was once Cassirer's teacher in Berlin.

Section 2 compares Cassirer's response to the controversy between mechanism, vitalism, and organicism, with that of three philosophers associated with the Vienna Circle and/or logical empiricism – Rudolph Carnap, Philipp Frank, and Ernst Nagel. Vitalism and organicism are both positions that express the autonomy of biology, as opposed to a mechanism consistent with the proposal that the natural sciences be unified under a physicalist standard where observation and linguistic report in biology are carried out under the same terms as in the sciences of non-living nature. As we see in Sect. 3, Cassirer's rejection of physicalism, and support for the autonomy of biology, are unsurprising given the high rank of Goethe in Cassirer's pantheon, and Cassirer's serious engagement with Kant's third *Critique.*

Section 3 also addresses the question of why Cassirer favours organicism over vitalism. Again, I show that the origins of this position can be found early in Cassirer's professional life – in the uptake of holistic principles on display in the 1910 book *Substance and Function.* The attachment to holism is persistent, and the story of the rise of holism in contemporary science is again told with approval in some of Cassirer's last publications. The more unexpected finding is the convergence of Cassirer and Heidegger in their specific arguments against Driesch who, it is claimed, did not sufficiently differentiate the causal mode of explanation proper to the physical sciences from the a-causal mode that begins with the sui-generis *phenomenon of life.*

The role, for Cassirer's philosophy of biology, of the notion of the basic phenomenon of life, brings up the question of whether the supposedly most "primitive" symbolic form, the *expressive function* has an ineliminable role to play in the mature natural science of biology. I argue in Sect. 4 that an equivocation on this very question is a symptom of a deeper tension between a radical pluralism and Enlightenment rationalism that each find various outlets within Cassirer's extensive reflections on scientific thought.

---

[4] See Schnädelbach (1984, 139).

[5] Peterson (2016, 6). Bergson (1907/1944, 48) is careful not to align his philosophy with Hans Driesch – see Posteraro, Chapter "Vitalism and the Problem of Individuation: Another Look at Bergson's *Élan Vital*", this volume.

2   Cassirer's Engagement with *Lebensphilosophie*

One could even say, as Cassirer did of Aristotle (Krois 2004, 282), that the "centre of gravity" of Cassirer's philosophy is in the examination of living and not inorganic phenomena. This claim appears arresting, if not straightforwardly false, under consideration of the fact that the bulk of Cassirer's output in the history and philosophy of science was allocated to the physical sciences. Cassirer wrote three monographs on the philosophy of modern physics but none on contemporary biology. The depiction of the advance of knowledge offered in the *Philosophy of Symbolic Forms* vol. 3 is the story of the development of mathematical physics. Correspondingly, the secondary literature on Cassirer's philosophy of science has made physics central to his thought. However, a different picture is obtained as soon as one considers that symbolic forms are *living forms*[6] – that Cassirer's mobile version of Kantianism is one in which the forms (erstwhile categories) of human thought grow and develop out of one another in an organic fashion; indeed, that the forms of the human mind are an outgrowth from a more comprehensive living order. Furthermore, it is the naturalist, Goethe, who contributes to the notion of living form, and whose influence on Cassirer's philosophy is pervasive. Cassirer is, in his own way, a philosopher of life, even though his writing places itself in deliberate opposition to the current that runs from Nietzsche and Schopenhauer to Bergson and Ludwig Klages.

We can follow Schnädelbach in describing *Lebensphilosophie* as, "a philosophical position which makes into the foundation and criterion of everything something which essentially stands *opposed* to rationality, reasons, concepts or the Idea – life as something irrational" (1984, 141).[7] Since this is a chapter on vitalism, I will concentrate first on a thinker who has been classified both as a vitalist and a philosopher of life, and with whom Cassirer directly engages – Henri Bergson. On the vitalism side, Bergson maintains that a fundamental vital impulse (*élan vital*) courses through the development of organic beings, rendering them unknowable by our discursive intellect – a mode of thought which must fall back on mechanistic explanations aiming at practical outcomes. Yet life is susceptible to a different form of intuitive knowing, whose basis is the insight disclosed in our own experience, since we are all living beings with an inner life. In particular, the true nature of

---

[6] The notion of "lebendige Gestalt" is discussed by Möckel (2005, 138–9), with reference to Cassirer's 1916 book, *Freiheit und Form*; and is deployed in *Feeling and Form* by Langer (1953), a book dedicated to the memory of Cassirer. A key feature of living forms is that they are unified *wholes,* not mere aggregates of parts (see Sect. 3). For Cassirer, both biological and cultural forms are wholes in this sense (Ferrari 1996, 109–10).

[7] Krois and Verene (1996, xi) point out that Cassirer associates the term *Lebensphilosophie* with a broader range of philosophers than is common now, applying it to the work of Schopenhauer, Kierkegaard and Nietzsche, as well as Heidegger. See Skidelsky (2009, chapter 7) on Cassirer's response to *Lebensphilosophie*. Of course, Cassirer does not share the anti-rationalist tendencies of this movement.

temporality, *durée*, is revealed to intuition but never to the intellect.[8] The anti-intellectual tenor of this last point is, alternatively, something praised and condemned in Bergson's philosophy.[9] At the same time, Cassirer argues, the recourse to a notion of immediate intuition of metaphysical truth puts Bergson in the ancient tradition of metaphysicians (Cassirer 1923/1955, 112). Indeed, Cassirer took *Lebensphilosophie* to be the modern manifestation of metaphysics (Krois and Verene 1996, xi).

Cassirer's rejection, but partial incorporation, of Bergsonian ideas, brings to light two of the major impulses that give coherence to the philosophy of symbolic forms: the renunciation of the metaphysician's ambition of *immediate* knowledge, in favour of the examination of symbolic mediacy; and the attempt to resolve the apparent conflict between human culture ("Geist") and the realm of natural life ("Leben") – a clash much lamented within the *Lebensphilosophie* of Weimar Germany. Schnädelbach (1984, 148) writes that,

> It was above all the influence of Henri Bergson's philosophy of the *élan vital* ….. which introduced temporality as a fundamental dimension into the 'Absolute' of life-philosophy. The Heracliteanism of the ontology of life-philosophy was thus given a quasi-epistemological justification. It is astonishing to what degree even those positions which explicitly attacked life-philosophy were affected by it. In the neo-Kantians, Kant's talk of the 'manifold' of sense became a 'heterogeneous continuuum', which was founded in the 'immediacy and irrational intuitiveness' of 'experienced life'.[10]

While we should not attribute these particular pieces of "neo-Kantian" terminology to Cassirer, the general claim is apt: that Cassirer's shaping of an alternative to the metaphysical philosophy of life utilised tropes of temporality and vitality congenial to that tradition.

On the first of these impulses – the renunciation of aspiration to immediate knowledge – we may take the introduction to *Philosophy of Symbolic Forms* vol. 3 as a long argument to the point that the "paradise of immediacy is closed to philosophy" (Cassirer 1929/1957, 40).[11] This piece seeks to reveal the inherent flaws within various iterations of the philosopher's quest for immediate knowledge, including the attempts of Berkeley and Mach to "return to the primal stratum of sensation and its pure facticity" (p. 25). Bergson's proclamation, from the "Introduction to Metaphysics" that, "La métaphysique est la science qui pretend se passer des symboles" (quoted, Cassirer 1929/1957, 36) is the culminating point in this account – it

---

[8] This summary is based on "An Introduction to Metaphysics" (Bergson 1903/1912) and *Creative Evolution* (Bergson 1907/1944). See essays in Burwick and Douglass (1992) on Bergson and the vitalism debate.

[9] Compare James (1909/1936) with Russell (1912).

[10] Indeed, one comes across the rhetoric of *Lebensphilosophie* in unexpected places. The final sentence of the "manifesto" of the Vienna Circle is: *"Die wissenschaftliche Weltauffassung dient dem Leben und das Leben nimmt sie auf* [The scientific world-conception serves life, and life elevates it]" (Carnap et al. 1929).

[11] Cf. Cassirer (1923/1955, 113): "To philosophy, which finds its fulfilment only in the sharpness of the concept and in the clarity of 'discursive' thought, the paradise of mysticism, the paradise of pure immediacy, is closed." See discussion in Ferrari (1995, 814).

is "perhaps the most radical rejection of the value and justification of symbolic formation in the whole history of metaphysics" (p. 36).

The Kantian roots of the replacement of a philosophical methodology resting in intuition, in favour of the examination of the forms of cognition and culture, are unmistakable.[12] It is just as important to see how Cassirer's matching of the Goethean to the Kantian allows him also to make use of the tropes of temporality and living-ness. In contrast to Simmel's essay on Goethe and Kant,[13] Cassirer's writing on these two figures emphasises their points of similarity over their differences. Both, for instance, show how one can and must be satisfied with the mediacy of our knowledge of nature:

> And the Kantian modesty was also quite congenial to his [Goethe's] thought. He was satis-
> fied with the 'colored reflection,'[14] and was convinced that in this colored reflection we
> possess life itself. 'We live amidst derivative phenomena,' he says, 'and do not know how
> to arrive at the ultimate question.'[15] This negation of 'absolute' knowledge meant therefore
> no loss to him, and it set no determinate limits to his way of inquiry. (Cassirer 1945a, 96)[16]

In addition, it is Goethe who points the way to a cognition of life that is not the deathly, abstract one which repulsed the post-Kantian Romantic philosophers, but is a cognition, through the medium of symbols, nonetheless (Möckel 2005, 82–3).[17] Goethe's scientific methodology depends on an adequacy of perceptual capacities of the researcher (perception of "facts" being, for Goethe, also a mode of theorization)[18] to the metamorphising forms of nature. Nature, for Goethe, is not the formless churn of the (stereotyped) Heraclitean flux, but a continual turnover of

---

[12] Cassirer (1923/1955, 113) points out Kant's betrayal of his own programme when he entertains the notion of the "intellectus archetypus" in the third *Critique*.

[13] Simmel (1906/1916) – and see remarks in Sect. 4 below.

[14] Hadot (2006, 258–9) relates that Faust, near the beginning of Part 2 of the play, "is forced to turn his back to the sun that blinds him, but he looks in ecstasy at the waterfall, where he sees the light of the day-star reflected in a rainbow: 'In the colored reflections we have life.'"

[15] Cassirer refers to Goethe *Maximen und Reflexionen* no. 1208.

[16] Cf. Cassirer (1945b, 76): "Kant was able to exert this influence on Goethe because at bottom the two agreed about dogmatic metaphysics."

[17] Cassirer (1930/1949) tells us that contemporary *Lebensphilosophie* is the direct descendent of the philosophy of romanticism, e.g. Schelling. Of additional relevance here is Cassirer's presenta-tion of Schelling's philosophy of nature in *The Problem of Knowledge* vol. 3, which Cassirer praises to the extent that it overlaps with Goethe's (1920, 23–270).

[18] "The highest thing would be to recognize that all fact is already theory" (Goethe *Maximen und Reflexionen* no. 575; quoted in Cassirer 1929/1957, 25 and 1996, 193).

forms that we can aspire to apprehend "if we ourselves remain mobile and supple".[19] As Hadot (2006, 254) relates, "Form" for Goethe " is not *Gestalt*, an immobile configuration, but *Bildung,* formation or growth."

Bergson contrasts the immediate intuition of life, with the mere "intellection" of it through the distorting medium of *rigid,* lifeless concepts:

> This empty and immobile space which is merely conceived, never perceived, has the value of a symbol only. How could you ever manufacture reality by manipulating symbols?
>
> But the symbol in this case responds to the most inveterate habits of our thought. We place ourselves as a rule in immobility, in which we find a point of support for practical purposes, and with this immobility we try to reconstruct motion. We only obtain in this way a clumsy imitation, a counterfeit of real movement, but this imitation is much more useful in life than the intuition of the thing itself would be. Now our mind has an irresistible tendency to consider that idea clearest which is most often useful to it. That is why immobility seems to it clearer than mobility, and rest anterior to movement. (Bergson 1903/1912, 52–3)

Cassirer, on the other hand, offers us "mobile" symbols, living forms – and concepts with open-ended extensions that do not elide the individuality of particulars but somehow set them in order with respect to one another.[20] There is a debt to Goethe, and to Georg Simmel (Ferrari 1996, 98–99).

Also Goethean is the hope that the relation of mind to life is not inherently adversarial; but I suspect that on this second impulse there is a more direct line of influence from Simmel. The case for the enmity between the intellectual "Geist" and life, or soul, is implicit in Bergson, and explicit in Klages, whose central work is entitled *Der Geist als Widersacher der Seele* (The Spirit[21] as Adversary of the Soul). Klages contends that, "life and spirit are two completely primary and essentially opposed powers, which can be reduced neither to each other nor to any third term" (1929, p. vii, quoted in Schnädelbach 1984, 149). [22] In contrast, Cassirer's position is that we should take Geist – understood here as the formative capacity of human thought and culture – as somehow having grown out of life, and therefore not inherently opposed to life. This claim is subject to extended treatment in the papers published as volume 4 of the *Philosophy of Symbolic Forms* (Cassirer 1996).

---

[19] Goethe *Zur Morphologie: Die Absicht eingeleitet*, quoted in Hadot (2006, 254). The holistic quality of human knowledge is a point at which Cassirer himself alludes to a happy alignment of mind and nature:

> Knowledge is 'organic' insofar as every part is conditioned by the whole and can be made 'understandable' only by reference to the whole. It cannot be composed of pieces, of elements, except to the extent that each part already carries in itself the 'form' of the whole (Cassirer 1996, 193).

[20] I discuss the significance of Cassirer's account of concept formation elsewhere (Chirimuuta 2020a, section 3).

[21] I follow the conventional translation of "Geist" as "spirit", though it must be appreciated that the German term does not have the "spiritualist" connotations of the English word.

[22] For Cassirer's discussions of Klages' writings see (1929/1957, 80–81; 1930/1949; 1996, 24ff).

Simmel was one of Cassirer's lecturers during undergraduate studies in Berlin, prior to Cassirer's induction into the Marburg school, and later a colleague at Berlin (Krois 1996). Simmel's last book, published in 1918, was a treatise on the philosophy of life entitled *Lebenanschauung* (The View of Life). The first two chapters are analysed in Cassirer's "'Geist' and 'Life'" text, the unpublished conclusion to the *Phenomenology of Knowledge*. Simmel argues that "transcendence is immanent in life" – that there is an inherent tendency for vital processes to reach beyond their present into their past and future. This is the basis for the "turn towards the idea" – life's negation of its own subjective drive for self-preservation through the positing of the objective ideals of human culture. The human mind, for Simmel, is an outgrowth of life but one whose operations can be tragically destructive of life (Simmel 1918/2010, 61). Cassirer's disagreement with Simmel, whose picture he mostly incorporates, boils down to a disavowal of its pessimistic implications, the prognosis of ever more vast separation between the mind and life.[23] The "turn to the idea", Cassirer instead proposes, "cannot be described as life bidding itself farewell in order to go forth into something foreign and distant from itself; rather, life must be seen as returning to itself, it 'comes into itself' in the medium of the symbolic forms" (1996, 19). It is interesting also to note that Simmel's *Lebenanschauung* exerted a strong influence on Heidegger during the genesis of *Being and Time* (Levine and Silver 2010, xxvi–xxvii), given that Cassirer and Heidegger presented quite similar criticisms of Driesch's vitalism (see Sect. 3).

3   The Rejection of Physicalism

Cassirer maintained cordial relations with members of the Vienna and Berlin circles of scientific philosophy, such as Rudolph Carnap and Hans Reichenbach, and was appreciative of their work (Krois 2000; Mormann 2016). That said, Cassirer's longest, published discussion of Carnap strikes a note of trenchant disagreement over the doctrine of physicalism presented by Carnap in his 1931 paper on the "physical language as the universal language of science" (Cassirer 1942/1961, 96–99). It appears in a late book, the *Zur Logik der Kulturwissenschaften* (Logic of the Cultural Sciences). This may seem a little far from our topic of vitalism but, as we will see, their stances towards physicalism condition these two philosophers' views regarding the autonomy of biology.

The physicalism developed by Carnap and Otto Neurath in the 1930s was not the ontological thesis prominent today in analytic philosophy, but the claim that all meaningful statements about the world can be stated in the physical language of objects bearing measurable properties, located in space and time. It follows that the languages of the various sciences (biology, sociology, psychology) are reducible to

---

[23] See Skidelsky (2003) and Geßner (1996) for comparison of the ideas of Simmel and Cassirer, especially regarding cultural pessimism.

physical language (Carnap 1931/1995, 66) and that with this single language in place, the division amongst the sciences disappears, such that a true *unity of science* is established (Carnap 1931/1995, 96). There is an immediate tension between the doctrine of the unity of science, and views that assert the autonomy of biology, social sciences and psychology. Whereas mechanism is consistent with physicalism and the unity of science, proponents of vitalism and organicism assert that biology is in some sense autonomous from the physical sciences – that it relies on its own laws or concepts, lacking correspondence to ones stateable in the physical language.[24]

The project of the philosophy of symbolic forms is tied to a pluralism about the ways that objectification and knowledge-making occur. The different departments of human thought and endeavour make use of sui generis symbolic forms, which amounts to their having different "languages" – in the broad sense of symbolic systems. This is not consistent with the elevation of the physical as a universal language (cf. Krois 2004, 280). Thus Cassirer argues for the autonomy of the human and biological sciences, and not for the unity of science in its physicalist formulation. In the next section I will account for why Cassirer opts for organicism/holism over the vitalist construal of the autonomy of biology. In the remainder of this section I will review arguments against both vitalism and holism, set out by three logical empiricists (Carnap, Philipp Frank, and Ernst Nagel) as these offer an important contrast class against which we will compare Cassirer's case.[25]

In the "manifesto" of the Vienna Circle, published collaboratively in 1929, we encounter a rejection of Driesch and Reinke's vitalism on the grounds that it is "metaphysical" (Carnap et al. 1929). It is asserted, however, that there is an "empirically graspable kernel" to vitalism, which is the claim that the laws applicable to the processes of organic nature are not reducible to physical laws – in other words, the thesis of the autonomy of biology. Carnap (1931/1995, 68) informs us that the, "Viennese circle is of the opinion that biological research in its present form is not adequate to answer the question" of reducibility, though the tentative expectation is that reduction will occur (p. 69). But, importantly, Carnap takes the view that the irreducibility of biological laws is *not inconsistent* with physicalism. Physicalism is only threatened by biology having its own *concepts* that cannot be defined in terms of "physical determinations". However, Carnap hastens to add, any such concepts would be metaphysical and 'nonsensical', in any case (p. 70). Thus Carnap rules out vitalism and organicism on the same grounds – both of these doctrines posit biological concepts, e.g. *entelechy* for vitalism, the concept of the 'whole' and 'organism' for organicism, which evade non-metaphysical construal.

Philipp Frank wrote at greater length than Carnap on foundational issues in biology. In a 1908 publication "Mechanismus oder Vitalismus?", Frank searches for a

---

[24] E.g. Driesch (1908, 143) writes, "What we have proved to be true has always been called *vitalism*, and so it may be called in our days again. But if you think a new and less ambitious term to be better for it, let us style it the doctrine of the *autonomy of life*.....[I]n our phrase autonomy is to signify the *being subjected* to laws peculiar to the phenomena in question."

[25] Chen (2019) provides a more comprehensive account of logical empiricist arguments against vitalism.

precise formulation of this question, especially in the light of the neo-vitalism of Driesch. Frank is even-handed in this paper: he criticizes both litigants for adding to the confusion, while taking the choice between mechanism and vitalism to be currently under-determined by empirical evidence. Only its track record of high heuristic value recommends mechanism over vitalism (Frank 1908, 408). The chapter, "Causality, Finalism and Vitalism" in the 1932 *Causal Law* book takes quite a different tone. Whereas the 1908 paper emphasizes the analogy between explanations positing vital "constants", and ordinary quantitative explanations in physics, the later book puts weight on the finalistic and spiritualistic aspects of Driesch's vitalism, which Frank takes to be a lightly veiled version of "ancient animism" (Frank 1932/1998, 112). Frank is here no more sympathetic to the organicist and holist alternatives to mechanism and vitalism, in spite of the fact that two organicist biologists, Ludwig von Bertalanffy and John Henry Woodger, were associated with the Vienna Circle (Hofer 2002). Frank's view is that the notion of the "whole" (beyond mere aggregate of parts) has no cognitive content, but "states something meaningful only about the emotional and volitional attitude of the claimant" (Frank 1932/1998, 129).[26] As we will see in the next section, this is a point at which Cassirer turns out to be in agreement with Heidegger, and misaligned with the logical empiricists, in spite of his apparently greater kinship with the philosophy of science of the latter group.

More involved arguments against the claim of organicists to be offering a meaningful alternative to both mechanism and vitalism are presented by Nagel (1951a, 1961). I mention this because it pertains to Nagel's fairly unsympathetic review of the *Problem of Knowledge* vol. 4[27] – the book that includes Cassirer's most extensive writing on the history and philosophy of biology, with organicism presented as a dialectical synthesis superior to both mechanism and vitalism (Cassirer 1950, 212). Nagel diagnoses Cassirer's misplaced acceptance of organicism as due to a tendency to "overrate Goethe as a thinker" (1951b, 149) and to be too much in thrall to the Sage of Königsberg:

> [Cassirer] finds merit in holism because it confirms in a significant manner Kant's conception of biological form as a heuristic rule. Opinions on the importance of holism as a philosophy of biology vary even among professional students of living organisms; but in this reviewer's judgment Cassirer certainly exaggerated its virtues. (Nagel, 1951b, 150)

In the next section I basically concur that the reach of Kant and Goethe is not to be discounted in the explanation of Cassirer's adherence to holistic approaches to

---

[26] Compare the 1934 Prague congress paper of Schlick:

> he presented an entire lecture, 'On the Concept of the Totality,' in which he claimed that while the distinction between totalities and aggregates might be linguistic or pragmatic, it was not a substantive distinction: there was no whole over and beyond the sum of parts. (Galison 1990, 744)

[27] Nagel seems to find Cassirer a wooly-headed thinker, who tries to mask his deficit in logical sharpness with an excess of scholarliness (1951b, 151).

biology. But we will also see that in this Cassirer is also prompted by concerns he shared with the phenomenologists.

In his critical discussion of Carnap (1931/1995), Cassirer writes, of the problem posed by physicalism, that its solution, "can be attained only by a phenomenological analysis." The task, Cassirer goes on to say, is "to understand each sort of language in its uniqueness – the language of science, the language of art, of religion, etc. We must seek to determine what each contributes to the building up of a 'common world.'" (1942/1961, 97). He gives the example of the contemplation of a painting which may be taken, under the determined attitude of the physicalist, as being merely a canvas with flecks of coloured paint on it. However, this would be to exclude from phenomenological examination, without justification, the "meaning" of the work, "which is not absorbed by what is merely physical, but is embodied upon and within it; it is the factor common to all that content which we designate as 'culture.'" (p. 98). We will see that an analogous consideration is at play in Cassirer's philosophy of biology: to apprehend living organisms in only physicalist terms is to *lose the phenomenon* – to neglect the way that living beings are first manifest to us in experience – in favour of a theoretically-motivated, unfaithful reconstruction.[28]

4   Holism Over Vitalism

It is not to be discounted that Cassirer had close personal connections with two prominent scientists associated with holism: the neurologist Kurt Goldstein, who was his cousin, and the biologist Jakob von Uexküll, a colleague at the University of Hamburg.[29] However, the case presented in *The Problem of Knowledge* vol. 4,[30] for the validity of organicism over vitalism makes more appeal to the authority of Goethe and Kant (of the *Critique of the Power of Judgement*) than to any contemporary figure. For instance, Cassirer criticizes Driesch for not, as claimed, staying true to the insights of Kant, but treating purposiveness as a fundamental power instead of a "point of view" (Cassirer 1950, 197–8); and he praises von Uexküll for following Goethe in his recognition that *morphology* (the study of form, structure) is the proper domain of biology as an autonomous science (pp. 203–5).[31]

The "Vitalism" chapter in *The Problem of Knowledge* vol. 4 starts with some recounting of the longer history of vitalism in the nineteenth century. More discussion is then allocated to Driesch's urchin experiments and the arguments for vitalism based on them. Like Carnap, Cassirer does seem to be bothered by the way that

---

[28] Cf. Merleau-Ponty (1942/1967, 126–7), quoted at the start of this paper.

[29] The intellectual connections with Goldstein and von Uexküll are examined in Chirimuuta (2020a) and Stjernfelt (2011), respectively.

[30] First published in 1950 (in English translation), but drafted around 1940.

[31] It should be noted that von Uexküll is often classified as a vitalist – as Cassirer (1950, 199) does. Stjernfelt (2011) observes that Cassirer tends to overplay the differences between von Uexküll and Driesch.

the positing of *entelechy* – something neither existing nor acting within space – takes Driesch beyond science and into metaphysics.[32] But unlike Carnap, Cassirer praises Driesch for isolating the singular problem posed by biological knowledge, as distinct from the science of the physical world:

> however one may regard his theory or disagree with its metaphysical implications, there is no denying that through his experiments and the questions that he raised in connection with them, Driesch contributed greatly to defining the characteristic methodological *principle* and *problem* of biology. (Cassirer 1950, 197)

According to Cassirer, von Uexküll's methodology provides a surer basis for the autonomy of biology, than Driesch's, because the starting point is anatomy rather than physiology (1950, 199). As he puts it, "[w]hile Driesch in his conception of entelechy wanted to demonstrate a specific autonomy of function, Uexküll started from the autonomy of *form*" (p. 200). This means that the biologist can avoid confrontation with the dynamical-causal explanations proper to physics, and take the study of biological structure as its own independent territory. This being possible because form, here, is not of the material sort studied in physics, but something more analogous to the ideal figure in geometry:

> Structure is not a material thing: it is the unity of immaterial relationships among the parts of an animal body. Just as plane geometry is the science not of the material triangles drawn on a blackboard with chalk but of the immaterial relationships between the three angles and three sides of a closed figure. (von Uexküll, quoted in Cassirer 1950, 200)

Uexküll's positing of a biological concept, form, not definable in terms of physical determinations raises his biology above the bar set by Carnap for scientific autonomy; of course it would also raise Carnapian suspicions of being hopelessly metaphysical – but those are not worries shared by Cassirer over this notion.

That Cassirer is open minded about this concept of form is probably due to his consistent endorsement of holistic notions across the sciences. As we saw in Sect. 2, Frank and Schlick shared an incredulity towards claims that organisms or social groups comprise wholes which hold some ontological or explanatory status independently of their parts. In contrast, we find in Cassirer (1910/1923, 333–4) an early endorsement of the claim of Gestalt psychology that, "[n]ot only the parts as such, but also their whole complex always produces definite effects upon our feeling and presentations", an endorsement reiterated much later in the *Logic of the Cultural Sciences* (1942/1961:170–2). In a late paper, "Structuralism in Modern Linguistics", Cassirer describes how physics, like psychology, broke away from the "classical tradition" and availed itself of the notion of the field which has ontological priority over material parts:

---

[32] "With these propositions [regarding the entelechy] Driesch wanted to establish biology as an independent fundamental science. But as we read them we get the impression that he has gone far beyond anything that *science* could establish and prove" (Cassirer 1950, 196).

The 'entelechy' is abstracted from the realm of spatial existence; it is something not sensible but supersensible. (Cassirer 1950, 198)

> The electro-magnetic field – in the sense of Faraday and Maxwell – is no aggregate of mate-
> rial points. We may, and must, indeed, speak of parts of the field; but these parts have no
> separate existence. The electron is, to use the term of Hermann Weyl, no element of the
> field; it is, rather, an outgrowth of the field ('eine Ausgeburt des Felds'). It is embedded in
> the field and exists only under the general structural conditions of the field. (Cassirer
> 1945b, 101)

Given that the concept of form most relevant to organicist biology is the notion of
the *whole organism* whose maintenance physiological processes appear to be
directed towards,[33] it is not surprising that Cassirer treats it as irreducible to collec-
tions of component parts.

Indeed, Cassirer takes the notion of wholeness in biology to be a metaphysically
non-committal replacement for the old ideas of teleology and purpose:

> To employ a teleological method in the study of living organisms means only that we exam-
> ine the processes of life so as to discover to what extent the character of preserving whole-
> ness manifests itself. (Cassirer 1950, 213)
> The expression 'wholeness' has the advantage of being completely free from hypothe-
> sis. It contains nothing psychic, nor does it assert that the events of life must always proceed
> in such a way as to achieve the highest degree of purposiveness. (Cassirer 1950, 213–4)

The other advantage of the version of autonomous biology that is based in the structural concept of the whole is that it allows for a division of labour between structural
and causal explanation, with the latter being left to physics.[34] Here the acausal
notion of the whole, or form, replaces the *final cause.* Two passages illustrate this
move, the first with reference to Uexküll, the second building up to an appreciative
discussion of Bertalanffy's 1932 book, *Theoretische Biologie*:

> In Uexküll's opinion physics is entirely correct in seeking to explain all connections in the
> world in terms solely of causality, but quite wrong if that means banishing every other mode
> of thinking from science. 'For causality is not the only rule at our disposal for construing
> the world.'" (Cassirer 1950, 203)
> Modern biology has not followed Driesch. But neither has it reverted to the 'machine
> theory of life.' It has avoided both extremes while concerning itself more and more with the
> purely methodological significance of the problem. For it, the primary question is not
> whether organic forms can be *explained* by means of purely mechanical forces; instead the
> emphasis falls on the fact that organic forms cannot be fully *described* through purely
> causal concepts. In demonstrating this it invokes the category of 'wholeness.' (Cassirer
> 1942/1961, 168)

The second passage is part of an argument that the cultural sciences (humanities)
must take the lead from biology in seeking their autonomy from the physical

---

[33] E.g. "There is no sense in trying to deny the maintenance of wholes in organic phenomena; the
correct procedure is first to discover this maintenance and then to explain it" (Bertalanffy, quoted
in Cassirer 1942/1961, 170).

[34] And we might add, to the special sciences when constructed in a physicalist manner – such as the
mechanistic biology of Loeb and others.

sciences via the study of sui generis forms.[35] The doctrine of physicalism, and the Unity of Science movement of which it was a part, was an attempt to dispense with the much discussed division between human and natural sciences (*Geisteswissenschaften* and *Naturwissenschaften*) (Carnap 1931/1995, 31–32).[36] It is noteworthy that Cassirer, in order to resist physicalism, places biology, alongside the humanities, as a science of form, shifting it from its traditional place alongside the natural sciences of the non-living world.

We now come to the comparison of Cassirer and Heidegger on the matter of vitalism. Both of these philosophers drew heavily from von Uexküll's notions of the organic, animal life and the formation of experiential worlds.[37] Both are critical of Driesch's vitalism. Where it goes wrong, they say, is in offering explanations of life that hypostatize vital forces conceptualized in the manner of physical forces – by treating life as explicable via causal principles, but of a different sort from causal-mechanical laws. This could not be more different from Frank's (1908) account, which takes no issue with Driesch treating vital forces in the manner of physical ones, but does not find a compelling enough case to be made for positing them. Driesch's physics-inspired methodology is a positive attribute, for Frank, not a liability; though subsequently, Frank (1932/1998) expressed objections to Driesch's attempts to emulate the pattern of physical explanation.

Cassirer draws from the third *Critique* the lesson that biological explanation must go beyond causation. Kant, he writes,

> insisted that these causal laws alone cannot enable us to *become even acquainted with that special realm of the phenomena* confronting us in organic nature, to say nothing of completely 'explaining' them. Here is the role of that other principle of order, which we call purposiveness. (Cassirer 1950, 198 emphasis added)[38]

Hence Driesch is criticised for his divergence from Kant. Likewise, Heidegger objects to the rush for causal explanation of distinctive properties of organisms, namely, self-production, self-regulation and self-renewal:

> The facts clearly do not allow us to doubt what has been said. What we have said also gives us a pointer to the peculiarity proper to the organism as against the machine, and thus also

---

[35] Cassirer (1942/1961, 172): "The 'scientific' study of the humanities can abandon itself to consideration of *its own forms* and *its own structures* with greater freedom and less restraint than heretofore as a consequence of the fact that the other fields of knowledge have become more mindful of their own very real problems of form. The logic of research is able now to assign each of these problems to its rightful place. Form-analysis and causal analysis are now seen to be orientations which, instead of contending with each other, complement each other, and which necessarily require each other in all knowledge."

[36] Cassirer (1942/1961, 88–92) discusses versions of the distinction, including the proposals of Windelband and Rickert.

[37] See Part 2 of Heidegger (1995) *Fundamental Concepts of Metaphysics*, derived from a lecture course of 1929–30. Krois (2004, 279) maintains that it was Cassirer who introduced Heidegger to Uexküll's work during the 1929 Davos lectures and debate. See also Gordon (2010) on Davos and the discussion of Uexküll.

[38] See Cassirer (1918/1981, chap. 6) for his interpretation of the *Critique of the Power of Judgement*, discussed by Ferrari (1996, chap. 3).

to the way in which organs belong to the organism as against the way in which machine components belong to the machine. And yet this pointer is still dangerous because it can, and repeatedly does, lead to the following conclusion: If the organism possesses this capacity for self-production, self-regulation and self-renewal, then the organism must contain an effective agency and power of its own, an entelechy and a vital agent which effects all this (a 'natural factor'). But this conception simply eliminates the problem, i.e., it no longer allows one to arise. Thus the real problem which is involved in determining the essence of life cannot even be seen because life is now handed over to some causal factor. (Heidegger 1995, 222–3)

What both find problematic about the encroachment of causal explanation is that it threatens appreciation of the unique *phenomena* of life. As Cassirer relates from Kant (in the quotation from the previous paragraph), the exclusive causal treatment "cannot enable us to become even acquainted" with those phenomena. This is how Heidegger makes a similar point, but without any explicit reference to Kant:

Thus Driesch was driven by his experiments to adopt his biological theory, known as a neovitalism, which is characterized by the appeal to a certain force or entelechy. This theory is repudiated in large measure by biology today. As far as biological problems are concerned, vitalism is just as dangerous as mechanism. While the latter does not allow the question of purposive behaviour to arise, vitalism tries to solve the problem too hastily. But the task is to recognize the full import of this purposive striving before appealing to some force which, moreover, explains nothing. (Heidegger 1995, 262)

As mentioned at the end of Sect. 2, a feature of phenomenologists', like Merleau-Ponty and Heidegger's, writings on biology is a concern not to "lose the phenomenon" in the midst of in-apt physicalistic descriptions that do not relate the features of living beings as they occur to us, irreducibly, in experience. It is telling that Cassirer concurs with them on this point, and that he takes it to be a lesson learned from Goethe:

The fundamental reality, the Urphänomen, in the sense of Goethe, the ultimate phenomenon may, indeed, be designated by the term 'life.' This phenomenon is accessible to everyone; but it is 'incomprehensible' in the sense that it admits no definition, no abstract theoretical explanation. We cannot explain it, if explanation means the reduction of an unknown to a better-known fact, for there is no better-known fact. (Cassirer 1942/1979, 193–4)

While not attempting "explanation", in drafts written around this time Cassirer (1996, Part II, chap. 2) himself deploys the notion of "basis phenomena" beginning, of course, with an analysis of Goethe.[39]

---

[39] Cassirer (1996, 137) states, relatedly, that basis phenomena are "the 'originär-gebenden' [primordially giving] intentions in Husserl's sense." See also note 43 below on this text.

5   An Interpretative Puzzle

In this final section I spell out one implication of my examination of Cassirer's response to vitalism for broader issues in the understanding of his philosophical project. The larger matter is the way that Cassirer attempts to uphold a progressivist narrative about the advance of science through the increasing mathematization of natural phenomena, while at the same time acknowledging concerns about the depersonalised character of the scientific worldview which attends this develop-ment. Cassirer's support for the autonomy of biology is not, strictly speaking, con-sistent with his progressivism. This leads to an unacknowledged tension within Cassirer's writings about the relative status of the non-scientific ("expressive") and physical-quantitative modes of apprehending nature, as I will now show.

In the second chapter of *Logic of the Cultural Sciences* Cassirer contrasts "per-ception of things" with "perception of expression", finding here the basic grounds for the distinction between natural and human sciences. In *expression-perception* we observe the world "as if it were something 'like ourselves'" – a bearer of subjec-tivity (1942/1961, 93), lacking causal determinacy (p. 94). In *thing-perception,* "we observe it [the world] as a completely spacial [sic] object and as the sum total of temporal transformations which complete themselves in this object" (p. 93); it is prerequisite for theoretical explanation and causal discovery (p. 94). Given that Cassirer rests the autonomy of biology, like that of the humanities, on the idea of an acausal knowing of form or structure (see 1942/1961, chap. 4), should we infer that for Cassirer, expression-perception is requisite for observation in the science of biology, for apprehension of the basic phenomenon of livingness, as well as for the appreciation of cultural objects like works of art? The interpretative puzzle is that Cassirer does not give us an unambiguous answer to this question.

There are passages that support an affirmative answer in Part 1 of *Philosophy of Symbolic Forms* vol. 3, where the notion of expression perception gets its first lengthy exposition.[40] For example:

> the reality we apprehend is in its original form is not a reality of a determinate world of things, originating apart from us; rather it is the certainty of a *living efficacy* that we experi-ence. Yet this access to reality is given us not by the datum of sensation but only in the origi-nal phenomenon of expression and expressive understanding. (Cassirer 1929/1957, 73 emphasis added)

And also he writes of the "expressive function", which is dominant mindset of the "mythical world" that it,

> can never wholly enter into this form [of physical causal relations] and never be submerged in it – for if it did so, not only would the mythical world of demons and gods disappear, but the fundamental phenomenon of 'livingness' as such [das Grundphänomen des 'Lebendingen' überhaupt] would vanish. Thus we see that the basic motif of consciousness

---

[40]Cassirer's debt to Klages in this text, and to another philosopher of life, Max Scheler, is not to be discounted.

which we have recognized as the actual organon of the mythical world intervenes at a decisive point in the structure of empirical reality (Cassirer 1929/1957, 88)

But a problem with this first interpretation – that the biologist's apprehension of the basic phenomenon of life depends on an employment of expression-perception – appears in a third passage from the same exposition of the expressive function:

> there is a kind of *experience* of reality which is *situated wholly outside this form of scientific explanation and interpretation* [naturwissenschaftlichen Erklärung und Deutung]. It is present wherever the being that is apprehended in perception confronts us not as a reality of things, of mere objects, but as a kind of presence of living subjects [Art des Daseins lebendiger Subjekte]. (Cassirer 1929/1957, 62 emphasis added)

The issue is that this "perception of life" is reported as lying "wholly outside…. scientific explanation and interpretation."

This lays ground for a negative answer to our question, one supported by some passages in the *Logic of Cultural Science*, where Cassirer describes the increasing de-personalisation of the worldview (dare we say, disenchantment) that has accompanied the development of natural science. He writes that,

> Natural science, as such, should and must be free to determine its way to the attainment of this goal. Not only does it seek increasingly to suppress all that is 'personal' it strives toward a conception of the world from which the 'personal' has been eliminated. It achieves its true aim only by disregarding the world of self and other. (Cassirer 1942/1961, 103–4)

Lest we think that an exception might be made for biology, Cassirer adds that,

> Even biology must not hold back; even for it, the dominance of 'vitalism' appears to have come to an end. Thus life is not only expelled from the inorganic, it is also banished from organic nature. Even the organism is subject to the laws of mechanics, the laws of pressure and impact, and without qualification.
>
> All attempts to oppose this radical 'devitalization' of nature with metaphysical arguments have not only miscarried but have compromised the very cause they have sought to serve. (1942/1961, 104)

And he goes on to describe the failed project of Gustav Fechner to resist this tendency. It is significant here that "vitalism" is equated with the attempt to retain a personalized view of organic nature, and mechanism is depicted as the inevitable outcome of scientific development.[41] The connection between vitalism and personalization makes sense of the linkage between the perception of subjectivity and the perception of life that is on display in the passages quoted above.

---

[41] Cassirer (1942/1961, 105–6) writes, "the mechanical theory has never tired of reducing to 'tropisms' all the phenomena in which Fechner sought to find proof of psychic life in plants; these can themselves be explained by means of known physical and chemical forces. According to the mechanical theory, heliotropism, geotropism, and phototropism are sufficient to account for the processes of plant life. The modern founders of tropism theory have not hesitated to extend the theory to animal life. In doing so they have regarded themselves as having established strict empirical proof for Descartes' theory that animals are automatons." And on this last point reference is made to Loeb's *Dynamics of Living Matter.*

Yet curiously, later in the same book, organicism is championed as the alternative to both mechanism and vitalism, and a path to the autonomy of biology (p. 168, quoted above).

Given that we find passages *within the same text* that support either the affirmative or negative answer to our question, we cannot put the problem down to Cassirer changing his mind over time. A plausible explanation is that the equivocation over the role of expression-perception in the natural science of biology is symptomatic of an underlying and unacknowledged tension in Cassirer's philosophy. There are two fundamental commitments in Cassirer's worldview that seem to jostle for supremacy. On the one hand he is a deep pluralist, endorsing the validity of a multiplicity of a symbolic worlds, each generating objects on their own conceptual terms; on the other he is a proponent of the commonplace narrative, associated with Enlightenment ideals of rationality, of mono-directional human progress towards the form of thought perfected in the exact sciences. As Carus (2007) relates, the philosophy of Carnap is one late occurrence of this Enlightenment project. The grand narrative structure of the *Phenomenology of Knowledge* is, likewise, in keeping with this tradition, where human thought is described as emerging from a "primitive" state, governed by mythological ideas, towards an "advanced" mindset that allows for the achievements of mathematical physics.

On the monotonic, progressivist narrative there should be no special role for "primitive" expression perception in biology, insofar as it is a 'mature' natural science. Yet without a distinct mode of perception for living beings, Cassirer has no principled grounds to reject a physicalism that denies the autonomy of biology. Cassirer's deep pluralism does allow for an autonomy of biology, and a view about the irreducible nature of living phenomena that is close to that of phenomenologists such as Heidegger. But unlike the phenomenologists, Cassirer declines to package these ideas with a generalised criticism of scientism within philosophy and wider culture.[42]

Therefore, it appears not that Cassirer is an *intermediate* figure between Carnap and Heidegger (as Friedman 2000 would have it), but that his philosophy is liable, like the Necker cube, to an aspect switch, depending on whether his pluralism or his progressivism is seen to be standing in the foreground. When the pluralism juts out, one finds a Cassirer closer to the phenomenologists;[43] when the progressivism projects forward, the philosophy of the Vienna Circle is not far away. This chapter begins with a quotation from Simmel who describes the modern condition as one

---

[42] In another essay I discuss the organicism of Kurt Goldstein in relation to the criticism of certain methods of abstraction within science, presented by Husserl and Merleau-Ponty (Chirimuuta 2020b).

[43] I say "closer" but not "identical" to. In Heidegger's review of the second volume of *Philosophy of Symbolic Forms*, on mythical thought, we find the criticism that Cassirer, adhering to the neo-Kantian epistemological interpretation of Kant, is not sufficiently aware of the depth of the difference (the "abyss") between the mythical and other modes of understanding

an interpretation of the mythical understanding of Being is much more labyrinthine and abysmal than is suggested by Cassirer's presentation. (Heidegger 1928/1997, 189) experiencing a kind of nervous oscillation between worldviews, alternately "mechanistic" (and, we may add, physicalist) or "vitalistic", Goethean and "artistic". Cassirer's vacillation on the role of the expressive function in the modern science of biology seems the perfect example of this:[44] it is not mere logical inconsistency, or lack of systematicity on Cassirer's part, but suggestive of something more deeply felt in Cassirer's response to the problem posed to philosophy by the apparent directionality, the semblance of a telos, manifest in the historical trend named modernity.

Acknowledgements I am very grateful to Bohang Chen, Philip Honenberger, and Charles Wolfe for their encouragement and comments on this work.

References

Bergson, Henri. 1903/1912. *An Introduction to Metaphysics*. Trans.T. E. Hulme. New York: G. P. Putnam's Sons.
————. 1907/1944. *Creative Evolution*. New York: Random House.
Burwick, Frederick, and Paul Douglass, eds. 1992. *The Crisis in Modernism: Bergson and the vitalist controversy*. Cambridge: Cambridge Uinversity Press.
Carnap, Rudolf. 1931/1995. *The Unity of Science*. Bristol: Thoemmes Press.
Carnap, Rudolf, Hans Hahn, and Otto Neurath. 1929. *Wissenschaftliche Weltauffassung – Der Wiener Kreis*. Wien: Wolf.
Carus, A.W. 2007. *Carnap and Twentieth Century Thought: Explication as Enlightenment*. Cambridge: Cambridge University Press.
Cassirer, Ernst. 1920. *Das Erkenntnisproblem in der Philosophie und Wissenschaft der neuren Zeit: die nachkantischen Systeme*. Vol. 3. Berlin: Verlag Bruno Cassirer.
————. 1910/1923. *Substance and Function, and Einstein's Theory of Relativity*. Trans. William Curtis Swabey and Marie Collins Swabey. Chicago: Open Court.
————. 1945a. *Rousseau, Kant and Goethe*. Trans. James Gutmann, Paul O. Kristeller, and John H. Randall Jr. Princeton: Princeton University Press.
————. 1945b. Structuralism in Modern Linguistics. *WORD* 1 (2): 99–120.
————. 1930/1949. 'Spirit' and 'Life' in Contemporary Philosophy. In *The Library of Living Philosophers, Volume VI: The Philosophy of Ernst Cassirer*, ed. Paul Arthur Schlipp, 855–880. Menasha: George Banta Publishing Company.
————. 1950. *The Problem of Knowledge: Philosophy, Science, and History since Hegel*. Trans. William H. Woglom. New Haven: Yale University Press.
————. 1923/1955. *The Philosophy of Symbolic Forms, Volume 1: Language*. Trans. Ralph Manheim. New Haven: Yale University Press.

---

[44] In the unpublished text "On Basis Phenomena" we have an indication that Cassirer was self-conscious about attempting to reconcile Simmel's two "worldviews". The key passage is Cassirer (1996, 136), which begins with the conflict between Goethe's resting place in "primary phenomena" and the demands of critical, reflective thought, dwells next on the Geist/Leben "opposition", and ends with the question,

Can we preserve respect for the primary phenomena, without acting in opposition to the critical spirit, without becoming guilty of sinning against the mind, which occurs when we deny its original right – its autonomy – so that we treat it as something foreign, an intruder (*l'intrus*)?

————. 1929/1957. *The Philosophy of Symbolic Forms, Volume 3: The Phenomenology of Knowledge*. New Haven: Yale University Press.

————. 1942/1961. *The Logic of the Humanities*. Trans. Clarence Smith Howe. New Have: Yale University Press.

————. 1942/1979. Language and Art II. In *Symbol Myth and Culture: Essays and Lectures of Ernst Cassirer 1935-1945*, ed. Donald Philip Verene, 166–195. New Haven: Yale University Press.

————. 1918/1981. *Kant's Life and Thought*. Trans. James Haden. New Haven: Yale University Press.

————. 1996. *The Philosophy of Symbolic Forms, Volume 4: The Metaphysics of Symbolic Forms*. Trans. J.M. Krois. New Haven: Yale University Press.

Chen, Bohang. 2019. Revisiting the Logical Empiricist Criticisms of Vitalism. *Transversal* 7: 1–17.

Chirimuuta, M. 2020a. Cassirer and Goldstein on Abstraction and the Autonomy of Biology. *HOPOS: The Journal of the International Society for the History of Philosophy of Science*. https://doi.org/10.1086/710181.

————. 2020b. The Reflex Machine and the Cybernetic Brain: The Critique of Abstraction and its Application to Computationalism. *Perspectives on Science* 28 (3): 421–457.

Driesch, Hans. 1908. *The Science and Philosophy of the Organism*. London: Adam and Charles Black.

Ferrari, Massimo. 1995. 'Metafisica delle Forme Simboliche': Note su Cassirer inedito. *Rivista di Storia della Filosofia* 50 (4): 809–837.

————. 1996. *Ernst Cassirer. Dalla scuola di Marburgo alla filosofia della cultura*. Florence: L.S. Olschki.

Frank, Phillipp. 1908. Mechanismus oder Vitalismus. Versuch einer präzisen Formulierung der Fragestellung. *Annalen der Naturphilosophie* 7: 393–409.

————. 1932/1998. *The Law of Causality and its Limits*. Trans. Marie Neurath and Robert S. Cohen. Dordrecht: Kluwer Academic Publishers.

Friedman, M. 2000. *A Parting of the Ways*. Chicago: Open Court.

Galison, P. 1990. Aufbau/Bauhaus: Logical Positivism and Architectural Modernism. *Critical Inquiry* 16 (4): 709–752.

Geßner, Willfried. 1996. Tragödie oder Schauspiel? Cassirers Kritik an Simmels Kulturkritik. *Simmel Newsletter* 6 (1): 57–72.

Gordon, P.E. 2010. *Continental Divide*. Cambridge, MA: Harvard University Press.

Hadot, Pierre. 2006. *The Veil of Isis: An Essay on the History of the Idea of Nature*. Trans. Michael Chase. Cambridge, MA: Belknap Press of Harvard University Press.

Heidegger, Martin. 1995. *The Fundamental Concepts of Metaphysics*. Trans. William McNeill and Nicholas Walker. Bloomington: Indiana University Press.

————. 1928/1997. Review of Ernst Cassirer, Philosophy of Symbolic Forms vol. 2: Mythical Thought. In *Kant and the Problem of Metaphysics*, ed. Richard Taft, 180–207 (Appendix II). Bloomington: Indiana University Press.

Hofer, Veronika. 2002. Philosophy of Biology around the Vienna Circle: Ludwig von Bertalanffy, Joseph Henry Woodger and Philipp Frank. In *History of Philosophy of Science: New Trends and Perspectives*, ed. Michael Heidelberger and Friedrich Stadler, 325–333. Dordrecht: Kluwer Academic.

James, William. 1909/1936. Bergson and his Critique of Intellectualism. In *Essays in Radical Empiricism – A Pluralistic Universe*, ed. Ralph Barton Perry. New York: Longmans, Green and Co.

Klages, Ludwig. 1929. *Der Geist als Widersacher der Seele*. Vol. I. Leipzig.

Krois, J.M. 1996. Ten Theses on Cassirer's Late Reception of Simmel's Thought. *Simmel Newsletter* 6 (1): 73–78.

————. 2000. Ernst Cassirer und der Wiener Kreis. In *Elemente moderner Wissenschaftstheorie*, ed. Friedrich Stadler. Vienna/New York: Springer.

————. 2004. Ernst Cassirer's Philosophy of Biology. *Sign Systems Studies* 32 (1–2): 277–295.

Krois, John Michael, and Donald Philip Verene. 1996. Introduction. In *The Philosophy of Symbolic Forms, Volume 4: The Metaphysics of Symbolic Forms*, ed. John Michael Krois and Donald Philip Verene. New Haven: Yale University Press.

Langer, Susanne. 1953. *Feeling and Form: A Theory of Art*. New York: Charles Scribner's Sons.

Levine, Donald N., and Daniel Silver. 2010. Introduction. In *The View of Life: Four Metaphysical Essays*, ed. Donald N. Levine, Daniel Silver, and Georg Simmel. Chicago: University of Chicago Press.

Merleau-Ponty, Maurice. 1942/1967. *The Structure of Behaviour*. Trans. Alden L. Fisher. Boston: Beacon Press.

Möckel, Christian. 2005. *Das Urphänomen des Lebens: Ersnt Cassirers Lebensbegriff*. Hamburg: Felix Meiner.

Mormann, Thomas. 2016. Wissenschaftliche Philosophie im Exil: Cassirer und der Wiener Kreis nach 1933. In *Husserl, Cassirer, Schlick*, ed. Matthias Neuber. Dordrecht: Springer.

Nagel, Ernst. 1951a. Mechanistic Explanation and Organismic Biology. *Philosophy and Phenomenological Research* 11 (3): 327–338.

———. 1951b. Review of The Problem of Knowledge. Philosophy, Science, and History since Hegel, by E. Cassirer. *Journal of Philosophy* 48 (5): 147–151.

———. 1961. *The Structure of Science: Problems in the Logic of Scientific Explanation*. New York: Harcourt, Brace & World.

Paci, E. 1973. La presa di coscienza della biologia in Cassirer. In *Idee per una enciclopedia fenomenologica*, ed. E. Paci, 456–464. Milan: Bompiani.

Peterson, Erik L. 2016. *The Life Organic: The Theoretical Biology Club and the Roots of Epigenetics*. Pittsburgh: Pittsburgh University Press.

Russell, Betrand. 1912. The Philosophy of Bergson. *The Monist* 22: 321–347.

Schnädelbach, Herbert. 1984. *Philosophy in Germany, 1831-1933*. Trans. Eric Matthews. Cambridge: Cambridge University Press.

Simmel, Georg. 1906/1916. *Kant und Goethe. Zur Geschichte der modernen Weltanschauung*. Trans. Josef Bleicher. Vol. https://generation-online.org/p/fp_simmel1.htm. Berlin: Kurt Wolff Verlag.

———. 1918/2010. *The View of Life: Four Metaphysical Chapters*. Trans. John A. Y. Andrews and Donald N. Levine. Chicago: Chicago University Press.

Skidelsky, E. 2003. From Epistemology to Cultural Criticism: Georg Simmel and Ernst Cassirer. *History of European Ideas* 29: 365–381.

———. 2009. *Ernst Cassirer: The Last Philosopher of Culture*. Princeton: Princeton University Press.

Stjernfelt, Frederik. 2011. Simple Animals and Complex Biology: Von Uexküll's Two-Fold Influence on Cassirer's Philosophy. *Synthese* 179: 169–186.

Canguilhem and the Greeks: Vitalism Between History and Philosophy

Brooke Holmes

Abstract In this essay, I examine the role of ancient Greek medicine and philosophy in Georges Canguilhem's analysis of vitalism at the intersection of history and philosophy in his essay "Aspects of Vitalism" in light of larger questions about the historicity of "life" as a concept in the history and philosophy of science and contemporary biopolitical theory. Vitalism, for Canguilhem, is not a proper object of the history of science. But nor is it a philosophy that exists outside of historical time. I show how Canguilhem embeds vitalism both historically and trans-historically by threading each of its three "aspects" in the essay through ancient Greece. Canguilhem distinguishes his own understanding of both life and vitalism from that of the "classical" vitalists of the eighteenth century by refusing to read ancient Greece as romantically naïve or pre-technological and instead locating a dialectic between vitalism and mechanism already in antiquity. I argue for a critical re-reading of Canguilhem's own conjunction of vitalism and Hellenism that resists its figuration of ancient Greece as the place where the human qua species first comes to take itself as an object of knowledge. I instead propose reading ancient Greek medical and philosophical texts that are read and reread in debates about the nature of human life

I'm very grateful to Charles Wolfe for all the excellent feedback on multiple versions of this paper. Thanks are due as well to Christopher Donohue and the other participants and audience members in the vitalism workshop at the National Human Genome Research Institute in July 2019, as well as to the participants and audience members at the "Ancient Holisms" conference at University College London, and above all to Chiara Thumiger, who, together with an anonymous reviewer, also offered invaluable comments on the written version; and to the audience at SUNY-Buffalo in November 2019, especially Kalliopi Nikolopoulou and Rodolphe Gasché. The audience at the virtual version of this talk in the "Critical Antiquities" network helped with final revisions, and I thank Tristan Bradshaw and Ben Brown for the invitation to present in such a stimulating setting. I also want to thank Colin Webster for his incisive questions on an earlier version, which much improved the paper; Stefanos Geroulanos for generous and helpful comments on the final version; and Paul Eberwine for careful editorial assistance. All remaining errors and shortcomings are, of course, my own.

B. Holmes (✉)
Department of Classics, Princeton University, Princeton, NJ, USA
e-mail: bholmes@princeton.edu

© The Author(s) 2023
C. Donohue, C. T. Wolfe (eds.), *Vitalism and Its Legacy in Twentieth Century Life Sciences and Philosophy*, History, Philosophy and Theory of the Life Sciences 29, https://doi.org/10.1007/978-3-031-12604-8_7

and the life of Nature over millennia as part of a milieu that shapes how contemporary thinkers theorize life in the interest of human flourishing.

The Greeks haunt life as an object of historical and philosophical inquiry. The field named biology—a science (*logos*) of *bios*, one of the Greek concepts of life—has been around for a little more than two centuries. But the prefix *bio–* has been proliferating in continental philosophy, critical theory, and political theory since the turn of the millennium under the auspices of biopolitics. Biopolitics is not synonymous with a history or philosophy of biology, even less so with the cluster of philosophical positions on "life" associated with vitalism. Yet the intensification of theoretical concern with life as an object of political, economic, and social control over the past few decades has produced a series of dominant narratives about the history of "life" as a concept deployed with material effects within networks of power. Within these narratives, both historiographical and philosophical, the Greeks tend to be positioned in one of two ways.

The first strategy consigns the Greeks to obsolescence in the face of modernity's appropriation of life as an object of scientific and medical knowledge and state management. Michel Foucault famously declared in 1966 in *The Order of Things* that life as an object of knowledge does not exist prior to the nineteenth century; only, rather, "living beings, which were viewed through a grid of knowledge constituted by natural history."[1] Historians of the life sciences have pointed to the emergence of biology as an autonomous academic discipline around 1800 as a critical turning point in the modern conceptualization of life. Even more important to the periodization of "life" for biopolitical theory is Foucault's claim in the final section of the first volume of the *History of Sexuality* that a society's "threshold of modernity" has been reached "when the life of the species is wagered on its own political strategies." "For millennia," Foucault continues, "man remained what he was for Aristotle: a living animal with the additional capacity for a political existence; modern man is an animal whose politics places his existence as a living being in question."[2] Foucault himself marked the form of power emergent on the threshold of modernity with the neologism "biopower," a term that is now largely superseded by biopolitics. Biopolitics is thus, on this version of a Foucauldian line, essentially modern. And biopolitics, in turn, defines modernity.

The historicity of "life" in relationship to the Greeks in Foucault is complicated by his late work on *bios*.[3] But it is not Foucault's Greeks who have been primarily responsible for keeping Greek "life" alive in biopolitical theory for the past two decades. Far more consequential has been Giorgio Agamben's ubiquitous claim that ancient Greek life is divided at its core into *bios* and *zōē*. On Agamben's analysis, *bios* defines life in the robust sense of the life lived by the proper political subject, whereas it falls to *zōē* to designate what he calls "natural life" or "bare life." Far from lying outside of politics, *zōē* marks a state of exception within the political,

---

[1] Foucault 1970: 139.

[2] Foucault 1978: 143.

[3] Holmes 2019.

where life is wholly vulnerable to the power of the sovereign. Agamben thus uses *bios* and *zōē* not only to articulate the foundational structure of Western politics but to establish politics itself "as the truly fundamental structure of Western metaphysics insofar as it occupies the threshold on which the relation between the living being and the *logos* is realized."[4] Agamben has his own understanding of a biopolitical modernity in the Nazi camps. Yet he invests a semantic split in Greek "life" with far-reaching structural–conceptual implications for metaphysics and politics in "the West." In so doing, he enacts another familiar way of conjugating "the Greeks" with life. They are there at the origins of the West and at the foundation of Man.

Here, then, are the choices on offer for the relationship of "the Greeks" and life. Either Greek life is obsolescent; or it survives, ghostly but powerful. Either modernity should be read in terms of autogenesis, or we should tell an ancient origin story. From one perspective, I find the choice easy enough: the ancient Greeks matter to how we historicize "life." More specifically, I think that the history of the concepts of life, nature, and the body elaborated in ancient Greek texts produces, over the course of multiple readings in multiple reception communities not limited to Western Europe, philosophically interesting puzzles that we are still grappling with.

Yet from another perspective, this choice commits us all too easily to an origin story that brings with it problems bound up in the entwined histories of modern vitalism, Philhellenism, and European theories of race in the nineteenth and twentieth centuries. Donna V. Jones has shown how constitutive racialism is of philosophies of life in this period; Alexander Weheliye has critiqued the dominant theories of biopolitics for misconstruing "how profoundly race and racism shape the modern idea of the human."[5] In conjunction with these critiques, I suggest we should also think about the ways in which vitalism meets Philhellenism, in particular, not only in explicit arguments for a racial biopolitics committed to purity and health, but also on the terrain of universal humanism.[6] It is in line with this commitment that I reject Agamben's account of *bios* and *zōē* not only on philological grounds, but also philosophical ones.[7] But it cannot just be a question of correcting an error. For doing so would leave intact the move Agamben makes—namely, the plotting of "bare life" as *zōē* at the primal scene of Aristotle's *Politics* as a performance of linguistic and hermeneutic expertise that claims for itself the petrifying authority of "the classical." For Agamben's philological practice traffics in a species of revealed truth that secures the totalizing reification of "the Western tradition." If we are to think more critically with both "the Greeks" and the concepts of "life" in which they are entangled, we—historians and philosophers alike—need something better than ancestral,

---

[4] Agamben 1998: 8.

[5] Jones 2010; Weheliye 2014: 4.

[6] On Philhellenism and a racialized biopolitics of health and purity, see Holmes 2019, with further references. Martin Bernal's arguments about the formation of modern Classics in the crucible of European racism in the first volume of *Black Athena* (Bernal 1987) have become newly urgent in disciplines focused on the study of the ancient Greek and Roman world: see Rankine 2019.

[7] For more detailed critique, see esp. Brill 2019: 7–16. See also Dubreuil 2006; Finlayson 2010; Holmes 2019; Miller 2020.

mystified Greeks and self-generated moderns. I hope this essay makes a small contribution to this larger critical project and the question of how to imagine the relationship of ancient Greek texts addressed to the question of "life" and contemporary problems in vitalism and biopolitics.

The challenge is that history at a scale that encompasses Greek antiquity while also trying to answer to the present is always at once philosophical and mythical. The mythic aspect of "the Greeks" is one reason why they do not go away. It is not the only reason, though. For the history of (what in standard academic terms is called) science, medicine, and philosophy unfolds for millennia along axes of reading and rereading ancient Greek texts. These texts are recursively embedded in radically diasporic multitudes of thought that cannot be neatly described as Western or European but span the globe. This multiplicity is another reason why the Greeks do not go away.

The challenge that I have just outlined is especially acute when the question concerns *life* as an object of philosophical and scientific knowledge within historical time. One of the most challenging and exciting twentieth-century engagements with this question is the vitalism of the historian and philosopher of science Georges Canguilhem. For Canguilhem, the blurred zone of passage between history and philosophy is vitalism. In this chapter, I aim to show that this passage is facilitated by "the Greeks" in Canguilhem's work in ways that have been neither fully articulated nor critically examined. I pursue this claim through a close reading of the role of the Greeks in his influential and difficult paper "Aspects of Vitalism," first presented as one of three lectures at the Collège philosophique in Paris in 1946–1947 and then published in 1952 in Canguilhem's second book, *The Knowledge of Life*.[8]

Although the role the Greeks play for Canguilhem across his writings is more complex than what I can show for "Aspects of Vitalism," the essay is important insofar as it is the only one that Canguilhem devoted specifically to vitalism.[9] Moreover, the essay's complicated staging of the Greeks at the intersection of philosophy, history, and myth sheds considerable light on why antiquity matters deeply to Canguilhem's larger vitalist project. Pressing Canguilhem's use of the Greeks in "Aspects of Vitalism" thus brings to the surface difficult but persistent questions about the science of life as a privileged expression of human nature versus a set of circumscribed practices of knowledge associated with "the West"; the origins of these practices; the work of racism, colonialism, and fascism in the stories that have been told about these origins; life as a "natural" object of systematic conceptualization in science and philosophy; and, finally, the temporality of biopolitics and the historical status of "life" as an object of control figured as *technē*. In closing, I suggest that by critically interrogating the way in which the Greeks mediate between

---

[8] The other two lectures, "Machine and Organism" and "The Living and Its Milieu" (Canguilhem 2008: 75–97 and 98–120, respectively), should be read alongside "Aspects of Vitalism."

[9] For the observation that "Aspects" is the only article that Canguilhem devoted exclusively to vitalism, see Etxeberria and Wolfe 2018: 58. On Canguilhem's vitalism, see further Geroulanos and Meyers 2012: 4–5, emphasizing Canguilhem's "negative vitalism"; Wolfe and Wong 2015; Wolfe 2017; Geroulanos 2018: 79–90

history and philosophy for Canguilhem in light of his approach to history as an expression of life, we can get clearer about what is at stake in the stories we choose to tell about life as an object of philosophy across historical time and work to craft them more critically and constructively.

***

Born in France in 1904, Canguilhem was educated as a philosopher, a historian of science, and a physician. In 1955 he took over from Gaston Bachelard as Director of the Institut d'Histoire des Sciences et Techniques at the Sorbonne, which he led for over 15 years in addition to supervising the examinations and training for philosophy. For both institutional and intellectual reasons, Canguilhem exercised a formative impact on a number of postwar French philosophers, including Louis Althusser, Jacques Lacan, Pierre Bourdieu, Jacques Derrida, Etienne Balibar, and Gilles Deleuze. But his relationship with Foucault was especially important.

In his definitive essay on Canguilhem's legacy, Foucault identified the marking of discontinuities as critical to Canguilhem's method in the history of science.[10] Canguilhem's views on method were themselves shaped by Bachelard's historical epistemology, which understood science as proceeding along a path pitted with "epistemological obstacles" to be surmounted.[11] These obstacles decisively fragment time's continuum, creating a *history* of science. In his fidelity to Bachelard's method Canguilhem was unsparing of his criticism of those historians of science who would seek out precursors—that is, those figures who ostensibly begin projects that are only finished at a later moment in history, at which point genuine progress is achieved. The precursor, Canguilhem argues in his essay "The Object of the History of Sciences" (2005), is a "false object" of the history of science that endangers its status *as historical*.[12] In making a historical figure a precursor, the historian cuts him off from the very conditions (social, cultural, intellectual, material) that determine his horizon of possibility, thus recasting time as empty space that can be traversed first one way, then another.[13] The ancient Greeks—Aristarchus, Hippocrates—function in the essay as paradigmatic cases of what Canguilhem calls, after J. T. Clark, "the precursor virus." We smile, he says, at those who invoke them to explain Copernicus or Harvey. We know better.

In insisting on the irreversibility of history, Canguilhem echoes the most important tenet of his best-known work, *The Normal and the Pathological*. All organisms, he argues there, follow a path of irreversible development that is coextensive with life itself. Health is always a dynamic state, changing in relationship to disease and

---

[10] Foucault 1991: 13–15. On the necessity of discontinuity in "effective" history, see Foucault 1977: 154.

[11] On Bachelard's method, see Rheinberger 2010: 19–34.

[12] Canguilhem 2005: 206.

[13] Canguilhem 2005: 205–206. By contrast, Canguilhem does allow anticipation, but "the facts that authorize it and the paths to its conclusion must be of the same order as those that impart to a theory its truly transitional reach" (2008: 38–39).

individuated by the organism's unique, contingent passage through time and by its relationship to what Canguilhem calls its "milieu."[14] The relationship between the life of the organism and science as it unfolds historically is not merely one of anal-ogy for Canguilhem.[15] For the discontinuities within the sciences express what Canguilhem sees as a fundamental axiom about life: life has a creative relationship to error, which is itself a product of the organism's ongoing interactions with its environment. The endless but irreversible "auto-correction" process of the sciences, the history *in* science, must be read, then, through Canguilhem's understanding of the relationship between knowledge and life.[16] On this understanding, knowledge is not opposed to life. It is, rather, a fundamental aspect of life's situatedness in its milieu. The transhistorical collective of scientists performs a continuous testing of life as an object of knowledge in a process that, because of the very nature of life as it is enacted by living subjects, is necessarily marked by discontinuities, ruptures, and shifts of knowledge. The "knowledge of life"—this is the title of the book in which "Aspects of Vitalism" appears—is at once knowledge that is made by life (and so belongs to it as the knowing subject) and knowledge of life insofar as life is understood as an object of knowledge.[17]

Canguilhem's approach to the history of the life sciences is therefore shaped by his understanding of the relationship between knowledge and life. But here we arrive at a paradox. On the one hand, Canguilhem insists on discontinuity within the history of science. It is because time is fragmented that the history of science is a history of discontinuity. On the other hand, by understanding the knowledge of life as not only the knowledge that life science gains of life but also the knowledge that belongs to life, we seem to have moved beyond the boundaries of history (though not of time), precisely because discontinuity is now read through a theory of life itself. In light of the problem, it is tempting to differentiate formally between the orientation of the philosopher and the work of the historian as two separable modes of inquiry. But the situation of Canguilhem's method is more complex, difficult,

---

[14] Canguilhem 1991. On the milieu, see also Canguilhem 2008: 98–120.

[15] As Foucault 1991: 21–24 observes.

[16] There are, more precisely, two processes of knowledge production and auto-correction that Canguilhem sharply distinguishes: science (as historical process) and the history of science. Each has its own object. Nevertheless, like science, the history of science is also subject to an ongoing reworking of its stories about the past due to its entanglement with the shifts of science in its own present, as Rheinberger 2005: 191–93 emphasizes.

[17] There is therefore more continuity between the subjective genitive and the objective genitive for Canguilhem than there is for Bergson, who drew a harder line between intuition and scientific knowledge: see Sholl 2012: 115–16; Wolfe and Wong 2015: 68–71.

and, in many ways, irresolvable.[18] Canguilhem himself insisted he was doing both together in a practice he called "epistemological history."[19] History and philosophy are stubbornly implicated in one another in Canguilhem's work. Their knotting can be seen most clearly at the dense core of Canguilhem's philosophical commitment, on which his claims about knowledge and life are founded—namely his vitalism.

Canguilhem makes it clear at the start of "Aspects of Vitalism" that vitalism attracts him as a *philosophical* position. He argues there that vitalism differs from other objects that a historian of science might take up, such as phlogiston or geocentrism. For unlike those erstwhile scientific objects and theories, vitalism can never be overcome. Rather, vitalism is "a certain orientation of biological thought," one that, "whatever the limited historical resonance of the name given to it, will be seen to have a significance greater than just that of a stage in biology's development."[20] It is "a permanent exigency of life within the living" and not a method; an ethics, not a theory.[21] Its timeless counterpart is mechanism. If vitalism resists all attempts to disprove it, it's because it's not something that *can* be disproven.

Yet Canguilhem still inscribes vitalism *within* history, in two important ways. First, vitalism is a label that has been claimed by or applied to particular thinkers in the history of the life sciences and medicine: the seventeenth-century Belgian "chemical" physician Jan Baptist van Helmont, for example, or the nineteenth-century "father" of histology Xavier Bichat, or the twentieth-century theorist of the organism as a whole, Kurt Goldstein. It is true that the meaning of vitalism in this context is constrained by its "limited historical resonance." But the very recurrence of vitalism as a historical phenomenon still points to something crucial about its larger meaning. Second, vitalism occupies a privileged moment within history because it is the product of a historical process through which life becomes an object of knowledge. "A philosophy that asks science for clarifications of concepts," Canguilhem writes, "cannot remain uninterested in the construction of this very science."[22] The founding of a science of life, in Canguilhem's view, transforms the

---

[18] Méthot, for example, pushes back against decomposing "historical epistemology" into its constituent parts: "in my view, the real interest of the concept of historical epistemology is precisely that *it indistinctly links the history and philosophy of science together in a single, coherent project*" (2013: 117, emphasis original). See also Wolfe and Wong 2015: "in a very real sense one cannot distinguish between a historical claim and a philosophical claim in Canguilhem's 'history of vitalism' or 'vitalism' (72); Wolfe 2017: 35–41. The entanglement of history and philosophy suggests a more complicated relationship between discontinuity and continuity in Canguilhem's thinking about the history of science. I thank Stefanos Geroulanos for drawing attention to this point.

[19] Méthot 2013: 116–17.

[20] Canguilhem 2008: 60. He there strongly contrasts a scientific view on vitalism from a philosophical one.

[21] Canguilhem 2008: 63. For this reason, vitalism escapes the standard games of truth, as Foucault suggests when he describes vitalism as an "indicator" in the history of the biology: on the one hand, a "theoretical indicator of problems to be solved" and, on the other hand, a "critical indicator of reductions to be avoided" (1991: 18).

[22] Canguilhem 2008: 60.

nature of what humans qua organisms generate in their interaction with their world. It is only within this new realm of activity that vitalism itself makes sense.

It is at this point in the argument that Canguilhem turns to the ancient Greeks. They surface as the critical point of passage at which Man's natural tendency, qua living being, to produce knowledge of life produces life as a formal, "scientific" object of knowledge. Aristotle occupies a crucial place here, but so does Hippocrates. The importance of the Greeks does not mean that Canguilhem neglects the boundary of modernity—far from it. To the extent that the boundary is operative, it separates a history of the philosophy of the living from a history of biology proper (in "Knowledge of Life," the Greeks appear in the section labeled "philosophy," not "history," which comprises only the essay on cell theory).[23] Yet because vitalism recurs within the history of the life sciences across a field of discontinuities, it points to the entanglement of the philosophy of life within the timebound "construction" of the life sciences. This entanglement points to the transhistorical persistence of the Greeks, blurring any strict boundary between antiquity and modernity.

The final sentence of "Aspects of Vitalism" reads, "In the end, to do justice to vitalism is simply to give life back to it."[24] Canguilhem implies that the work of the historian who thinks vitalism *in time* is instrumental to making this gift. But what kind of historical time do the Greeks, in fact, occupy? What does it actually mean to historicize—an act we usually associate with contingency and denaturalization—an orientation towards life that Canguilhem describes as "the expression of the confidence the living being has in life, of the self-identity of life within the living human being conscious of living?" In the essay itself, Canguilhem isolates out three aspects of vitalism: first, the *vitality* of vitalism; second, its *fecundity*; and third, its *honesty*. Each of these aspects offers not only a different facet of vitalism; each also offers a different perspective on antiquity, which recurs throughout the essay in a discontinuous yet insistent fashion, thereby performing, as it were, its own vitality. I turn now to examine each of these three aspects in detail.

***

Canguilhem begins his inquiry by suspending the question of whether vitalism is true from a scientific perspective. Instead he asserts that it is true "historically." The fact that it keeps making comebacks is its "vitality," the first aspect of vitalism Canguilhem discusses. But precisely because it eludes any one historical period,

---

[23] At Canguilhem 1989: 546, Canguihem adopts Foucault's position on the invention of a modern concept of life. This essay on "*vie*" comes more than four decades after "Aspects," but it isn't incompatible with the distinction already operative in the division of sections in *La connaissance de la vie*, which distinguishes between life as an object of philosophy (=vitalism) and as an object of a modern science. As in the later Foucault (e.g., Foucault 2017: 39–42 on paganism as an object of historical study bound up in romanticism), Canguilhem's periodization is itself shaped by a sense of how later receptions of Greek vitalism frame our reading of Greek "life," without assuming that the latter is only a product of modern projection.

[24] Canguilhem 2008: 74.

vitalism's vitality is a problem for the philosopher rather than for the historian. Nevertheless, even if we are within the remit of philosophy, we are not talking about an expanse of time coextensive with life itself. Rather, we are talking about the beginnings of *a* history, the history of biological theory, which "reveals itself to be a thinking that throughout its history has been divided and oscillating."[25] These oscillations move between not only what Canguilhem calls Discontinuity and Continuity (on the problem of the succession of forms) but also Preformation and Epigenesis (on the problem of development), and Atomicity and Totality (on the problem of individuality). Each of these oscillations, Canguilhem proposes, expresses a dialectical process that life produces in thought once it has emerged as an object of a science of life. What this means is that the "life" at stake in vitalism is always already defined by a science of life. And as Canguilhem makes immediately clear in this essay, this science of life is given as Greek. It's the "Hippocratic" expression of trust in *vis medicatrix naturae*, the idea, that is, that nature possesses its own healing power, or Platonist-Aristotelian-Galenic views on universal sympathy.

Yet from another perspective, the life expressed in (Greek) vitalism precedes it. For vitalism turns out to express a more basic, "instinctual" distrust of the power of technology (*technique*) over life. In other words, it translates something internal to life itself, described by Canguilhem as "the expression of the confidence the living being has in life, of the self-identity of life within the living human being conscious of living"; he goes on to describe this, as we have seen, as "a permanent exigency of life in the living (*une exigence permanente de la vie dans le vivant*), the self-identity of life immanent to the living."[26] Life, in other words, always expresses a need, a demand (connotations of the French word *exigence*). But it is also confident in its fundamentally agonistic relationship to its milieu. This agonism with the external world will eventually take shape as vitalism's counterpart, mechanism. But before mechanism, the living human being has an attitude towards life that Canguilhem glosses as the "cunning of reason." The phrase calls Hegel to mind. But Canguilhem is also playing with the sense of "stratagem" present in the Greek word for "machine," *mēchanē*, as we will see further. Suffice to say here that mechanism "as a scientific method and as a philosophy" is already *implicit* in the human use of tools.[27]

The human *tout court* is constituted here by a fundamental push and pull. Canguilhem emphasizes the point by turning to the Czech biologist and historian of science Emanuel Rádl for his claims about the two ways in which the human being considers nature. Either "Man" sees himself as its (or her, in keeping with the unmarked but gendered Man and the feminization of Nature) child, feeling a sentiment of belonging and subordination to Nature; or Nature is held before him as a

---

[25] Canguilhem 2008: 61.

[26] Canguilhem 2008: 61.

[27] Canguilhem 2008: 63.

"foreign, indefinable object."[28] Part of the vitality of vitalism is therefore secured by a polarized definition of the human being in relationship to Nature: kinship and sympathy (this is the position of the vitalist) or by a sense of distance and mastery (mechanism). The polarity of human nature produces the permanent oscillation of biological theory. This means that vitalism's "vitality" ultimately points to the irrepressibility of life's faith in itself. Nevertheless, this faith comes to be mediated in human beings by historically situated forms of vitalism. And as we have seen, the first of these forms is Greek. Life latches onto history in the guise of vitalism through Greek names: first Hippocrates, then Plato, Aristotle, and Galen.

In truth, the nature of historical emergence as it takes shape in the essay is more complicated than the account I have just offered for reasons that challenge any simple linearity in the history of vitalism. Although Hippocrates is the first to be named in the essay, Canguilhem introduces him in the context of a vitalist tradition already defined by the eighteenth-century Montpellier physician Paul-Joseph Barthez. The guiding principle of this tradition is the "Hippocratic" principle of *vis medicatrix naturae*. For Canguilhem, the principle represents "the biology of physicians skeptical of the constraining power of remedies" and distrustful "of the power of technique over life." By introducing Hippocrates via an eighteenth-century "spirit of Hippocratism," Canguilhem initially frames it as a reaction to mechanism and, more specifically, Cartesianism.

Yet the appeal to a *vis medicatrix naturae* can be traced back to the Hippocratic Corpus and, more specifically, to a passage from *Epidemics* VI on "untaught nature" that was taken up by Galen as foundational for the philosophy of nature in the texts that he ascribed to Hippocrates: "The body's nature," the passage runs, "is the physician in disease. Nature finds the way for itself, not from thought."[29] This Hippocratic vitalism thus "gets behind" Cartesian mechanism to claim historical priority. But what kind of priority should we accord ancient vitalism, if it is, by definition, *already* a translation of something else, namely, life's faith in itself? What mechanism motivates *this* translation, so far removed from the Cartesian moment?

The difficulties inherent in such a question will become clearer when Canguilhem turns to the second aspect of vitalism. But they are already evident in his invocation of Plato, Aristotle, and Galen, names given as examples of the attitude towards Nature that Rádl had defined in terms of sympathy and in opposition to an attitude that views Nature as a "foreign, indefinable object" to be mastered. Rádl's theory is a philosophical anthropology: there are two timeless options. But Canguilhem frames the sympathetic attitude of Plato, Aristotle, and Galen as *already vitalist*,

---

[28] Canguilhem 2008: 63.

[29] The text where the idea of "untaught nature" first appears, *Epidemics* VI, is dated to the mid fourth-century BCE (see [Hippocrates] *Epidemics* VI 5.1 [Littré 5.314]). For Galen's reception of the idea of "untaught nature" and his pitting of the atomists against continuum theorists, see esp. *On the Natural Faculties* 1.12 (2.26–30 Kühn) and Jouanna 2012. The concept of *vis medicatrix naturae* makes a complicated comeback in debates in biology in the 1920s and 1930s, which form the backdrop to Canguilhem's engagement with it: see Geroulanos 2018: 161–206, 292–315, esp. 298–302.

already inside a science of life. What this means is that their vitalism has to have been shaped by the prior emergence of the machine or technique as an obstacle to life.

It is worth emphasizing that, for Canguilhem, *all* life is agonistic vis-à-vis its milieu and creative in its strategies for survival. The use of technology is "a universal biological phenomenon," as he writes in "Machine and Organism."[30] Fundamentally, technology is not opposed to life. Rather, it is an extension of life into the milieu. For human beings, tools and machines are the expression of this extension of life.[31] At the same time, it is in humans' use of tools that we find the seeds of mechan-*ism*. Here I suggest that Canguilhem's double translation of the Greek word *mēchanē* as "machine" and "cunning" shows its own cunning. In a basic sense, the *mēchanē* is always double-sided. On the one hand, as "machine" or "tool," it's an extension of (human) life into the milieu. On the other hand, as "cunning," it represents a disposition towards its object, that is, Nature—namely, as a foreign, hostile object to be hunted, captured, and controlled, precisely because Nature is seen as the source of obstacles to (a) life.

This double-sided quality of the *mēchanē* is doubled again in Canguilhem's choice of the Greek lexeme, which he allies with the figure of Ulysses. With *mēchanē*, we also capture a relationship that is at once "permanent" and historical. It is in the use of tools among humans that we find the seeds of mechanism. But the Greek of the lexeme, together with the appeal to Ulysses, also signals the *historical* potential of the dialectic between Man and Nature held within the Greek term *mēchanē*. And this is crucial, because it is only within a historical process by which Nature is objectified as an object of control that vitalism can express the opposing sentiment—namely, that Man is one with Nature. So the *mēchanē* expresses not just the doublet "machine/cunning," but the beginnings of a historical process through which the use of machines produces an attitude of mastery vis-à-vis Nature—a trust in tools and forms of epistemic capture, and a corresponding definition of the world they act on as inert and passive—that will eventually transmute into mechanism and trigger, in turn, the formation of vitalism.[32] In short, the *mēchanē* enables the transition from the universal nature of Man into the history of science on the quasi-mythic terrain of ancient Greece. Canguilhem does not explicitly locate the Greeks within a historical process here. Rather, the *mēchanē* and Ulysses hover in the space of mythic time, on the cusp of the transformation of life into vitalism. They stand behind vitalism proper. When we turn to the second aspect of vitalism, however, we do find the Greeks being invoked in historical terms through Canguilhem's critique

---

[30] Canguilhem 2008: 96.

[31] The machine is a "work of man" and an organ "of the human species" (Canguilhem 2008: 76, 87).

[32] "But is one not then justified in concluding that the theory of the living machine is a human ruse that, if taken literally, would nullify the living? If the animal is nothing more than a machine, and the same holds for the whole of nature, why is so much human effort expended in order to reduce them to that?" (2008: 63). The very effort required of mechanism to corral its object becomes a symptom of its misprision of that object.

of modern vitalism's "return to antiquity." I turn now to consider how Canguilhem's imagination of the Greeks as historical actors produces his own "unclassical" vitalism.

***

Canguilhem calls the second aspect of vitalism its "fecundity." What Canguilhem wants to address with "fecundity" is whether vitalism is *productive* within the history of biology, or whether it is an obstacle to progress that has to be discarded for biology to make real discoveries. After all, vitalism could be vital and still sterile. It could persist not because it is rooted in some truth about life but because it is an especially tenacious and pernicious illusion. At the start of the essay, Canguilhem had set aside questions of truth. He now shifts back into the mode of historical epistemology to reframe the question of vitalism's vitality in terms of Bachelard's "epistemological frontier." Is this vitality, he asks, only a symptom of mechanism's historically contingent limits? If that were true, then vitalism merely marks the (ever-shrinking) territory that mechanism has yet to conquer. To argue otherwise would only be to refuse mechanism "the time it needs to complete its project."[33] Canguilhem's position here looks even harder to defend today, with the advances in biotechnology casting life as "no more ontologically interesting than stardust," as Jones writes.[34] But as Jones goes on to argue, vitalism is tenacious. It is unlikely that Canguilhem would have ceded its ground even today.

Canguilhem takes different tacks in responding to the charge that vitalism is a scientific dead-end. He first lists the contributions of vitalists to the history of biology. But he quickly runs into the problem that whatever the import of these contributions to biology, the concepts produced to explain them (e.g., vital principle, entelechy, hormé) seem to indicate neither progress nor discovery. Rather, in their (re)generation of Greek concepts, they seem to signal regression: "vitalism's fecundity appears at first glance to be all the more contestable in that…it always presents itself as a return to antiquity," whether one considers the return to Aristotle against Descartes in van Helmont or in the Montpellier vitalists, or the return of Renaissance humanists to the Plato of the *Timaeus* over an Aristotle who had become "overly rationalized" in medieval scholasticism.[35] The return to the Greeks only seems to confirm vitalism's sterility. Canguilhem thus, like an organism in the face of an obstacle, changes course. His new strategy pivots on a split in vitalism. On one side is what Canguilhem calls "classical vitalism," by which he is referring primarily to the vitalists of the eighteenth century who become conflated with "Hippocrates." On the other is his own vitalism. He elucidates this split by differentiating two ways of "returning" to the Greeks.

---

[33] Canguilhem 2008: 69.

[34] Jones 2010: 3.

[35] Canguilhem 2008: 66.

The return to antiquity, Canguilhem argues, need not be about revalorizing concepts that are worn out. Rather it is about going back to the sources, fueled by "a nostalgia for intuitions ontologically more original and closer to their object."[36] What are these intuitions? Not vitalism itself, at least as it is defined according to the "Hippocratic" spirit of the Montpellier vitalists. Rather, these intuitions pertain to a dialectic within the science of life as it is played out in historical time *already in antiquity*. Here Canguilhem proposes mechanism as a philosophical position historically realized in fifth-century Greece: Aristotle's vitalism responds to Democritus' mechanism; Plato's finalism responds to Anaxagoras' mechanism.[37] Canguilhem is spelling out what was only hinted at earlier. Insofar as the first named vitalists are Greek, they are responding to the emergence of historical mechanists like Democritus and Anaxagoras in a space that was already opened up by a mythic *mēchanē* in the discussion of vitalism's first aspect.

Canguilhem's decision to name Greek mechanists at this point turns out to be critical for his own definition of vitalism. For he is subtly but surely taking his distance at this point from what he calls "the vitalist's eye," a product of eighteenth-century Hippocratism.[38] The vitalist's eye offers only a naïve vision of the past as pre-technological and pre-logical, "a vision of life anterior to tools and language, that is, to instruments created by man to extend and consolidate life."[39] Canguilhem aims to see both the history of (ancient) vitalism and life itself differently.

We can now better understand Canguilhem's earlier conflation of "Hippocrates" with *vis medicatrix naturae*. He is not interested in going back to the writings of the Hippocratic Corpus as a site for the very dialectic that interests him. His lack of concern makes little sense from a historical or even philosophical perspective. After all, the theorization of medicine as a *technē* is a guiding thread of the writings in the "Hippocratic" Corpus, and medicine, in turn, was the paradigmatic *technē*, especially vis-à-vis- the mastery of (human) nature, in the fifth and fourth centuries BCE.[40] Indeed, the principle of *vis medicatrix naturae* appears late in the writings of the Hippocratic Corpus, perhaps as a response to the dominance of "technical"

---

[36] Canguilhem 2008: 66.

[37] Cf. Canguilhem 2008: 79–80, where there is a relationship of mechanism and vitalism within Aristotle, because he has a notion of the automaton. This claim enables Aristotle to precede and also subsume Cartesian mechanism (85). On Canguilhem's shifting engagement with Descartes in the 1930s and 1940s, see Geroulanos 2018: 83–87.

[38] Canguilhem 2008: 67.

[39] Canguilhem 2008: 67.

[40] There is plenty of evidence of "technological optimism," e.g., [Hippocrates] *On the Technē* 11 (Littré 6.18–20 = 237, 6–14 Jouanna). See also, e.g., [Hippocrates] *On Ancient Medicine* 24 (Littré 1.636 = 153, 16–19 Jouanna), though this text, *On Ancient Medicine*, is more complicated in its famous ("vitalist") argument in Chap. 9 that the practice of medicine will always be probabilistic. See Lonie 1981, which reads strands of mechanism and vitalism in the Hippocratic Corpus through the lens of Friedrich Hoffmann's iatromechanism in a way that both accepts the anachronism of the categories and does not reduce them to Hoffmann's terms. See also Holmes 2014 on the problems of positing a vitalist "sympathy" in the Hippocratic Corpus; Holmes 2010: 192–227 for medicine as a model *technē*.

intervention in the earlier writings of the Corpus, much as Canguilhem's account here would predict. In other words, an argument can easily be made, appropriating Canguilhem's notion of a dialectic within biological theory, that the principle of "untaught nature" is itself a response to certain strains of technological optimism, conditioned by the valorization of *technē* as the source of medicine's power and authority, in fifth- and fourth-century BCE medical writing. But if Canguilhem is unconcerned with whether Hippocratic "vitalism" might itself already be embedded in the dialectic that he identifies elsewhere, it is because he is using the name of Hippocrates to label the "emic" expression of classical vitalism. He does this even as he recognizes that this expression is properly an effect of the reception of "Hippocrates" from Galen to the revival of Hippocratism in the Montpellier vitalists.

In other words, Canguilhem uses Hippocrates strategically to name the eighteenth-century vitalist's refusal to cede *any* ground to mechanism. It is because the so-called "Hippocratic spirit" believes nature has no need of technique—in later writings, Canguilhem treats "Hippocratic" medicine as non-interventionist, even "contemplative"—that it can produce the naïve vision of pre-technological Man.[41] In negotiating the apparent obsolescence of vitalism, Canguilhem thus appropriates Hippocrates to designate both the refusal to acknowledge the entanglement of life and technique *and* a false origin for later vitalists. These vitalists mistakenly use Hippocrates to prove the possibility of medicine before or without technique, rather than recognizing that vitalism and mechanism go hand in hand already in classical antiquity.

The wrong kind of vitalist "return to antiquity" is therefore implicated in the "classical"—that is, eighteenth-century—vitalist's misunderstanding of the relationship between life and technique. Canguilhem accordingly needs to correct both aspects of the classical vitalist's error. When he looks to the relationship between Democritus and Aristotle for philosophical clarity on the problem of vitalism, he is looking to a different antiquity. This antiquity is viewed not from within the classical eighteenth-century vitalist perspective, which posits Hippocrates as a purer, pre-technological figure. Rather, Canguilhem is adopting the perspective of a different vitalism, what we could call an "unclassical" vitalism, a vitalism that is nevertheless still oriented towards classical Greek antiquity. Let us see what this looks like for Canguilhem.

Canguilhem's own vitalism takes shape in his last defense of vitalism's fecundity, This defense requires the outright rejection of classical eighteenth-century vitalism. Discussing Bichat, he identifies classical vitalism's "philosophically inexcusable fault" as its designation of an "empire within an empire" (that is, Spinoza's

---

[41] See Canguilhem 1994: 130–31 where he opposes contemplative ancient medicine to activist modern medicine (though he admits this is an oversimplification). To the extent he commits to this view, there would be no true dialectic between vitalism and mechanism until the modern period, at least in medicine. In later essays, Canguilhem explicitly rejects the "Hippocratic" dictum of the "healing power of nature" (*vis medicatrix naturae*) under the influence of his resistance to the anti-medicine and anti-psychiatry movements of the late 1960s and early 1970s: see Canguilhem 2012: 25–33, with Geroulanos and Meyers 2012: 13–15.

imperium in imperio) in order to secure the specificity of the biological against the laws of physics.[42] Far from erring in its ambitions, classical vitalism is too modest, Canguilhem declares. His argument builds on his philosophy of the organism and is developed at greater length in the lectures that followed "Aspects," "Machine and Organism" and "The Living and Its Milieu." His strategy throughout these lectures is to nest mechanism within vitalism. "To live," he writes in "The Living and Its Milieu," "is to radiate: it is to organize the milieu from and around a center of reference, which cannot itself be referred to without losing its original meaning."[43] The very behavior of "man as technician and scientist" therefore demands an explanation that privileges life.[44] In this way Canguilhem claims a priority for life vis-à-vis reason or mechanism that is at once logical and historical. In place of the classical vitalist's naïve vision of pre-technological Man, he offers a vision of Man whose use of tools and concepts is as rooted in life as is the vitalist resistance to technique.

Nevertheless, vitalism itself is a specific translation of the permanent exigency of life, enmeshed in the analogous mechanist "translation" of a more fundamental *mēchanē*. As such, it falls back into history, and history returns us to Greek origins. Canguilhem attributes an even more powerful historical priority to the ancients, with the promise of originality and purity such priority entails, in his explication of the third and last aspect of vitalism: its honesty.

***

Canguilhem's defense of the "honesty" of vitalism shows him grappling with a charge against vitalism that is more damning than that of obsolescence or error— namely, the charge that it is politically reactionary, counterrevolutionary, or something even more toxic: complicit in the horrors of the Holocaust through the Nazi appropriation of vitalist biology for a racist biopolitics.[45] These charges were not abstract to Canguilhem, who fought in the French Resistance. It is worth remembering, too, that the lectures date to 1946–1947. But he refuses to accept the charges against vitalism on the grounds that politics borrows from biology what it has already lent it. The blame for Nazi biology, he insists, lies with the moral and philosophical failures of Nazi biologists.

If you are going to be persuaded of vitalism's innocence, these arguments would seem sufficient. But Canguilhem goes a step further, seeking to exculpate vitalism due to the innocence of its origins: "If we look for vitalism's meaning in its origins and for its purity at its sources, we will not be tempted to reproach Hippocrates or

---

[42] Canguilhem 2008: 70.

[43] Canguilhem 2008: 113–14.

[44] Canguilhem 2008: 71.

[45] On the political problem of vitalism in the years after the war, see Bianco and de Beistegui 2015: 3–5. The charges against vitalism arising from Nazi appropriations have not gone away. See, e.g., Braidotti 2010, repudiating the "new vitalism" of Jane Bennett and others. Cf. Weinstein and Colebrook 2017 on "critical life studies."

the Renaissance humanist for the dishonesty of their vitalism."[46] Why does Canguilhem feel compelled to make this move? We could read it as a final encounter between the historian's understanding of vitalism and the philosopher's. For Canguilhem goes on to acknowledge that it is in part true to see behind the twentieth-century resurgence of vitalism the workings of political economy and, more specifically, bourgeois society's crisis of confidence in capitalism—a reading of vitalism's attraction that could no doubt be developed with even greater force in the early twenty-first century. But Canguilhem is generally impatient with externalist explanation. So he sets aside historical materialism to return to the space in Greek antiquity where history transmutes into an expression of life itself—and, indeed, an expression of life's resistance to technique: "the rebirths of vitalism translate, perhaps in discontinuous fashion, life's permanent distrust of the mechanization of life."[47]

Yet the nostalgic longing for the "purity" of vitalism's sources is itself implicated in a different kind of history in which the Greeks represent an innocent origin. History in the mode of nostalgic mythmaking feels decidedly impure in light of another -ism that became complicit in the formation of Nazi ideology—namely, German Philhellenism in its attachment to a lost Greek homeland.[48] Does politics borrow from Philhellenism what it has already lent to it? Do these transactions take place across a sharp border? Has the love of the Greeks ever been innocent? Surely, as Nietzsche at his most critical would insist, the answer must be "no."[49] If we need a critical vitalism, can we pursue this end without also thinking critically about the Greeks? So what can we say about Canguilhem's Greeks? What can we do with them?

By this point, I hope to have demonstrated how thoroughly the Greeks permeate Canguilhem's reading of vitalism as at once historical and philosophical in "Aspects of Vitalism." They occupy consequential roles in each of the three aspects of vitalism: vitality, fecundity, honesty. I have tried to show how, on Canguilhem's account, the Greeks produce the first historical articulation of vitalism. Hippocratism is thus the form in which life asserts itself, and keeps asserting itself, within the repeatedly renegotiated terms of life science. So as long as we remain within a science of life, vitalism as recursion has to be read vis-à-vis a Hippocratic origin if its vitality is to be fully appreciated. But the "Hippocratic" vitalism of the eighteenth century also failed to comprehend the dialectic between vitalism and mechanism out of which it was created. This ancient dialectic, Canguilhem argues, can and must therefore be read in terms of his *own* vitalism, which repairs vitalism's ancient history in order to repair its philosophy. Within Canguilhem's terms, the oscillation between vitalism and mechanism in biological theory is part of a much longer, continuous, and

---

[46] Canguilhem 2008: 73.

[47] Canguilhem 2008: 73.

[48] See Goldhill 2000; Fleming 2012: 109–17, with further references; Chapoutot 2016.

[49] See Porter 2019.

indeed non-human story of life's exercise of technique. But it is also the result of a critical *rupture* in the relationship of (human) life to knowledge. Canguilhem's own theory encompasses the philosophical position that life by its very nature distrusts mechanization, even as it uses tools to evade obstacles. It also encompasses the historical position that once Greek vitalism emerges it becomes the template for life's knowing, conscious rebellion against science that is retraced in vitalism's later rebirths. In the third part of the essay, this template is revisited as vitalism's innocent origin.

Greece thus mediates between history and philosophy in "Aspects of Vitalism." It enables Canguilhem to fold his philosophical argument about the relation of (human) life and knowledge into history as the expression of a recurrent but still timebound phenomenon: vitalism. It is therefore impossible to grapple with the entwinement of history and philosophy in Canguilhem without also thinking about how he approaches the history *of* philosophy in ancient Greece. At the same time, Canguilhem is not exactly doing the history of philosophy or biology in "Aspects of Vitalism." If he invokes the agonism between Democritus and Aristotle, it is not with the aim of analyzing or even documenting a philosophical debate. Rather, he surgically deploys that dynamism as a historical fact against the naïve vitalist's view of history and in the service of his own theory of life and its history. Canguilhem can do this because he takes for granted the place of classical Greece as the spatiotemporal zone where life consequentially crosses into history as the object of both mechanistic control and vitalist care.

By attending to the function of the Greeks in "Aspects," we can see that the essay is, at a fundamental level, addressed to the question of *what it means to historicize a permanent exigency of life*. It is worth recalling the warped temporality created by Canguilhem's introduction of Hippocrates via the Montpellier vitalists. The blurring between "Hippocrates" and "the Hippocratic spirit" implies that Canguilhem is well aware that his ancients are mediated by post-Cartesian vitalism as it flourished in France in the eighteenth and early nineteenth centuries, even as he reads the Latin *vis medicatrix naturae* as the Greek of "Hippocrates." He is committed to the view that organismic time is irreversible. But still, the Greeks cannot be contained within this modern reception. If they're imbricated in Canguilhem's "history of the present," to borrow a term of Foucault's, it is precisely because they exceed modernity.[50] What exceeds the eighteenth-century reception is something like the "life" that fuels Canguilhem's own vitalism. Philosophy and history are mixed up because the Greeks are inside and outside of history.

I will try to be more precise. We could take up ancient Greek vitalism and its reception from a historicist perspective. The history of the life sciences has been shaped by various "returns" to "the Greeks," and especially Hippocrates, Plato, Aristotle, and Galen. Problems concerning the mastery of the nature of living things

---

[50] Geroulanos and Meyers 2012: 12–13 (the phrase first appears in Foucault's *Discipline and Punish* [1975]).

and the care of human life and bodies are knit into this history, recursively and dialectically articulated through textual transmission and scenes of experimentation
situated in institutions and communities like translation movements patronized by
elites and/or the state, medical faculties in universities, and commercial laboratories. The proper work of historicism here would seem to be to stress contingency,
the sense that the history of life might have been otherwise, in defiance of Whig
history and the teleology of progress. It would be to emphasize, too, the very density
and messiness of a thickening around Greek texts, which decenter the proprietary
claims on the Greeks qua ancestors exercised by the figure of vitalist return.

But Canguilhem's transhistorical account of vitalism does something different,
cutting through the complexity of thick history. The essay appears, we can recall, in
the "philosophy" section of *Knowledge of Life*, not the history section. Why
"Aspects" belongs in the section on philosophy presumably has something to do
with what it means to be working as a historian of life within a milieu. Insofar as
Canguilhem locates himself within the tradition of philosophical biology, he knows
as a historian that the making of that tradition informs his organization of the world
and his understanding of his own experience, as much as experimental work in biology informs his thinking.[51] But this tradition does not just offer life as an object of
knowledge. It offers a choice of orientation: mechanism *or* vitalism. Not just as a
philosopher, but as someone who is, simply, alive, Canguilhem sees himself as obli-
gated to decide between these options, even as he creatively reinterprets what that
decision entails—that is, even as he "chooses" vitalism but affirms a version of
vitalism that differs in fundamental ways from classical vitalism. It is this space of
choice or *orientation towards*, I suggest, that Canguilhem is trying to hold open
when he describes vitalism as an ethics. Ethics in this sense expresses his position
that being alive always entails the organization of the world in terms of values. The
force of vitalism may then be derived from the idea that however contingent or pro-
vincial or distant its historical emergence was, in the world shaped by its emergence
or, more precisely, by an embedded dialectic of vitalism and mechanism, vitalism is
choiceworthy.

Is it possible to read Canguilhem's return to the Greeks in the same spirit?
Canguilhem himself, as I have argued, assumes the conjunction of vitalism and
the backwards-glance towards the Greeks. His intervention in classical vitalism is
not to sever the relationship with the Greeks. Rather, in remaking it through a
reading of the mechanism/vitalism dialectic in antiquity, he implies that to be a
vitalist is to always go back to the Greeks. That return is part of what secures the
vitality of vitalism itself as over and above modernity (and especially as pre-Car-
tesian). There is a real sense here that mechanism is aligned with the myth of the
autogenesis of modernity. For insofar as mechanism expresses a faith in the autogenesis of biological science and technology, it denies the relationship of the

---

[51] Thus it is true his views of life as dynamic and productive of its own norms is informed by
Darwin or genetics and ongoing work by biologists (Sholl 2012: 122; Etxeberria and Wolfe 2018),
but it is also informed by his stance as a historian, by which he reads this scientific information as
operating within a regime of truth that is historicized at different levels.

history of life science to a philosophy of life (vitalism), as well as the relationship of a philosophy of life to life itself. But the choice that I outlined at the start of my paper is precisely about whether the Greeks are a necessary part of this history, the history of life as an object of knowledge. That question implies others. Are the Greeks irrelevant to what we call modernity or, say, biopolitics? Are they irrelevant to any work of orienting *towards* life, away from the biopolitical, however it is understood?

In this chapter, I have tried to open up some space to pause and assess Canguilhem's complex conflation of Hellenism and vitalism. One problem with the conflation is that it reenforces the figuration of the Greeks as the origin of science and philosophy. The problem is familiar. But I would point specifically to how the origin-function of the Greeks in Canguilhem enacts the idea that, prior to the birth of philosophy in Greece, Man only had "technique," understood as the most primitive form of *mēchanē*; the idea that Greece is where Man first thinks himself.[52] Another problem arises from the very elision of Hellenism in modern genealogies of the biopolitical that draw a sharp line between obsolescent views on life and a science of life. What these genealogies have too often overlooked is the Greeks' persistent role in figuring the life that escapes modernity, the fuller life that is at once lost and able to be regained.[53] We need to think more critically about the surplus of life that modern thinkers have invested in "the Greeks," who are regularly taken together with "the savage," "the primitive," and other highly racialized non-Western Others to figure a vitality lost to those who identify as European.[54] Canguilhem's desire for a purer vitalism at the origins of Greek antiquity thus cannot be disentangled from a deeply raced and sexed paradigm of Man aligned with "the Greeks". Here the choice is not between ancients and moderns but between what kinds of *longue durée* stories we tell.

In grappling with this choice, perhaps we could think about Canguilhem's strategy to capture mechanism with vitalism. Remember that, on Canguilhem's revised vitalism, mechanism is only the hypertrophy of a technique that begins as an extension of life. Life and technique are inseparable. By analogy with this claim, could

---

[52] For the opposition of technique and science as a means of differentiating pre-Greek and Greece science, see Rochberg 1992: 550–51, citing a passage from William Wightman's 1951 *Growth of Scientific Ideas*: "Greek thought differed from all that had gone before in respect both of *generality* and *rigour*. In *general* thinking, which opens up possibilities undreamed of by the severely 'practical man,' we pass from *percepts*—things to which we can point, like triangular fields—to *concepts*, creations of the process known as *abstraction*...Science is, among other things, the generalization of perceptual experience by means of adequate concepts." See further Rochberg 2016: 38–58. Cf. the way this position persists at Sholl 2012: 119: "With the rise of scientific knowledge, thought is not that which merges with life or appears as a continuation of vital process, as is technique. Rather, thought objectifies life, turning it into something to be studied."

[53] See Holmes 2019. Esposito 2008, for example, ignores the role of Philhellenism in Nietzsche's vitalism (78–109) and Nazi biology. A critical awareness of the function of the Greeks would complicate his appropriation of Canguilhem's vitalism at Esposito 2008: 189–91.

[54] See Payne 2018 on the Greeks in this role. See also Jones 2010, esp. 102–28 on vitalism, "the primitive mind," and racial memory in Bergson.

the choice between ancients and moderns be equally illusory? That to the extent the ancient Greeks set at least some of the terms within which "we moderns" make choices about life, or politics, they are always already determining the way we orient towards life, or towards our mechanisms, our technologies? These orientations, the argument would go, unfold within the terms of a life that precedes them, and so shapes the very nature of the choices, even when those choices are to affirm modernity, just as life conditions even the choice of mechanism.

The analogy has conceptual force. It also has problems. By framing the claim of the Greeks on the moderns in these terms, I think we can also see more clearly the challenges and limits of Canguilhem's organicism at the level of historiographical method; that is, problems that go beyond his "naïve" or romantic return to the Greeks as pure. Here it is a question of what the organicism of vitalism itself does for the unity of biological theory or a philosophy of life. For in extending vitalism back to the Greeks, Canguilhem suggests that each manifestation belongs to the same life. Admittedly, the life that Canguilhem has in mind has been freed, we could say, from the unity of classic holism, where the parts are always parts of a larger whole governed by a structure that secures oneness despite multiplicity. Instead of an organism governed by the teleological normativity of Aristotle's biology, Canguilhem seeks an organism individuated in time and creatively remade through its encounters with a milieu. Nevertheless, even if we read the history of a philosophy of life through Canguilhem's revised model of life itself, we have still imposed the unity of *a* life on that history, suggesting a coherence, boundedness, and inner determination that reifies and indeed naturalizes the "Western" tradition in biological terms.

Instead, what if we were to work through the organicism of Canguilhem's history of a philosophy of life by thinking of the ancient Greeks not as an origin point from which a life has unfolded, but as themselves part of the milieu? At a minimum this would resist the recurrence of the Greeks in a philosophy of life science as an index of life itself. It would resist, too, the very definition of the Greeks as the site where Man takes life as an object of systemic knowledge and practice. Instead, it would attend to the very materiality of ancient Greek texts about life, and nature, and body, and soul—and their readings and re-readings—as part of the world inhabited by later thinkers struggling to define the flourishing of human life is as well as the nature of the cosmos. It would take seriously the choice-making of thinkers situated in a milieu, a collective of thinkers, whose boundaries are, at present, uncertain.

In other words, whether we read the Greeks as part of the vitalist project becomes a decision informed by what we think will contribute to our flourishing at a particular historical moment. Part of the challenge here is navigating between reading the Greeks as inextricable from this milieu, on the one hand, and, on the other hand, framing a "return to the Greeks" as a choice on the order of choosing vitalism. Another way to say this is that Hellenism is woven together with vitalism to an extent that we cannot choose to undo their historical entwinement. But getting clearer about how they work together might help us make choices about how to engage with vitalism, the Greeks, and the problems, historical and philosophical, posed by "life" within communities that are not reified by nation nor by race nor by the phantasm of Europe but are, rather, dynamic and heterogeneous and heteroglossic. It is within a milieu where lives are lived together under circumstances at once radically contingent and deeply informed by entrenched structures of inequity, exploitation, and violence that we make decisions about which pasts to read with, and on what terms.[55] In the end, Canguilhem's philosophico-historical perspective on vitalism asks us to think about the impossibility of neutrality within a landscape that is not of our own making but is, nevertheless, remade, however incrementally, by our choices of which pasts to value, and how, and why.[56] This perspective can help us see how the Greeks in "Aspects of Vitalism" function as at once historically consequential texts, myths, and objects of elective affinity. By diagnosing the different facets of classicism's continuity in the essay, we may gain a better sense of how reading the history and philosophy of vitalism through the Greeks is always an ethico-political choice made in a particular milieu, at a particular moment.

Bibliography

Agamben, Giorgio. 1998. *Homo Sacer: Sovereign Power and Bare Life.* trans. Daniel Heller-Roazen. Stanford: Stanford University Press.

Bernal, Martin. 1987. *Black Athena: The Afroasiatic Roots of Classical Civilization, Vol. 1 The Fabrication of Ancient Greece 1785-1985.* New Brunswick: Rutgers University Press.

Bianco, Giuseppe, and Miguel de Beistegui. 2015. Introduction: Life: Who Cares? In *The Care of Life: Transdisciplinary Perspectives in Bioethics and Biopolitics,* ed. Miguel de Beistegui, Giuseppe Bianco, and Marjorie Gracieuse, 1–21. Lanham: Rowman and Littlefield.

Braidotti, Rosi. 2010. The Politics of 'Life Itself' and New Ways of Dying. In *New Materialisms: Ontology, Agency, and Politics,* ed. Diana Coole and Samantha Frost, 201–220. Durham: Durham University Press.

Brill, Sara. 2019. *Aristotle on the Concept of Shared Life.* Oxford: Oxford University Press.

Canguilhem, Georges. 1989. Vie. *Encyclopedia Universalis* 23: 530–546.

———. 1991. *The Normal and the Pathological.* trans. Carolyn Fawcett in collaboration with Robert S. Cohen. New York: Zone.

———. 1994. A Vital Rationalist. In *Selected Writings from Georges Canguilhem.* New York: Zone.

———. 2005. The Object of the History of Sciences. In *Continental Philosophy of Science,* ed. Gary Gutting, 198–207. Malden, MA: Blackwell.

———. 2008. *Knowledge of Life.* trans. Stefanos Geroulanos and Daniela Ginsburg. New York: Fordham University Press.

———. 2012. *Writings on Medicine.* trans. with an Introduction by Stefanos Geroulanos and Todd Meyers. New York: Fordham University Press.

Chapoutot, Johann. 2016. *Greeks, Romans, Germans: How the Nazis Usurped Europe's Classical Past.* trans. Richard R. Nybakken. Berkeley: University of California Press.

---

[55] These decisions do not need to dictate exclusive affiliations. I would point to Francesca Rochberg's claim of a "certain kinship" between knowledge and methods of knowing developed in the cuneiform world of the last two millennia BCE and those (methods) of later periods (2016: 3). Rochberg engages with recent work on Amazonian ontologies to argue that we should not seek a concept of nature in cuneiform sources as the grounds of our kinship with them in a history of science. Rather, we should imagine kinship along other lines.

[56] For analysis of "classical" value, see Porter 2006; The Postclassicisms Collective 2019: 8–18.

Dubreuil, Laurent. 2006. Leaving Politics: *Bios, Zōē*, Life. *Diacritics* 36 (2): 83–98.

Esposito, Roberto. 2008. *Bios: Biopolitics and Philosophy*. trans. Timothy Campbell. Minneapolis: University of Minnesota Press.

Etxeberria, Arantza, and Charles T. Wolfe. 2018. Canguilhem and the Logic of Life. *Transversal: International Journal for the Historiography of Science* 4: 47–63.

Finlayson, James Gordon. 2010. 'Bare Life' and Politics in Agamben's Reading of Aristotle. *The Review of Politics* 72: 97–126.

Fleming, Katie. 2012. Odysseus and Enlightenment: Horkheimer and Adorno's 'Dialektik der Aufklärung'. *International Journal of the Classical Tradition* 19 (2): 107–128.

Foucault, Michel. 1970. *The Order of Things: An Archaeology of the Human Sciences*. New York: Vintage Books.

———. 1977. Nietzsche, Genealogy, History. In *Language, Counter-Memory, Practice: Selected Essays and Interviews*, 139–164. Ithaca: Cornell University Press.

———. 1978. *The History of Sexuality. Vol. 1: An Introduction*. trans. R. Hurley. New York: Vintage Books.

———. 1991. Introduction. In Georges Canguilhem, *The Normal and the Pathological*, 7–24. New York: Zone.

———. 2017. *Subjectivity and Truth, Lectures at the Collège de France 1980–1981*. ed. Arnold I. Davidson, trans. Graham Burchell. New York: Palgrave Macmillan.

Geroulanos, Stefanos. 2018. *Transparency in Postwar France: A Critical History of the Present*. Stanford: Stanford University Press.

Geroulanos, Stefanos and Todd Meyers. 2012. Introduction: Georges Canguilhem's Critique of Medical Reason. In George Canguilhem, *Writings on Medicine*. trans. with an Introduction by Stefanos Geroulanos and Todd Meyers, pp. 1–24. New York: Fordham University Press.

Goldhill, Simon. 2000. Whose Antiquity? Whose Modernity?: The Rainbow Bridges of Exile. *Antike und Abendland* 46: 1–20.

Holmes, Brooke. 2010. *The Symptom and the Subject: The Emergence of the Physical Body in Ancient Greece*. Princeton: Princeton University Press.

———. 2014. Proto-Sympathy in the Hippocratic Corpus. In *Hippocrate et les hippocratismes: médecine, religion, société: Actes du XIVᵉ Colloque International Hippocratique*, ed. Jacques Jouanna and Michel Zink, 123–138. Paris: Académie des Inscriptions et Belles-Lettres.

———. 2019. *Bios. Political Concepts: A Critical Lexicon* 5. http://www.politicalconcepts.org/bios-brooke-holmes/.

Jones, Donna V. 2010. *The Racial Discourses of Life Philosophy: Négritude, Vitalism, and Modernity*. New York: Columbia University Press.

Jouanna, Jacques. 2012. Galen's Concept of Nature. In *Greek Medicine from Hippocrates to Galen: Selected Papers*, 287–311. Leiden: Brill.

Lonie, I.M. 1981. Hippocrates the Iatromechanist. *Medical History* 25: 113–150.

Méthot, Pierre-Olivier. 2013. On the Genealogy of Concepts and Experimental Practices: Rethinking Georges Canguilhem's Historical Epistemology. *Studies in History and Philosophy of Science* 44: 112–123.

Miller, Paul Allen. 2020. Against Agamben: Or Living Your Life, *Zōē* Versus *Bios* in the Late Foucault. In *Biotheory: Life and Death under Capitalism*, ed. Jeffrey R. Di Leo and Peter Hitchcock, 23–41. London: Routledge.

Payne, Mark. 2018. *Hontology: Depressive Anthropology and the Shame of Life*. Alres Ford: Zero Books.

Porter, J.I. 2006. Introduction: What is 'Classical' About Classical Antiquity. In *Classical Pasts: The Classical Traditions of Greece and Rome*, ed. J.I. Porter, 1–65. Princeton: Princeton University Press.

———. 2019. Nietzsche's Untimely Antiquity. In *The New Cambridge Companion to Antiquity*, ed. T. Stern, 49–71. Cambridge: Cambridge University Press.

Rankine, Patrice D. 2019. The Classics, Race, and Community-Engaged or Public Scholarship. *American Journal of Philology* 140: 345–359.

Rheinberger, Hans-Jörg. 2005. Reassessing the Historical Epistemology of Georges Canguilhem. In *Continental Philosophy of Science*, ed. Gary Gutting, 187–197. Malden, MA: Blackwell.

———. 2010. *On Historicizing Epistemology: An Essay.* trans. David Fernbach. Stanford: Stanford University Press.

Rochberg, Francesca. 1992. The Cultures of Ancient Science: Some Historical Reflections; Introduction. *Isis* 83 (4): 547–553.

———. 2016. *Before Nature: Cuneiform Knowledge and the History of Science.* Chicago: University of Chicago Press.

Sholl, Jonathan. 2012. The Knowledge of Life in Canguilhem's Critical Naturalism. *Pli* 23: 107–127.

The Postclassicisms Collective. 2019. *Postclassicisms.* Chicago: University of Chicago Press.

Weheliye, Alexander G. 2014. *Habeas Viscus: Racializing Assemblages, Biopolitics, and Black Feminist Theories of the Human.* Durham: Duke University Press.

Weinstein, Jami, and Claire Colebrook. 2017. Introduction: Critical Life Studies and the Problems of Inhuman Rites and Posthumous Life. In *Posthumous Life: Theorizing Beyond the Posthuman*, ed. Jami Weinstein and Claire Colebrook, 1–14. New York: Columbia University Press.

Wolfe, Charles T. 2017. La biophilosophie de Georges Canguilhem. *Scienza & Filosofia* 17: 33–54.

Wolfe, Charles T., and Andy Wong. 2015. The Return of Vitalism: Canguilhem, Bergson and the Project of a Biophilosophy. In *The Care of Life: Transdisciplinary Perspectives in Bioethics and Biopolitics*, ed. Miguel de Beistegui, Giuseppe Bianco, and Marjorie Gracieuse, 63–77. Lanham: Rowman and Littlefield.

Canguilhem and the Logic of Life

Arantza Etxeberria and Charles T. Wolfe

[L]a vie déconcerte la logique.
(Canguilhem 1977a, 1)
[T]o do biology, even with the aid of intelligence, we sometimes
need to feel like beasts ourselves.

(Canguilhem 2008a, xx)

Abstract We examine aspects of Canguilhem's philosophy of biology, concerning the knowledge of life and its consequences on science and vitalism. His concept of life stems from the idea of a living individual endowed with creative subjectivity and norms, a Kantian view which "disconcerts logic." In contrast, we examine two naturalistic perspectives in the 1970s exploring the logic of life (Jacob) and the logic of the living individual (Maturana and Varela). Canguilhem can be considered to be a precursor of the second view, but there are divergences; for example, unlike them, he does not dismiss vitalism, often referring to it in his work, and even at times describing himself as a vitalist. The reason may lie in their different views of science.

Keywords Canguilhem · Vitalism · Biology · Logic of life · Autopoetic/ heteropoetic · Analysis/synthesis · Living individual

A. Etxeberria
IAS Research for Life, Mind, and Society, Department of Philosophy,
University of the Basque Country, UPV-EHU, Donostia/San Sebastián, Spain
e-mail: arantza.etxeberria@ehu.eus

C. T. Wolfe (✉)
Département de Philosophie & ERRAPHIS, Université de Toulouse Jean-Jaurès,
Toulouse, France

C. Donohue, C. T. Wolfe (eds.), *Vitalism and Its Legacy in Twentieth Century Life Sciences and Philosophy*, History, Philosophy and Theory of the Life Sciences 29, https://doi.org/10.1007/978-3-031-12604-8_8

1   Introduction

In Canguilhem's philosophy, life disconcerts logic because of its intrinsically self-produced or "autopoetic"[1] nature, in contrast with mechanical devices. This is 'logic' in the sense of the method of scientific discovery, even that which Claude Bernard theorized for experimenting with organisms in vivo,[2] but also, 'logic' understood as a concept or scheme of the internal organization and functional integration that underlies the living state. Knowledge of life, the method, is challenging for biology. Experimental biology tends to consider living beings as machines, and the knowledge operation required for that leaves part of life aside; it cannot grasp life in full. In fact, although Canguilhem as a historian and philosopher of science has a high regard for biology,[3] he finds that something goes missing when scientific knowledge aims to understand life by means of an analysis of wholes in(to) parts, even if this may be the only way to proceed. The analysis/synthesis dichotomy is important in Canguilhem's view on life and vitality, as well as for his understanding of medicine, and of the pathological more generally. Life is not analyzable, he contends, i.e., life defies scientific methods because of its inherent plasticity and variability, which are evident in its interactive or relational capacity. The latter is particularly emphasized in his conceptualization of the milieu as forming a constitutive relation with the organism,[4] notably close to, and inspired by, von Uexküll's idea of *Umwelt*.

Canguilhem's understanding of life sets a high value on its capacity to establish its own multiple norms according to its relationship with the milieu, and change them within a range of potentialities to establish a new physiological order when required.[5] This entanglement of organism and milieu in a normative relation is the main distinctive feature of Canguilhem's vitalism. Not that he posits the matter of life as an ontological or metaphysical entity different from that of physicochemical systems. Canguilhem denies that vitalism is a metaphysics, and then adds immediately afterwards that it is "the recognition of the originality of the fact of life [*le fait*

---

[1] As we discuss below, Canguilhem writes "autopoetic", rather than "autopoietic".

[2] Bernard 1865, Coleman 1985.

[3] Canguilhem's relation to science itself and the knowledge of science has been questioned; for example, Gabel 2015 mentions Jacob's comment in the *Web Stories Video*, that Canguilhem told him that he would not have written much of what he did, had he read Jacob earlier. Although Jacob seems to have understood Canguilhem's remark at face value, we could always think he was just being polite and appreciative of the work of the scientist.

[4] In "The Living and its Milieu," he writes that "the relationship between the organism and the environment is the same as that between the parts and the whole of an organism" (Canguilhem 2008a, 111).

[5] "Man is only truly healthy when he is capable of multiple norms, when he is more than normal. The measure of health is a certain capacity to overcome organic crises in order to establish a new physiological order, different from the initial order" ("Le normal et le pathologique," in Canguilhem 1965, 167). See also Canguilhem 1972, 77, 155.

vital]."[6] What is this originality, then? It is not an ontological specificity (like a Drieschian entelechy), yet it is a feature which resists any 'logic of life'. Although he warns that there are intellectual dangers inherent in positing that living beings are like an empire within an empire (*imperium in imperio*, Canguilhem 1965, 95), he asserts that *Life itself* determines livings beings to act in interpretive, purposive, normative, vital ways. Life "disconcerts logic" (Canguilhem 1977a, 1). He does not reject biology's kind of knowledge as science either, but contends that being alive is the same as being synthetic, as opposed to analytic; and synthetic, like autopoetic, means that it is a system in a state of continuous creativity.

In this essay we examine Canguilhem's ideas concerning the knowledge of life and its consequences for science and vitalism. First, his concept of life, which stems from the idea of the living individual as endowed with creative subjectivity and norms; we will consider it as a Kantian view in that it shares Kant's challenge to a science of living beings (Sect. 2). Second, why life disconcerts logic. In order to explicate this, we examine two different perspectives. One is the evolutionary and genetically based logic of life of works such as Jacob's (1973) (Sect. 3). The other is the organizational dynamic logic of the individual of the autopoietic school (Maturana and Varela 1973). Although Canguilhem seems to have preceded and influenced the latter, there are divergences. For example, unlike them, he does not dismiss vitalism (Sect. 4). Third, we explore his claim for vitalism connected with views about the role of analysis in the scientific knowledge of life and his characterization of life as synthesis (Sect. 5).

2 Canguilhem and the Life of an Organism

Since the second half of the nineteenth century there have been two distinctive styles in the study of biology. One of them, physiology, focuses on the living organism, and takes as its main topic the arrangement of parts or organs into an organized whole. The other style, that of evolutionary biology, is concerned with differential changes of traits in populations and lineages through – mainly – genealogical processes. The main exponents of each of these styles or traditions were, respectively, Claude Bernard (1865) and Charles Darwin (1859).[7]

A main philosophical topic of physiology is biological individuality, its delimitation and its cohesion: how a living individual maintains its integrity and organization through the causal interactions of its parts and the regulation of those interactions (Bernard 1878–1879; Pradeu 2016). This was the approach pursued experimentally by Claude Bernard's physiology, as it aimed to reach scientific status; some considered him as the Newton of medicine (see note 5 below). Physiology thus understood

---

[6] Canguilhem, "Le normal et le pathologique," in Canguilhem 1965, 156.

[7] As several authors have noted, Claude Bernard's physiological tradition had little interest in evolutionary or developmental biology, which it did not view as proper sciences (see Normandin 2007).

parted ways with the more observational approach of natural history and intended to be scientific. Canguilhem's philosophy of life operates within this physiological framework permeated by the antagonism between mechanist and vitalist views concerning the special status of living beings and the challenge thus posed for the scientific knowledge of life.

This subject matter is reminiscent of Kant's view of organized beings and scientific knowledge. Kant promoted the view of living beings as purposeful and self-organized in his 1790 *Critique of Judgment* (§ 65, AA 5, 374). There he established the grounds for understanding organized beings whose components are mutually dependent on each other and on the whole they generate. Being self-organized, they are very different from artifacts such as a watch, organised according to a designer's plan. But this understanding set a limit for science. Difficulties appear in the project of reconciling this with the conceptual framework Kant developed for natural sciences in his *Critique of Pure Reason* (1781/1787), founded in natural laws without purposiveness, in external causes, and in mechanical principles (Nuño de la Rosa and Etxeberria 2010). Kant did not think there can be a naturalist scientific explanation for living beings, such as there is one for physical systems. His declaration that there will not be a "Newton of the blade of grass"[8] is well known: "Indeed, so certain is it, that we may confidently assert that it is absurd for human beings even to entertain any thought of so doing or to hope that maybe another Newton may some day arise, to make intelligible to us even the genesis of but a blade of grass from natural laws that no design has ordered" (Kant 1790/1987 § 75, AA 5, 400).[9]

Kant's view of the organism as a self-organized system constitutes a challenge for the scientific analytic method. There have been attempts to reconcile teleology and mechanism, such as Lenoir (1982)'s who understood the Kantian tradition as a way to integrate self-organization and teleology within scientific biology. But those naturalizing efforts or scientific explanations of material self-organizing appear to be reductionistic (Moss and Newman 2016). In sum, the Kantian challenge is basically the problem of whether our knowledge of life or of living systems can be naturalized. Kant seems to hold that it cannot (and some praise him for this, while others

---

[8] Since then, there was a long controversy about who could be the scientific figure that would contradict Kant. According to Cassirer, for biologists like Haeckel, Darwin was the "Newton of the blade of grass," yet Roux rejected this (Cassirer 1950, 163). Others have mentioned Claude Bernard (Prochiantz 1990), and still others thoroughly agree with Kant (Nuño de la Rosa and Etxeberria 2010). We return to this topic in Sect. 4.

[9] This statement of Kant's is often quoted approvingly, a rare exception being Zammito 2006, who notes that Kant is neatly placing himself in the rearguard of scientific thought of his time concerning living entities. Our point here is simply to note the existence of this influential position according to which 'Life' is not reducible to a certain set of empirical (measurable, quantifiable) features. In that sense Canguilhem can be said to be a Kantian (see Brilman 2018 for an interesting development of this connection).

reproach him for just the same claim).[10] Canguilhem's approach to this, in some respects similar to Kant's (Brilman 2018), asks how we can know about living beings with the kind of knowledge developed to investigate inanimate beings and the production of technical devices.

Canguilhem's view of living systems as actively self-produced or autopoetic establishes the difference with respect to technological objects. He referred to living beings as "autopoétique" or "autopoetic" in "L'expérimentation dans la biologie animale," an essay on the experimental tradition started by Claude Bernard, originally delivered as a talk in 1951 and included in *La connaissance de la vie*. There he distinguishes the "heteropoetic" character of human technical activity in its interaction with the environment: "Man first experiences and experiments with biological activity in his relations of technical adaptation to the milieu. Such technique is heteropoetic, adjusted to the outside, and it takes from the outside its means, or the means to its means" (Canguilhem 2008a, 9). However, he contends that when in interaction with other living beings, experimenters become aware of the "autopoetic character of organic activity." The realization of this has been an achievement: "Only after a long series of obstacles surmounted and errors acknowledged did man come to suspect and recognize the autopoetic character of organic activity and to rectify progressively, in contact with biological phenomena, the guiding concepts of experimentation." Human action producing technology "presupposes a *minimal* logic – for the representation of the external real, which human technique modifies, determines the discursive, reasoned facet of the artisan's activity, and all the more so the engineer's." This does not work in the case of living entities because humans cannot produce them from the exterior, therefore "we must abandon this logic of human action if we are to understand living functions" (Canguilhem 2008a, 9).

Canguilhem's attention is focused on the kind of knowledge of or attitude towards living entities, in epistemological terms. The "autopoetic" character of living beings, in contrast with artefacts, refers to the kind of object of knowledge. Later Maturana and Varela (1973, 1980) will use a similar term (autopoiesis) to characterise the constitutive organization of living beings. This topic appears also in the analysis/synthesis opposition: Canguilhem insists that knowledge of living systems proceeds by analysis; to know living individuals, science or biology has to analyze them, while ontologically they are synthetic, as they dynamically produce themselves in an active and creative way. "The physiology of regulation (or homeostasis, as it has been called since Walter Bradford Cannon), together with cytologic morphology, enabled Bernard to treat the organism as a whole and to develop an analytic science of organic functions without brushing aside the fact that a living

---

[10] Respectively, Chen 2019 and Zammito 2006 and 2018. The extent to which Kant 'refuses' naturalization or perhaps just the integration of the life sciences in a mechanistic scientific project (on Kant's 'Newtonian' understanding of science), given that naturalization itself is a debated and non-transparent category, is somewhat controversial: for a different view from ours see Duchesneau 2018 (thanks to G. Bolduc for pointing this out), and the review of this book by one of us (Wolfe 2020).

thing is, in the true sense of the word, a *synthesis*" ("Le tout et la partie dans la pensée biologique," translated in Canguilhem 1994, 298).

To argue that living bodies are special, Canguilhem takes over Kurt Goldstein's chief holistic or organismic idea presented in his influential work *The Organism* (1934) – it is the organism as a totality, not a cluster of functions or organs, which acts and reacts as a unified approach to its environment and its challenges (Canguilhem 1972, 49) – and strips it of some of its more overtly metaphysical trappings. In Canguilhem's unique way of engaging with 'organisms' and the question of their uniqueness we find one of the curious features of Goldstein's account: the way in which he wavers or moves back and forth between a cautious, epistemological position (reminiscent of the Kantian regulative ideal in the third *Critique*) in which organisms are real and special *because of the way we cognitively constitute them*, and a bold, ontological position in which organisms are real because of basic, intrinsic features which are just there. However, this convenient distinction between the epistemological (projective, externally constitutive) vision of biological entities and the ontological vision (strong vitalist, 'rational metaphysics' as Kant might have said) is somewhat muddied when Canguilhem introduces a further vitalist twist, in "Aspects du vitalisme": that it might be an objective ('ontological') feature of living beings that they are interpretive beings, and especially that they need to regard *other* entities as being, like themselves, organismic, purposive, vital. We interpret Canguilhem as alluding to this need of being interactively immersed with other organisms to know what they are, when he writes in *La connaissance de la vie* that "We suspect that, to do mathematics, it would suffice that we be angels. But to do biology, even with the aid of intelligence, we sometimes need to feel like beasts ourselves" (Canguilhem 2008a, Introduction xx). There may also be an existentialist *parfum* in Canguilhem's reflections, as when he describes this interpretive stance as essentially a kind of fundamental existential attitude – not a 'fact' but a way of life, indeed a contemplative way of life. In any case, what is distinctive of his position, especially when we consider the core arguments of *The Normal and the Pathological*, is the presupposition that normativity is a power or capacity proper to *living* beings:

> We, on the other hand, think that the fact that a living man reacts to a lesion, infection, functional anarchy by means of a disease, expresses the fundamental fact that life is not indifferent to the conditions in which it is possible, that life is polarity and thereby even an unconscious position of value; in short, life is in fact a normative activity. Normative, in philosophy, means every judgment which evaluates or qualifies a fact in relation to a norm, but this mode of judgment is essentially subordinate to that which establishes norms. Normative, in the fullest sense of the word, is that which establishes norms. And it is in this sense that we plan to talk about biological normativity. (Canguilhem 1972, 126–127)

We find here an insistence that there is something unique about living entities that makes them creators of a certain world which they inhabit. Upon closer examination, this idea seems to contain some Nietzschean overtones (Foucault also pointed to this aspect in his mentor's work: Foucault 1991, 21), namely, the idea that values, norms and other higher-level constructs are in fact products of our vital instincts, so that life integrates rationality to itself through its normative activity. In a lecture in the problem of regulations in the organism and society, Canguilhem also insists that

> An organism is an entirely exceptional mode of being, because there is no real difference, properly speaking, between its existence and the rule or norm of its existence. From the time an organism exists, is alive, that organism is 'possible', i.e., it fulfils the ideal of an organism; the norm or rule of its being [*existence*] is given by its existence itself (Canguilhem 2002, 106–107).

Organisms have agency. Yet Canguilhem does not appeal to a disembodied, foundational subjectivity, as we might find in more anti-naturalistic trends in phenomenology; there is no pure ego contemplating the reality of the flesh like a sailor in a ship, for him. As regards the relevance of experience, it would seem that – despite their shared affinity for Goldstein – it is more than unlikely that Canguilhem would go as far as Merleau-Ponty, as we see when he reflects on the limitations of a conceptualization of the living body as "inaccessible to others, accessible only to its titular holder" (Canguilhem 2008b, 476).

Canguilhem's position on organic uniqueness and what he somewhat cryptically calls 'experience' is subtly yet significantly different:

> the classical vitalist grants that living beings belong to a physical environment, yet asserts that they are an exception to physical laws. This is the inexcusable philosophical mistake, in my view. *There can be no kingdom within a kingdom [empire dans un empire]*, or else there is no kingdom at all. There can only be one philosophy of empire, that which rejects division and imperialism. . . . One cannot defend the originality of biological phenomena and by extension, of biology, by delimiting a zone of indeterminacy, dissidence or heresy within an overall physicochemical environment of motion and inertia. *If we are to affirm the originality of the biological, it must be as a reign over the totality of experience, not over little islands of experience.* Ultimately, classical vitalism is (paradoxically) too modest, in its reluctance to universalize its conception of experience.[11]

'Classical' vitalism as described here is what one of us has termed substantival vitalism elsewhere (Wolfe 2011, 2015a). That is, a form of vitalism claiming that living beings are ontologically special, different from the rest of the physical world, and perhaps even unexplainably so. And Canguilhem's diagnosis of an "inexcusable philosophical mistake" is clear enough (whether we explicate this in Spinozist terms – no kingdom within a kingdom – or in physicalist terms – no gaps in the lawlike physical world; neither of these are to be confused with a more 'Gaian' sense of 'one world' in which life is coeval with this world). But what should we make, then, of his defence of the "originality of the biology," i.e. the autonomy of biology, as a "reign over the totality of experience"? What looks at first glance like metaphysical holism might instead be an 'attitudinal' conception, that is, a *point of view* on experience.

---

[11] Canguilhem, "Aspects du vitalisme," in Canguilhem 1965, 95, emphasis ours. To our knowledge, this unusually 'phenomenological'-sounding appeal to 'lived experience' rather than 'life' has not been pointed out in Canguilhem, with the exception perhaps of Paul Rabinow's comments on Canguilhem's "not-so-latent existentialism," in his introduction to the Canguilhem anthology *A Vital Rationalist* (Rabinow 1994, 18).

Canguilhem was aware and acceptant of the biology of his times, and paid attention both to the physiological perspective and to the evolutionary|molecular biology perspective. Yet he does not appear to be keen to develop what we could call a logic of life or the living; why? Taking into account Canguilhem's views on living individuals, we can now consider the reason why life disconcerts logic according to him. For this, we examine some views on the nature of life and organisms in the biology of the 1970s, such as François Jacob's and Humberto Maturana and Francisco Varela's, each proposing a particular proposal for a logic in biology. Canguilhem's ideas contrast with those of biologists of the time: we will specifically take into account Jacob's evolutionary perspective in *La logique du vivant* and Maturana and Varela's organizational perspective in *Autopoiesis and cognition*. Both books were originally written in the early 1970s (in languages different from English) and elaborate very different research programs to explore living organization.

3   The Logic of "Life at Large"

As mentioned earlier, two major traditions of biological thought emerged in the nineteenth century, which articulate and convey distinctively different intuitions concerning life. The evolutionary approach conceives of life as a whole and unique phenomenon, whereas the physiological approach is concerned with the organization of particular living individuals underlying the living state.

In the early 1970s, the two styles of biology generated different and opposed views on whether the organization of living organisms or the phenomena of reproduction and evolution of living entities was primary in biology. A previous clash between the two styles can be also found earlier in the twentieth century, when scientists and philosophers of the Theoretical Biology Club (such as Needham, Woodger, and Waddington, among others) developed a distinctive organicist framework in contrast to the then-emerging framework of molecular biology and the Modern Synthesis in evolutionary biology (Etxeberria and Umerez 2006; Peterson 2017).

François Jacob's *La logique du vivant* was a very important book in the 1970s in which the author, already a Nobel Prize winner and a widely recognized molecular biologist, made a remarkable attempt at reconstructing the history and philosophy of biology around the notion of biological organization. His "logic of life" stands for the then-prevailing genetic and evolutionary consensus that the most important feature of life is reproduction (and evolution), visible in the recent findings in molecular biology. Jacob attempted to reconcile the received tradition of continental European biological thought with views stemming from contemporary ideas on genetics and evolution. It is full of enthusiasm towards the notion of biological information and the logic of genetics of the 1960s and 1970s, which he understands to be the corollary of biological struggles to understand biological organization, through a model of life sympathetic to informational formalisms for genetic action.

This he shares with Canguilhem to a certain point, as is visible in the latter's 1966 additions to *Le normal et le pathologique*, displaying a real openness to genetics as a "nouvelle connaissance de la vie." (see Morange 2000; Loison 2018).

For Jacob, the special features of life appear in the evolutionary genetic framework, linked not to the properties of living beings such as studied by classical philosophy (e.g. Thomism) but to the new image made possible by the evolutionary science of the time. It enhances a collective view of life connected as a genealogical succession, rather than mechanistically explainable at the level of the individual living being:

> An organism is merely a transition, a stage between what was and what will be. . . . (Jacob 1973, 2). Everything in a living being is centred on reproduction (ibid, 4).
>
> Let us imagine an uninhabited world. We can conceive the establishment of systems possessing certain properties of life, such as the ability to react to certain stimuli, to assimilate, to breathe, or even to grow - but not to reproduce. Can they be called living systems? Each represents the fruit of long and laborious elaboration. Each birth is a unique event, without a morrow. Each occasion is an eternal recommencement. Always at the mercy of some local cataclysm, such organizations can have only an ephemeral existence. Moreover, their structure is rigidly fixed at the outset, incapable of change. If, on the contrary, there emerges a system capable of reproduction, even if only badly, slowly, and at great cost, that is a living system without any doubt. (ibid, 4–5)

Jacob distinguishes explicitly the two views of biology. According to his preferred perspective, evolutionary accounts consider the genealogical connection among living beings in the sense that living beings are not systems that arise and disappear due to their physicochemical properties, or at least not only because of them. As many of their capacities have been inherited from their ancestors, these systems, or part of them, have been *informed* by others:

> Much of the controversy and misunderstanding, particularly with regard to the finality of living beings, is caused by a confusion between these two attitudes. Each tries to establish a system of order in the living world. For one, it is the order which links beings to one another, sets up relationships and defines speciations. For the other, it is the order between the structures by which functions are determined, activities coordinated and the organism integrated. One considers living beings as the elements of a vast system embracing the whole earth. The other considers the system formed by each living being. One seeks to establish order between organisms; the other within each organism. The two kinds of order meet at the level of heredity, which constitutes the order of biological order, so to speak (Jacob 1973, 7–8).

Darwinian evolution implies two main ideas: that of the common ancestor, which entails that there is a genealogical connection among forms of life, often represented as the tree of life, and secondly, a claim about the causes of evolutionary processes, such as natural selection (thus understood as explanatory of key features of living beings). The *received view* of the Modern Synthesis reinterpreted both aspects, and by the 1970s many proposed genes as the main ontology, but critics of this view affirm, and Jacob was among them, that evolution by natural selection does not contribute to our knowledge of living organization if it is not by studying development and regulation. After the 2000s, new approaches in systemic and synthetic biology made clear the need to take into account organismic approaches both for molecular and evolutionary biology. Jacob made a great effort to integrate the new biology based on genetics and molecular biology with the organizational tradition (Jacob 1977). In fact, Jacob and Monod's distinction between structural and regulative genes has been identified by some as a source for the field of evo-devo (Morange 2017, 278). Jacob identifies the genetic program, and the determinism it embodies, as the fundamental element of the emerging new theory of the living. However, he also admits the importance of different levels of integration in the domain of life, called *integrons*, each of them being characterized by some independence with respect to lower ones.

Current systemic approaches search for a complementarity of the two perspectives. On the one hand, the study of evolution needs to include the mechanical causal processes taking place in development – in addition to population dynamics at various levels and contingent events – and processes responsible for organizations and entities that emerge in interactions, such as symbionts, ecosystems, etc. The *extended evolutionary synthesis* attempts to advance a perspective that would be encompassing and inclusive (Pigliucci and Müller 2010). On the other hand, research on the nature of living organization cannot rely only on formal, mathematical aspects; organizations need to be studied in the material domain, including historical evolutionary events. Jacob emphasizes how biology is an exploration of a logic of life beyond any logic of the living organism. From this perspective, biological knowledge is not concerned with individuality, finality or causal mechanisms; it is a science of living forms that appear and are transmitted in a contingent way.

In his review of François Jacob's evolutionary perspective, Canguilhem (1971) addresses the view of life taking place at the level of cells and the logic of reproduction, as disclosed by the genetics of the time. He there confronts the view that "in order to understand what we are as living beings, we must look to the chromosome, the gene, the DNA molecule. The biochemical study of the bacteria is the beginning of self-knowledge of oneself as a living being" (Canguilhem 1971, 23). This view seems to oblige one to reject finalism, and the centrality of individuality (see Brilman 2018 for a reading of Canguilhem that emphasizes how, in contrast, his philosophy of norms is or can accommodate a philosophy of individuality). In addition, Canguilhem treats the new playground of biological science, namely the informational perspective, as a new logic for understanding life and getting rid of vitalism in order to achieve full continuity of the vital and the inanimate. Canguilhem's review of Jacob's book praises his effort but does not give in, to the contrary.

Gabel quotes Jacob saying that only in the fifties did Canguilhem begin to take account of contemporary biological research, and contends that after that he gave up his vitalism. "Though he did not renounce his old positions –in fact he seems to have felt his philosophy to be consistent with the discoveries of genetics and molecular biology- he in fact moved away from both humanism and vitalism" (Gabel 2015, 82). We disagree, as in his review Canguilhem remains sceptical about the informational logic of life. Indeed, as Morange has maintained, molecular biology does bear out the continued relevance of some of Canguilhem's ideas (even if Canguilhem did not analyse developments in biology adequately). Also, philosophical positions may be modulated, but are not dictated by scientific facts, and this is evident in Canguilhem's case, as when he writes: "The execution of a program that is identified with its realization is a blunt fact, without cause or responsibility. The logic of life does not refer to any logician" (Canguilhem 1971, 23). In that sense, blind evolution is a change without history, as "evolution through natural selection is only history in its incidents, errors and rare events" (24). And at the end of the review, Canguilhem reflects on Jacob's much-quoted pronouncement that biological research no longer "'inquires into Life" ("On n'interroge plus la vie aujourd'hui dans les laboratoires"), i.e., that the concept of Life (and by extension any ontologically foundational clauses attached to work in the life sciences) no longer serves any purpose in such work.[12] With a curious kind of pathos that is however not 'Romantically anti-scientific', he observes that living beings "think they live" a life "outside of laboratories," not realizing (Canguilhem literally writes "not knowing") that in laboratories, "Life has lost its life with its secret."[13]

4   The Logic of the Living Individual

Very soon after the 1970s, and especially at the turn of the century, both in the philosophy of biology and in most biological disciplines there was a significant movement in search of systemic and organizational principles, as is made evident by current advances in systems biology, synthetic biology and the extended evolutionary synthesis. Historically there are (at least) two organizational traditions: the physiological one starting with Claude Bernard, to which autopoiesis (and most of the work on biological autonomy) belongs, and the developmental one which has led to structuralism and Evo-Devo (Etxeberria and Bich 2017; Etxeberria 2004). Both have connections to Kant's view of organisms in his *Critique of Judgment* in their ways of arguing holistically and/or mereologically, although they have kept quite apart during the twentieth century.

To Jacob's plea for a logic of life, Varela and Maturana respond with a new vindication of the centrality of the living individual as a foundation of biology, this time looking for a "logic" of the living individual. Maturana and Varela's notion of autopoiesis can be considered to be an answer to Jacob's picture that especially rejects the informational perspective in biology, a view shared by the Developmental Systems Theory in philosophy of biology, especially after Susan Oyama (1985).

---

[12] Jacob 1970, 320 (all translations are ours unless otherwise indicated). At the conceptual level, this corresponds to Edouard Machery's deliberately deflationary suggestion (Machery 2012) that we should give up seeking to provide definitions of life, as these are either folk concepts, or unresolvable with other competing definitions: namely, evolutionists, theoretical biologists, self-organization theorists, molecular biochemists and artificial life researchers cannot agree on a definition.

[13] Canguilhem 1971, 25. He adds that "it is outside laboratories that love, birth and death continue to present living beings – the children of order and chance – the immemorial figures of these questions that life science no longer asks of life.

Their narrative on the logic of the living individual, clearly influenced by Jacob's book, deliberately disputes many of his positions about the logic of life and the centrality of reproduction and evolution. In contrast, for the positive part, it often draws Canguilhem's views to contrast Jacob's informational stance. But their main claim goes far beyond Canguilhem's position and points to developments in biology that Canguilhem did not foresee, in particular in work in cybernetics and artificial systems that aimed to explore living phenomena through synthetic and systemic models and simulations. (see however Canguilhem's remarks on cybernetics in his essay on the emergence of the concept of biological regulation, in Canguilhem 1977b, 82). It is within this research field that Maturana and Varela's contributions flourished.

The autopoietic approach to the living belongs to the above-mentioned physiological systemic tradition focused on the problem of the relational unity of the living, additionally associated with Kant's understanding of organisms, Claude Bernard's concept of *milieu intérieur,* and the organicist tradition that considers life as organization – a tradition including Hans Jonas and Jean Piaget among others.[14] Other clear associations are with the cybernetic movement, especially with second-order cybernetics (on this relation see Bich and Etxeberria 2013).

The notion of autopoiesis was proposed by Maturana and Varela[15] (Varela 1979; Maturana and Varela 1973, 1980) to refer to the biological self-organization of individual living beings, in contrast with other properties of life that the biology of their time considered as primary (genes as DNA or informational properties). The basic idea of autopoiesis is self-production, as a relational dynamic of components that generates or brings forth a membrane or boundary, which constitutes the individual living being's identity as separated from the surroundings (Varela 1981). The autopoietic approach to life criticises evolutionary and molecular biology, and focuses in contrast on, autonomy and identity, aiming to naturalize them as marks of life, prior to reproduction or evolution. The autopoietic theorists claim that living organization has primacy with respect to the other phenomena associated with life (Etxeberria 2004). The status of individual identity, constituted by the system itself, and not by anything external (heteropoietic), is a central idea of this approach. As we noted, some of the distinctions they stress, for example that between autopoiesis and heteropoiesis, already appear in Canguilhem's *La connaissance de la vie.* The relations of the autopoietic unity and its surroundings cannot be understood in terms of input/output fixed interactions. Instead, non-specific perturbations are coupled with plastic behaviors of the system within the range of internal coherence.

In their initial writings, the authors embrace mechanism and criticize vitalism, in the name of their logic of self-production. This is important because, according to

---

[14] Canguilhem sits somewhat unsteadily here, given that he is less of a "naïve (ontological) organicist" than the rest.

[15] Both Francisco Varela and Humberto Maturana have separately claimed to have coined the term, referring to different sources for the invention. Here we suggest that they probably conceived the notion based on those passages of Canguilhem's *La connaissance de la vie* in which he describes living systems as autopoetic.

them, vitalism focuses on entities bearing properties, in contrast with the relational approach they defend in which properties appear as a result of relations among components (see Bich and Arnellos 2012, 79). Maturana writes that

> in a vitalistic explanation, the observer explicitly or implicitly assumes that the properties of the system, or the characteristics of the phenomenon to be explained, are to be found among the properties or among the characteristics of at least one of the components or processes that constitute the system or phenomenon. In a mechanistic explanation the relations between components are necessary; in a vitalistic explanation they are superfluous (Maturana 1978, 30).

For interactionist or ecological perspectives living beings cannot be fully accounted for in terms of *intrinsic* properties, but need to take into account *relational* properties arising from interactions between living constituents.[16]

One difference between Canguilhem's usage of the term 'autopoetic' and Maturana and Varela's account of autopoiesis may be that the latter intend to explore ways in which the autopoiesis of living systems can be explored in artificial models. This is not exactly like Canguilhem, who thinks that the autopoetic character of living beings is equivalent to their not being susceptible to be grasped by knowledge. Canguilhem starts his book *La connaissance de la vie* with the sentence: "Connaître c'est analyser" ("To know is to analyse") only to remind us quickly of the difficulties of grasping a true knowledge of what it means to be alive through analysis.

Although Maturana and Varela also recognise the difficulty of knowing life, "regarded as rationality's blind spot" (Brilman 2018, 40), they rely more confidently on the possibility of exploring the autopoietic organization through the construction of networks and other artificial systems that will allow for the exploration of important aspects of self-production. Today not everyone would accept that biology as a science proceeds merely by analysis. On the contrary, many fields including synthetic biology and, earlier, Artificial Life, have attempted to build synthetic models, systems or simulations by integrating knowledge from different sources and exploring their emergent and creative properties. As a result, the concept of scientific knowledge associated with many fields is far from the idea that the aim of models is to represent reality. Rheinberger (2015) has reflected on the nature of the different epistemological objects produced by science to explain life, and 'organizational' theories in the philosophy of biology (see Moreno and Mossio 2015; Bechtel 2007) have contributed to understand their epistemological and ontological properties.

Many of those systems can in some ways be creative or autopoetic as well, at least in that the complexity of their organization and their operation is opaque to

---

[16] We do not discuss Canguilhem's relation to ecology here for want of space but in his rather little-known essay "Qu'est-ce que l'écologie?" (1974) he expresses a rather cautious, at times deflationary attitude towards what one might term the more Romantic and/or political determinations of ecology, without thereby dismissing it out of hand. It is tempting, however, to see his reflections on organism and environment as lending themselves to an almost 'Gaian' type of understanding, according to which nothing that we know on Earth as natural can be understood as such without the intervention of life: biological phenomena permeate and intervene in the physics and chemistry of all the natural world, so that biology is fundamental.

rationality. Then to ask 'are organisms unique in the physical world? If so, why?' as an orienting question does not only affect issues of an ontological kind (what is life), or an epistemological kind (how can we know life?), but also highlights the presence of a specifically *relational* dimension.

In considering these matters, Hacking (1998, 202) highlights the relational character of Canguilhem's anti-Cartesian thinking about machines: "He takes all tools and machines to be extensions of the body, and part of life itself" (Hacking 1998, 202). In this sense, Canguilhem's approach stresses that there may be an aspect of life which cannot be grasped only by considering the system's properties (as when we question whether machines and living beings are or not similar). On the contrary, what is considered is a certain relation to the milieu (Gayon 1998; Etxeberria 2020). Something that is still opaque to logic.

5   Canguilhem's Claim for Vitalism

A main feature of vitalism in scientific research is to consider that living beings are in some sense different from inorganic or inert beings; this does not always have further ontological and methodological implications. Canguilhem's work appears to be among those defending that view, although he is critical of ... uncritical ontological vitalism. At the same time, Canguilhem appears more cautious than Jacob or other prominent figures who try to dissolve the problem of what life is into an evolutionary logic. This deflationary view underlies usual attempts to replace the definition of life with a list of living properties, such as those appearing in many biology textbooks. In contrast Canguilhem is suspicious of the rejection of vitalism in this way because many of the features that are associated with life, in contrast with those of inanimate systems do surreptitiously appear in normative concepts such as evolutionary advantage (1971, 24).

Canguilhem often refers to vitalism in his work, going as far as describing himself – playfully, yet not just playfully, given the circumstances – as a vitalist.[17] He acknowledges that vitalism is a position that is difficult to maintain. As Dominique Lecourt comments, "Canguilhem, a hero of the Resistance, clearly expresses the difficulty of presenting himself as a 'vitalist' in 1946-1947" (Lecourt 2011, 13) and he thus quotes this passage from "Aspects of Vitalism":

> Today, above all, the usage of vitalist biology by Nazi ideology, the mystification that consisted in using theories of *Ganzheit* to advocate against individualist, atomist, and mechanist liberalism and in favor of totalitarian forces and social forms, and the rather easy conversion of vitalist biologists to Nazism have served to confirm the accusation formulated

---

[17] For example in the Foreword to his book on the development of the notion of reflex: "Il nous importe peu d'être ou tenu pour vitaliste..." and when he presents the book itself as a "defense of vitalist biology" (Canguilhem 1977a, Avant-Propos, 1). Some years earlier, he had devoted one article exclusively to the topic, "Aspects du vitalisme" (originally a series of lectures given at the Collège Philosophique in Paris in 1946–1947), in Canguilhem 1965.

by positivist philosophers like Philipp Frank, as well as by Marxists (Marcel Prenant) that
it is a "reactionary biology" (Canguilhem 1965, 97, 2008a, 72)

At the same time, Canguilhem was comfortable identifying himself with the
equally problematic figure of Nietzsche, and his reference to Marxist criticisms of
vitalism should not be taken to mean outright agreement. In the same essay,
Canguilhem asserts from the outset that when the philosopher inquiries into biological life, she has little to expect or gain from "a biology fascinated by the prestige of the physicochemical sciences, reduced to the role of a satellite of these
sciences" (Canguilhem 1965, 83, 2008a, 59). What this entails for vitalism is that
it has a specifically *philosophical* place, whether it is scientifically 'validated' or
'refuted', and apart from its status as a scientific 'construction'. In this sense, as
Canguilhem suggests, *vitalism is not like geocentrism or phlogiston: it is not refutable in quite the same way.*[18]

To summarize these two dimensions of Canguilhem's thought, one could say that
on the one hand his vitalism is *heuristic*, a claim that living phenomena need to be
approached in a certain way in order to be understood; but on the other hand, it possesses a more *ontological*, Aristotelian dimension. Consider the example Canguilhem
had given in "Aspects du vitalisme": vitalism is not like (the theory of) phlogiston
or geocentrism. Now, faced with this 'fact' that vitalism is not like phlogiston, there
are two possible responses: it's not like phlogiston because it's *true* and thus one's
ontology needs to include it or it's not like phlogiston because it has this *heuristic
value*, or explanatory power.

For Canguilhem vitalism is a way to understand Life in a certain way in order not
to miss its essential spontaneity; historically, thinkers known as vitalists have had
what he calls "this vitalist confidence in the spontaneity of life."[19] In other words,
the philosopher in this position is almost inexorably led to a vitalist *positionnement*.
The type of questions she will have for biological science entails that the latter not
be conceived of in reductionist terms, although Canguilhem doesn't explicitly say if
a purely physicochemical perspective on biological entities is flawed ontologically,
or just methodologically. Nevertheless, this is a loaded, rather a prioristic conception of biological science, actually quite reminiscent of the holism of Goldstein,
who Canguilhem openly credits as a major influence.[20]

---

[18] Canguilhem, "Aspects du vitalisme," in Canguilhem 1965, 84; Canguilhem 2008a, 60. The
Medawars note that it is hard to devise an experiment to 'refute' vitalism (Medawar and
Medawar 1983).

[19] "Aspects du vitalisme," in Canguilhem 1965, 89.

[20] On Canguilhem and Goldstein, Gayon 1998, 309–310, and Métraux 2005 make some useful
observations (Métraux also reproduces a letter from Canguilhem to Goldstein); see also Wolfe
2015b. Gayon notes several further references to Goldstein in Canguilhem: Canguilhem 1965,
11–13, 24, 146; Canguilhem 2002, 347; Canguilhem 1977b, 138. Canguilhem (along with
Merleau-Ponty) played a key role in the introduction of Goldstein into France, through the translation of the Organism book (Goldstein 1934/1939), which Canguilhem initiated (the co-translator,
Jean Kuntz was his student) and also by translating Goldstein's article on the "problème épisté-
mologique de la biologie" together with his wife Simone.

But what sort of claim is the insistence on the originality of vital facts? Just because it is not naïve ontological vitalism does not mean it is vitalism without any ontology. As this is not an analysis of vitalism in general but of certain issues in the thought of Canguilhem, it may be worth rapidly clarifying this terminology. It seems that, in addition to a kind of 'de facto' vitalism of some life scientists who insist on the specificity of the systems they study, including in relation to the objects of other sciences such as chemistry and physics, there is a non-ontological vitalism, articulated in thinkers like Claude Bernard and at times in Xavier Bichat (Wolfe 2019), distinct from an *ontological* vitalism in that the latter will consider the difference between living and non-living beings, organisms and mechanisms, 'whole-person' analyses in medicine and molecular analyses, etc., as having ontological significance and/or as being ontologically grounded.

This sense of privacy, of inaccessible interiority, is a crucial feature of many defences of what organisms are and how they are different from machines, but as we mentioned earlier, one should rather think in relational terms. This raises the issue of the relation between Canguilhem and phenomenology.[21] That is, while mainstream biologists thought the problem with vitalism was its appeal to immaterial vital forces, or 'entelechies' that could not themselves be located anywhere in the spatiotemporal world, there may be a different, more philosophical problem with vitalism, in that it can become an appeal to a kind of foundationalist subjectivity, a Self, a Centre, whether this is equated with life itself (as in the old Aristotelian motif that 'the soul is life') or is seen as a precondition thereof. Interesting – and idiosyncratically – Canguilhem's way of renewing vitalism is neither that of the "classical" vitalist, in his terms (which matches the standard critical portrayal of the vitalist), nor that of the subjectivist.

Kurt Goldstein and Canguilhem were, we suggest, onto something when they insisted that rather than say what is unique about the biological, we look to the *observer*: to be an organism is to have a *point of view* on organisms; one which produces intelligibility, which reveals organisms as meaning-producing beings. Goldstein stressed a kind of 'standpoint' dimension in 'the organism' (in fact, typically the human patient), namely, the idea that we necessarily have 'points of view' on our environment and that such points of view enter into the basic definition of what it is to be such an organism. Canguilhem gave further inflection to this idea by speaking of how vitalism is not a mere scientific theory (true or false, refutable, experimental, etc.) but, crucially, something existential, what he calls an *exigence*:

> Vitalism expresses a permanent requirement or demand [*exigence*] of life in living beings, the self-identity of life which is immanent in living beings. This explains why mechanistic biologists and rationalist philosophers criticize vitalism for being nebulous and vague. It is normal, if vitalism is primarily a 'demand', that it is difficult to formulate it in a series of determinations ("Aspects du vitalisme," in Canguilhem 1965, 86).

---

[21] For a nice discussion which makes Canguilhem a phenomenologist see Gérard 2010; for an equally convincing reading which seeks to draw Canguilhem away from phenomenology, see Sholl 2012.

An *exigence* is not a vital 'fact' in a static sense but rather something processual and indeed agential. Other prominent recent figures like Varela also underline the uniqueness of the biological by rejecting that life can be characterized by providing some empirical criteria and vindicate the need for a concept of life that takes into account the self-producing activities of living systems. Yet he explicitly rejects vitalism and embraces naturalism. In this respect Weber and Varela differ from Kant, who believed that living organization cannot be explained scientifically: "Our immodest conclusion is that Kant, although foreseeing the impossibility of a purely mechanical, Newtonian account of life, nonetheless was wrong in denying the possibility of a coherent explanation of the organism. But this 'Newton of the Grassblade' was surely not Darwin." Instead, they maintain that it is the "convergence of philosophical and biological thinking" which offered "an objective account of biological individuality that joins in circle with the constitution of a subject" (Weber and Varela, 2002, 120–121). Thus, they think that the times are ripe for a naturalistic understanding of the living individual as autopoietic.

6   Conclusions

In this paper we have shown some of the problems Canguilhem faced in challenging the existence of a logic of life that can be known by science, in contrast to Jacob and Maturana and Varela, who are more confident than him, but with very different arguments. Some of Canguilhem's difficulties derive from embodiment, relations of the living with the milieu and with other living organisms, and his apparent sympathy for certain phenomenological approaches to the nature of life and living bodies, notably their 'existential' and 'attitudinal' dimensions (even though this definitely does not make him a phenomenologist), although he doesn't go all the way and literally appeal to the "truth of my body", as Merleau-Ponty did (Canguilhem 2008b, 475); his residual existentialism (with occasional overtones of anthropocentrism) may hold some lessons for present-day thinking about life.

Perhaps the difference between vitalism and organicism, given the Kantian difficulties for a science of the living,[22] lies in the difference between a complete skepticism (towards some vitalist positions, although most of them are caricatures) and the hope that science can advance, however partially or perspectively, in understanding at least some aspects of biological organization. Although it is clear that most vitalists were in agreement with this position, criticisms (like for example those of logical empiricists like Frank, although closer reading reveals important

---

[22] Aside from some of the difficulties mentioned earlier, the specific 'conceptual difficulty' lies in the way some non-reductionist programs like Varela's strongly invoke the Kantian pedigree, while somehow overlooking the fact that a core element of the Kantian concept of organism is that it cannot be the object of a causal-naturalist science.

nuances)[23] have built a straw-man of vitalism as a position that wholly rejects scientific understanding of life and embraces mysticism instead.

Canguilhem is not a vitalist according to this excessively partial picture, yet he also does not believe that life has a logic that can be grasped in fixed norms or regulations. And this not only because the norms are internal or internally produced and managed (like in autopoiesis), but also and more importantly because they are variable and their very organization may be contingent in some respects. We suggest that the recognition that some scientific models may have properties of the kind Canguilhem attributes to living beings – that is to say, they are also emergent, creative, and synthetic, and oblige scientists to interact with their products instead of just analysing or representing them – may be a landmark separating different views of science. Organicism tends to value these models as naturalistic, whereas vitalism as understood by and in Canguilhem, takes a step back, and stresses their relational nature.

Acknowledgments An earlier version of this paper was published online in *Transversal: International Journal for the Historiography of Science*, 4, special issue on Canguilhem (2018): 47–63. The authors wish to thank the editor of the journal for permission to reprint a revised version, Marina Brilman and Sebastjan Vörös for their helpful reading of earlier drafts of this paper, and Ghyslain Bolduc for his comments on the penultimate version.

AE acknowledges funding from Grants IT 1228-19 from the Basque Government) and PID2019-104576GB-I00 from the Ministry of Science and Innovation, Spain. CW was funded by the FWO (Flemish Research Council) and then by the European Union's Horizon 2020 Research and Innovation Programme (GA n. 725883 ERC-EarlyModernCosmology).

References

Bechtel, William. 2007. Biological Mechanisms: Organized to Maintain Autonomy. In *Systems Biology: Philosophical Foundations*, ed. F. Boogerd, F.J. Bruggeman, J.-H.S. Hofmeyr, and H.V. Westerhoff, 269–302. Amsterdam: Elsevier.

Bernard, Claude. 1865. *Introduction à l'étude de la médecine expérimentale*. Paris: J.-B. Baillière & Fils.

———. 1878–1879. *Leçons sur les phénomènes de la vie communs aux animaux et aux végétaux* (2 vols.). Paris: J.-B. Baillière.

Bich, Leonardo, and Argyris Arnellos. 2012. Autopoiesis, Autonomy, and Organizational Biology: Critical Remarks on 'Life After Ashby'. *Cybernetics & Human Knowing* 19 (4): 75–103.

Bich, Leonardo, and Arantza Etxeberria. 2013. Systems, Autopoietic. In *Encyclopedia of Systems Biology*, ed. W. Dubitzky, O. Wolkenhauer, K.-H. Cho, and H. Yokota, 2110–2113. New York: Springer.

Brilman, Marina. 2018. Canguilhem's Critique of Kant: Bringing Rationality Back to Life. *Theory, Culture & Society* 35 (2): 25–46.

Canguilhem, Georges. 1965. *La connaissance de la vie,* revised edition. Paris: Vrin (First published 1952).

———. 1971. Logique du vivant et histoire de la biologie. *Sciences* 71 (mars–avril): 20–25.

———. 1972. *Le Normal et le pathologique,* 3d revised edition. Paris: PUF (First published 1943).

---

[23] On such nuances see Chen 2018, 2019.

———. 1974. Qu'est-ce que l'écologie? *Dialogue* (mars): 37–44.

———. 1977a. *La formation du concept de réflexe aux XVIIe et XVIIIe siècles,* 2nd revised edition. Paris: Vrin (First published 1955).

———. 1977b. *Idéologie et rationalité dans l'histoire des sciences de la vie.* Paris: Vrin.

———. 1994. *A Vital Rationalist: Selected Writings from Georges Canguilhem.* Ed. François Delaporte and Trans. Arthur Goldhammer. New York: Zone Books.

———. 2002. Puissance et limites de la rationalité en médecine (1978). In *Études d'histoire et de philosophie des sciences concernant les vivants et la vie,* ed. Canguilhem, 392–411. Paris: Vrin.

———. 2008a. *Knowledge of Life.* Ed. Paola Marrati, Todd Meyers and Trans. Stefanos Geroulanos and Daniela Ginsburg. New York: Fordham University Press.

———. 2008b. Health: Crude Concept and Philosophical Question (translation of "La santé, concept vulgaire et question philosophique" (1988), by T. Meyers and S. Geroulanos). *Public Culture* 20 (3): 467–477.

Cassirer, Ernst. 1950. *The Problem of Knowledge: Philosophy, Science, and History Since Hegel.* Trans. W. Woglom and C. Hendel. New Haven: Yale University Press.

Chen, Bohang. 2018. A Non-metaphysical Evaluation of Vitalism in the Early Twentieth Century. *History and Philosophy of the Life Sciences* 40 (3): 50.

———. 2019. Revisiting the Logical Empiricist Criticisms of Vitalism. *Transversal: International Journal for the Historiography of Science* 7: 25–40.

Coleman, William. 1985. The Cognitive Basis of the Discipline: Claude Bernard on Physiology. *Isis* 76: 49–70.

Duchesneau, François. 2018. *Organisme et corps organique de Leibniz à Kant.* Paris: Vrin, coll. Mathesis.

Etxeberria, Arantza. 2004. Autopoiesis and Natural Drift: Genetic Information, Reproduction, and Evolution Revisited. *Artificial Life* 10 (3): 347–360.

Etxeberria, A. 2020. Regulation, Milieu, and Norms: Georges Canguilhem's Individual Organisms as Relations. In *Vital Norms. Canguilhem's the Normal and the Pathological in the Twenty-First Century,* ed. Pierre-Olivier Methot and Jonathan Sholl, 295–332. Paris: Hermann.

Etxeberria, Arantza, and Leonardo Bich. 2017. Auto-organización y autopoiesis. In *Diccionario Interdisciplinar Austral,* ed. C.E. Vanney, I. Silva, and J.F. Franck. http://dia.austral.edu.ar/Autoorganizaci%C3%B3n_y_autopoiesis.

Etxeberria, Arantza, and Jon Umerez. 2006. Organización y organismo en la Biología Teórica ¿Vuelta al organicismo? *Ludus Vitalis* 26: 3–38.

Foucault, Michel. 1991. Introduction. In G. Canguilhem, *The Normal and the Pathological.* Ed. Robert S. Cohen and Trans. Carolyn R. Fawcett, 7–24. New York: Zone Books.

Gabel, Isabel. 2015. *Biology and the Philosophy of History in Mid-Twentieth-Century France.* Doctoral Dissertation, Columbia University.

Gayon, Jean. 1998. The Concept of Individuality in Canguilhem's Philosophy of Biology. *Journal of the History of Biology* 31 (3): 305–325.

Gérard, Marie. 2010. Canguilhem, Erwin Straus et la phénoménologie: La question de l'organisme vivant. *Bulletin d'analyse Phénoménologique* 6 (2): 118–145. http://popups.ulg.ac.be/bap.htm.

Goldstein, Kurt. 1939. *The Organism: A Holistic Approach to Biology Derived from Pathological Data in Man.* New York: American Book (reprint, New York: Zone Books, 1995) (translation of *Der Aufbau des Organismus* (1934)).

Hacking, Ian. 1998. Canguilhem Amid the Cyborgs. *Economy and Society* 27 (2): 202–216.

Jacob, François. 1970. *La logique du vivant.* Paris: Gallimard.

———. 1973. *The Logic of Life. A History of Heredity.* Trans. B. Spillmann. New York: Pantheon Books.

———. 1977. Evolution and Tinkering. *Science* 196: 1161–1166.

Kant, Immanuel. (1790/1987). *Critique of Judgment.* Trans. W. Pluhar. Indianapolis: Hackett.

Lecourt, Dominique. 2011. La philosophie de la vie de Georges Canguilhem. In *Repenser le vitalisme: histoire et philosophie du vitalisme,* ed. Pascal Nouvel, 5–13. Paris: PUF.

Lenoir, Timothy. 1982. *The Strategy of Life: Teleology and Mechanics in Nineteenth-Century German Biology*. Dordrecht/Boston: D. Reidel.

Loison, Laurent. 2018. Un enthousiasme paradoxal? Georges Canguilhem et la biologie moléculaire (1966-1973). *Revue d'Histoire des Sciences* 71 (2): 271–300.

Machery, Edouard. 2012. Why I Stopped worrying About the Definition of Life... And Why You Should as Well. *Synthese* 185: 145–164.

Maturana, Humberto. 1978. Biology of language. In *Psychology and Biology of Language and Thought: Essays in Honor of Eric Lenneberg*, ed. G.A. Miller and E. Lenneberg, 27–63. New York: Academic.

Maturana, Humberto R., and Francisco J. Varela. 1973. *De máquinas y seres vivos. Autopoiesis, la organización de lo vivo*. Editorial Universitaria; new edition in Lumen, Santiago, 1994.

———. 1980. *Autopoiesis and Cognition: The Realization of the Living*. Dordrecht: Springer.

Medawar, P.B., and J.S. Medawar. 1983. Reductionism and Vitalism. In *From Aristotle to Zoos: A Philosophical Dictionary of Biology*, 227–232, 275–277. Cambridge, MA: Harvard University Press.

Métraux, Alexandre. 2005. Canguilhem als Architekt einer Philosophie des Lebenden. In *Maß und Eigensinn. Studien im Anschluß an Georges Canguilhem*, ed. C. Borck, V. Hess, and H. Schmidgen, 317–346. München: Wilhelm Fink.

Morange, Michel. 2000. Georges Canguilhem et la biologie du XXe siècle. *Revue d'histoire des sciences* 53 (1): 83–105.

———. 2017. Molecularizing Evolutionary Biology. In *The Darwinian Tradition in Context*, 271–288. Cham: Springer.

Moreno, Alvaro, and Matteo Mossio. 2015. *Biological Autonomy: A Philosophical and Theoretical Enquiry*. Dordrecht: Springer.

Moss, Lenny, and Stuart A. Newman. 2016. The Grassblade Beyond Newton: The Pragmatizing of Kant for Evolutionary-Developmental Biology. *Lebenswelt* 7: 94–111.

Normandin, Sebastian. 2007. Claude Bernard and *an Introduction to the Study of Experimental Medicine*: 'Physical Vitalism', Dialectic, and Epistemology. *Journal of the History of Medicine and Allied Sciences* 62 (4): 495–528.

Nuño de la Rosa, Laura, and Arantza Etxeberria. 2010. ¿Fue Darwin el 'Newton de la brizna de hierba'? La herencia de Kant en la teoría darwinista de la evolución. *Endoxa* 24: 185–216.

Oyama, Susan. 1985. *The Ontogeny of Information*. Cambridge: Cambridge University Press.

Peterson, Erik L. 2017. *The Life Organic: The Theoretical Biology Club and the Roots of Epigenetics*. Pittsburgh: University of Pittsburgh Press.

Pigliucci, Massimo, and Gerd Müller, eds. 2010. *Evolution-the Extended Synthesis*. Cambridge, MA: MIT Press.

Pradeu, T. 2016. Organisms or Biological Individuals? Combining Physiological and Evolutionary Individuality. *Biology and Philosophy* 31 (6): 797–817.

Prochiantz, Alain. 1990. *Claude Bernard. La révolution physiologique*. Paris: PUF.

Rabinow, Paul. 1994. Introduction. In *A Vital Rationalist: Selected Writings from Georges Canguilhem*, ed. François Delaporte, 11–22. New York: Zone Books.

Rheinberger, Hans-Jörg. 2015. Preparations, Models, and Simulations. *History and Philosophy of the Life Sciences* 36 (3): 321–334.

Sholl, Jonathan. 2012. The Knowledge of Life in Canguilhem's Critical Naturalism. *Pli* 23: 107–127.

Varela, Francisco J. 1979. *Principles of Biological Autonomy*. New York: Elsevier/North-Holland.

———. 1981. Autonomy and autopoiesis. In *Self-Organizing Systems: An Interdisciplinary Approach*, ed. G. Roth and H. Schwegler, 14–24. Frankfurt/New York: Campus Verlag.

Weber, Andreas, and Francisco J. Varela. 2002. Life after Kant: Natural Purposes and the Autopoietic Foundations of Biological Individuality. *Phenomenology and the Cognitive Sciences* 1: 97–125.

Wolfe, Charles T. 2011. From Substantival to Functional Vitalism and Beyond, or from Stahlian Animas to Canguilhemian Attitudes. *Eidos* 14: 212–235.

———. 2015a. Il fascino discreto del vitalismo settecentesco e le sue riproposizioni. In *Il libro della natura, vol. 1: Scienze e filosofia da Copernico a Darwin*, ed. Paolo Pecere, 273–299. Rome: Carocci.

———. 2015b. Was Canguilhem a Biochauvinist? Goldstein, Canguilhem and the Project of 'Biophilosophy'. In *Medicine and Society, New Continental Perspectives*, ed. Darian Meacham, 197–212. Dordrecht: Springer.

———. 2019. *La philosophie de la biologie avant la biologie: une histoire du vitalisme*. Paris: Classiques Garnier.

———. 2020. Review essay on F. Duchesneau, *Organisme et corps organique de Leibniz à Kant*. Paris: Vrin, 2018. *Studia Leibnitiana* 50: 254–258.

Zammito, John. 2006. Teleology Then and Now: The Question of Kant's Relevance for Contemporary Controversies over Function in Biology. *Studies in History and Philosophy of Biological and Biomedical Sciences* 37: 748–770.

———. 2018. *The Gestation of German Biology. Philosophy and Physiology from Stahl to Schelling*. Chicago: University of Chicago Press.

Is There Not a Truth of Vitalism? Vital Normativity in Canguilhem and Merleau-Ponty

Sebastjan Vörös

Abstract The paper investigates the phenomenon of vitalism through the lens of vital normativity as expounded by Maurice Merleau-Ponty and Georges Canguilhem. I argue that the two authors independently developed complementary critiques of the mechanical-behaviourist conception of life sciences, which culminated in a surprisingly similar notion of life construed as a normative (polarized) activity, i.e., an activity that is not indifferent to its own conditions of possibility. Such an alternative conception of life has far-reaching consequences for the epistemology of life sciences, for it requires it to reconsider not only its object of inquiry - the nature of (the relationship between) an organism and its environment -, but also, since scientists themselves are living beings, the nature of its epistemic practices. What I call the truth of (a specific variety of) vitalism is thus reflected not only in how life is cognized, but also in how life cognizes (itself). This last point is of particular philosophical importance, as it paves the way towards a more dynamic conception of reflection (tentatively called ouroboric thought), which takes seriously that we, as cognizers of life, at the same time live the lives of cognizers.

1 Historical Blindspot: Opportunities Missed, Opportunities Seized

In a sense, this paper is a reflection on a missed opportunity; an opportunity for a productive dialogue between two French thinkers who, working at the same time and in the same intellectual milieu, addressed similar issues and drew surprisingly similar conclusions, but who never, at least to my knowledge, engaged with each other's work directly. The two thinkers in question are Georges Canguilhem and Maurice Merleau-Ponty, and the topic they both grappled with - from two different,

S. Vörös (✉)
Faculty of Arts, University of Ljubljana, Ljubljana, Slovenia
e-mail: sebastjan.voros@ff.uni-lj.si

© The Author(s) 2023
C. Donohue, C. T. Wolfe (eds.), *Vitalism and Its Legacy in Twentieth Century Life Sciences and Philosophy*, History, Philosophy and Theory of the Life Sciences 29, https://doi.org/10.1007/978-3-031-12604-8_9

yet complementary perspectives - is vital normativity and its implications for philosophy and science.[1]

This surprising oversight has not gone completely unnoticed and has been acknowledged briefly by at least one of the two protagonists of our story. Thus, in his preface to the second edition of the *Essay on Some Problems Concerning the Normal and the Pathological* (published in 1950), Canguilhem points out that, during the inception of the *Essay* (first published in 1943), he could have profited by drawing on Merleau-Ponty's *The Structure of Behavior* (first published in 1942). However, as it was brought to his attention when the manuscript had already been in print, he could give it but a passing nod of recognition. Yet to his mind, such an omission is not necessarily something to be regretted since, as he puts it, he would much prefer "a convergence whose fortuitous character better emphasizes the value of intellectual necessity to an acquiescence, even fully sincere, in the view of others" (NP, 29–30).

In what follows, I will try to unearth and explore this 'fortuitous convergence', especially as it pertains to the topic of vital normativity and its relation to vitalism. I will draw mostly on the two texts mentioned above, although I will occasionally turn to some later works, particularly Canguilhem's *Knowledge of Life* (first published in 1952) and Merleau-Ponty's *Phenomenology of Perception* (first published in 1945). My approach will not be exegetical, but thematic: by drawing on both authors I will put forward a case for vital normativity which, as I will try to show, lends credence to a general impetus behind vitalist approaches. In so doing, I will, for the most part, skim over many an undoubtedly important difference between the two authors, leaving a more thorough comparative study to those better versed in such undertakings.

2   Dialectical Blindspot: An Immodestly Vitalist Proposal

Before I proceed to the main topic, however, there is one obvious hurdle I need to tackle. This hurdle is related to the fact that, while Canguilhem had been quite vocal in his defense of (some version of) vitalism,[2] Merleau-Ponty never seems to have ventured down the vitalist path; in fact, he was often openly dismissive of vitalism.[3] Doesn't this, in and of itself, undermine the project I have set out to achieve?

---

[1] Despite the thematic convergence of their investigative agendas, there is currently a surprising lack of comparative studies on the topic; a notable exception is Peña-Guzmán (2013).

[2] For a more in-depth analysis of Canguilhem's vitalism see Wolfe and Wong 2014.

[3] For instance: "We are upholding no species of vitalism whatsoever here. We do not mean that the analysis of the living body encounters a limit in irreducible vital forces." (SB, 151) And even more radically: "These remarks cannot serve to justify a vitalism, however - even the refined vitalism of Bergson. The relation of the vital élan to that which it produces is not conceivable, it is magical." (SB, 158)

Let us approach this from a somewhat skewed angle. Underneath the thematic polyphony of Merleau-Ponty's work one finds a common methodological thread. The latter is best illustrated by his tendency to flesh out his own views against the backdrop of the antagonistic approaches he is trying to supersede: realism vs. transcendentalism, empiricism vs. intellectualism, naturalism vs. vitalism, etc. Now, while on the surface level this may seem like a fairly trivial strategy - after all, hasn't such a contrastive approach become the gold standard of contemporary academic debate? - Merleau-Ponty supplements it with a unique dialectical twist.[4] For, instead of simply refuting the two views under scrutiny, he is bent on showing that, while ultimately *erroneous*, they are nevertheless *motivated* errors, "rest[ing] on an authentic phenomenon which philosophy has the function of making explicit" (SB, 216). In other words, while rejecting some aspects of each antagonistic pair, he retains others, and he does so by reintegrating them, in a broadly Hegelian fashion, into a more comprehensive whole, in light of which their meaning undergoes a substantial change. Thus, one finds explicit references to a "truth of dualism" (SB, 209), "truth of sociologism" (SB, 211), "truth of naturalism and realism" (SB, 224), even "truth of solipsism" (PP, 419), as well as claims to the effect that intellectualism/transcendentalism is "less false than abstract" (PP, 143), etc.

However, there is a glaring omission in Merleau-Ponty's strategy, for, although featuring in several of his dialectical arguments, he never raises the question of the truth *of vitalism*. This is all the more striking if we consider that not only is its antagonistic pair, naturalism, afforded ample treatment - the penultimate section of *The Structure of Behavior* even carries the title "Is There Not a Truth of Naturalism?" (SB, 201–221) - but also that, as mentioned, Merleau-Ponty speaks of the truth of sociologism *despite the fact* that sociologism plays a marginal role in his treatise. Is vitalism, that dreaded Urfiend of modern biological science, so flawed in his eyes as to exceed even the rehabilitative scope of the dialectical approach?

The way in which Merleau-Ponty normally speaks of vitalism, namely as a view propounding the existence of an ephemeral supplement, variously called "life force", "élan vital", "entelechy", etc., to the mechanistic processes found in living organisms, lends credence to such interpretation. Thus, by Merleau-Ponty's lights, both naturalism and vitalism subscribe to the "realist postulate" (SB, 46), i.e., they both view the organism as a bundle of interrelated causal mechanisms, with naturalism "juxtaposing separated mechanisms" and vitalism "subordinating them to an entelechy" (SB, 3). However, in his view, both of these conceptions are mistaken, since the proper object of biology is neither (pace naturalism) "the superposition of elementary reflexes" nor (pace vitalism) "the intervention of 'vital' force", but rather "an indecomposable structure of behavior" (SB, 46; more on this presently).

What Merleau-Ponty has in mind when he speaks of vitalism, then, seems to be the run-of-the-mill *substantive* variety (Wolfe 2011) which, in Canguilhem's words, accepts "the insertion of the living organism into a physical milieu to whose laws it constitutes exception" (AV, 69) and tries to underscore the originality of life "by

---

[4] For Merleau-Ponty's use of "dialectics" see Pollard 2016.

demarcating within the physico-chemical territory [...] enclaves of indetermination, zones of dissidence, or foyers of heresy" (ibid.). Canguilhem agrees with Merleau-Ponty that such 'vitalism of two empires' - the empire of physio-chemical processes and the empire of vital-biological forces - leads to philosophical blind alleys, for "there cannot be an empire within an empire without there being no longer any empire" (ibid.). Yet, as Canguilhem is quick to add, the classical substantivalist vitalism errs not in pushing its agenda too far - by, say, reverting to supernatural forces - but in *not going far enough*:

> If one is to assert the originality of the biological, this must be in terms of the originality of one realm over the whole of experience, and not over islets of experiences. In the end, classical vitalism sins, paradoxically, only in its excessive modesty, in its reluctance to universalize its conception of experience. (ibid.; cf. SB, 158)

There is, then, an *immodest* variety of vitalism, of which Canguilhem explicitly speaks and to which Merleau-Ponty merely alludes, a vitalism that is both less and more radical than its substantive cousin. It is *less* radical in that it does not posit epistemic and ontological cracks in the edifice of scientific knowledge; however, it is *more* radical in that, if taken seriously, it forces us to "'comprehend' matter within life, and the sciences of matter - which is science itself - within the activity of the living" (AV, 70). That is, vitalism, in this broader sense, is not a scientific theory, but a persistent demand to inquire into the *existential underpinnings* of scientific inquiry; it has to do not so much with centers of indetermination within the scientific framework as with decentering scientific determinations against the backdrop of vital normativity. And, as I will argue, it is precisely in this epistemic and existential immodesty that, as a germinating seed, the truth of vitalism lies.

3   Mechanical Blindspot: Vital Normativity

I begin with a brief exposition of Merleau-Ponty's and Canguilhem's critiques of the mechanicist-behaviourist account of life. The two approaches, although waging war against a common foe, tackle the issue from different perspectives, with Merleau-Ponty focusing primarily on the notion of behaviour and Canguilhem dealing with the notion of (un)health and (ab)normality. However, their critiques prove mutually enriching and lead to a strikingly similar understanding of life.

Merleau-Ponty starts his exploration of behaviour with a critical assessment of behaviourism, the prevalent school of thought in the biological and psychological sciences of the time (SB, 4–5). According to the behaviourist, the objective conception of animal and human comportment must be shorn of intentionality, meaning and normativity, and construed solely in terms of mechanical action,

> in which the cause and the effect are decomposable into real elements which have a one-to-one correspondence. In elementary actions, the dependence is uni-directional [...] and, even when one speaks of reciprocal action between two terms, it can be reduced to a series of uni-directional determinations. (SB, 160-161)

For the behaviourist, every behavioural act consists of a sequence of causally inter-related mechanical actions which take place in discrete physio-anatomical parts: an outside stimulus impinges on a specific sense organ (affector); this triggers a spe-cific neural pathway (reflex arc); finally, the stimulated pathway brings about the activation of a specific motor unit (effector). Thus, all behaviour is ultimately a summative (re-)action caused by an outside stimulus and mechanically mediated by the activity of independent reflex arcs: "The 'normal' activity of an organism is only the functioning of this apparatus constructed by nature; there are no genuine norms; there are only effects" (SB, 9).

However, drawing on the findings of numerous non-reductionist (holist) scien-tists of the time - primarily Goldstein,[5] but also Buytendijk, von Weizsäcker, Wertheimer, Köhler and Koffka -, Merleau-Ponty argues that behaviourism is both empirically unsubstantiated and conceptually moot. To begin with, it accords poorly with actual observations.[6] In words of Noah Brender:

> The 'elementary reflex' which was supposed to be the basic unit of behaviour [...] turned out to be largely mythical. [...] Instead, the effect of a given stimulus was found to vary according to the presence or absence of other stimuli, the history of the organism, and the activity it was engaged in. (Brender 2013, 251-2)

However, when confronted with such empirical discrepancies, the behaviourist, instead of eschewing his initial (mechanistic) presuppositions, tries to make obser-vational data fit theoretical predictions. It does this by introducing auxiliary, and fundamentally unobservable, processes (inhibition, shunting, etc.), "which are almost in contradiction with it, just as the Ptolemaic system revealed its inadequacy by the large number of *ad hoc* suppositions which became necessary in order to make it accord with the facts" (SB, 16).

An alternative approach, adhering more closely to observational data, portrays behaviour not as something that can be decomposed into a succession of reflexes but as something pertaining to *the organism as a whole*. Even more importantly, behav-iour, on this view, cannot be understood by studying processes that are passively triggered in a living being by external influences, but becomes intelligible only in light of *an active engagement of the organism with its environment*. For example, consider a predator observing its prey: "When the eye and the ear follow an animal in flight, it is impossible to say 'which started first' in the exchange of the stimuli." (ibid., 13) One could, of course, say that "the movements of the organism are always

---

[5] It could be claimed that, since both Canguilhem and Merleau-Ponty were influenced by Goldstein's organismic biology, we shouldn't be surprised to find their work cohere in so many respects. However, this is but partly, perhaps only trivially, true; for while it is undoubtedly the case that both drew heavily on Goldstein, it is also the case that each of them not only focused on dif-ferent aspects of Goldstein's work, but also expanded and modified them in substantial ways that merit further exploration. While there are some high-quality studies on the impact of Goldstein on each of the two authors individually - see, e.g., Smyth 2017 for Merleau-Ponty and Wolfe 2015 for Canguilhem -, a comprehensive comparative analysis of the two "Goldsteinians" remains yet to be written.

[6] For a comprehensive overview see Sheredos 2017, 194–7.

conditioned by external influences", i.e., that behaviour is the *effect* of stimulations; but one could just as rightfully say that these stimulations have "been made possible only by its preceding movements which have culminated in exposing the receptor organ to the external influences", i.e., that behaviour is the *cause* of all stimulations (ibid.).

The organism is not akin to a keyboard that reacts, in a predetermined way, to a set of external factors; instead, it actively participates, "by its proper manner of offering itself to actions from the outside", in the process of *selecting, and structuring of,* the stimuli to which it will be sensitive (SB, 13). This is why the organism does not respond to discrete physicochemical stimuli, as behaviourists would have it, but rather to their frequency, intensity, temporal sequence, etc., i.e., to the complexes of stimuli organized as *structured wholes* (SB, 8). Thus, by shaping and elaborating the stimuli, the organism carves out of the physical domain a unique *virtual domain* - a 'milieu' or an 'Umwelt' as a domain of structures that are *significant for it*: "One could say that the [milieu] emerges from the world through the being and the actualization of the organism [...] an organism can exist only if it succeeds in finding in the world an adequate [milieu] - in shaping a [milieu]" (SB, 13; quoting Goldstein, cf. O, 85).[7]

This last point is what, in Merleau-Ponty's opinion, distinguishes vital systems from *all* physical systems, even those that, as suggested by Gestalt theorists, act as organized wholes or as what in modern parlance would be called *complex dynamic systems*. For instance, Merleau-Ponty mentions the formation of a spherical soap bubble (SB, 131, 146) as an example of a physical process that is best described not in terms of causal interactions between discrete particles, but in terms of *dynamic field phenomena* adhering to the *law of minimum energy displacement and maximum level of stability* (SB, 36, 146). Thus, in the soap bubble,

> the external forces exerted on the surface of the bubble tend to compress it into a point; the pressure of the enclosed air on the other hand demands as large a volume as possible. The spherical structure which is realized represents the only possible solution to this problem of minimum and maximum. (SB, 146)

When exposed to external influences, dynamic physical systems (re)act as *integrated wholes*: they undergo concerted transformations of their co-determining 'parts' which collectively tend towards a mechanical-energetic equilibrium.

Now, the comportment of living beings can be said to resemble that of dynamic physical systems in that it also tends towards an equilibrium: every outside disturbance triggers imbalances in the organism's sensory parts, which are then compensated by changes in the motor parts, only to trigger further alterations in the sensory parts, and so on (SB, 36–7). However, Merleau-Ponty argues that, unlike physical systems, in which the equilibrium is obtained with regards to *real* (i.e.,

---

[7] In fact, as both Canguilhem and Merleau-Ponty emphasize, organisms respond to discrete physico-chemical stimuli only in pathological and laboratory conditions, i.e., in conditions that are *abnormal* for the living being. In such cases, the organism can be "momentarily reduced to the conditions of the physical [cause-effect] system" (SB, 150); however, this is not a mode of being that is *proper to the organism* (more on this below).

physico-chemical) conditions, vital systems obtain the equilibrium "with respect to *virtual* conditions which the system itself brings into existence" (SB, 145; my emphasis). Put differently, the equilibrium that the vital system tends towards is not the mechanical-energetic equilibrium, but rather the "vital equilibrium" (SB, 147), which depends on the *organisation, history and activities of the organism*: "[T]he [vital] structure, instead of procuring release from the forces with which it is penetrated through the pressure of external conditions, executes a work beyond its proper limits and constitutes a proper milieu for itself" (SB, 145–6). Thus, whereas actions exercised by the physical system always have "the effect of *reducing a state of tension*, that is to say, of *advancing the system toward rest*" (SB, 145; my emphasis), living beings tend not towards rest, i.e., towards the simplest energetic state with regards to their physico-chemical surrounding, but rather towards the maintenance of a *level of tension that is characteristic for them*. This level of tension is simplest not with regards to physico-chemical factors but *with regards to the organism's characteristic engagements with its milieu* (SB, 146–7).

According to Merleau-Ponty, then, the comportment of living beings needs to be situated into the virtual domain of significance, which they construct by submitting the external stimuli to their "*descriptive norms*" (SB, 28; my emphasis). Hence, unlike physical systems, whose 'behaviour' can be encapsulated in mathematically expressible laws, "organic structures are understood only by *a norm*, by a certain type of transitive action which characterizes the individual" (SB, 148; my emphasis). However, although integral to his argument, Merleau-Ponty never elucidates what he means by "norm"; luckily, Canguilhem's Goldstein-inspired explorations into the nature of health/disease and normal/pathological provide a much-needed supplement to this interpretative hiatus.

As already mentioned, Canguilhem approaches a similar topic from a different angle, focusing primarily on the question of whether life sciences (medicine in particular) can be couched exclusively in objectivist terms. Now, what exactly would objective science of health and normality amount to? For one thing, it would construe health and normality as *descriptive (statistical) facts* represented by "a canonical collection of functional constants" (NP, 122): body temperature, pulse rate, blood pressure, etc. Analogously, disease and abnormality would be conceived as deviations from the established statistical averages (NP, 151): they would denote *quantitative privations*, i.e., divergences from physiological constants, instead of qualitative alterations, i.e., changes in the manner of being (NP, 42). Consequently, scientific medicine would require that pathology (science of disease/abnormality) be grounded in physiology (science of health/normality), with the latter studying normal (i.e., statistically average) and the former studying abnormal (i.e., statistically divergent) physiological processes.

According to Canguilhem, the ideal of objective medicine harkens back to the grand philosophical narrative, developed by the protagonists of what has been later termed scientific revolution, whose ultimate goal was the all-encompassing mechanization of nature (NP, 128). However, any approach that tries to account for vital behaviour in mechanical terms faces a serious challenge, for it needs to explain why physical systems are indifferent to their movements, while this is decidedly *not* the case with living systems:

> In establishing the science of movement on the principle of inertia, modern mechanics in effect made the distinction between natural and violent movements absurd, as inertia is precisely an indifference with respect to directions and variations in movement. Life is far removed from such an indifference to the conditions which are made for it; life is polarity. The simplest biological nutritive system of assimilation and excretion expresses a polarity. (ibid.)

This last point is crucial. Whenever we try to develop a strictly descriptive science of health and normality, we implicitly presuppose that certain states are deemed (un)desirable by, and thus normative for, the living being (NP, 222). Put differently, the reason why, say, a given physiological state is considered pathological is not because it statistically diverges from certain physiological averages, but because the living being itself shows us, by means of its behaviour, that these statistical divergences are *significant for it*. So, instead of being descriptive facts, health/disease and normality/pathology turn out to be *normative concepts*: "[M]edicine exists as the art of life because the human being himself [or, as he later adds, living beings in general] call[s] certain dreaded states or behaviors pathological (hence requiring avoidance or correction) relative to the dynamic polarity of life" (NP, 126).

By characterizing life as *dynamic polarity*, Canguilhem points to what he feels is "the fundamental fact that life is not indifferent to the conditions in which it is possible" (NP, 127): the organism's engagement with its surrounding is never neutral but involves "preference and exclusion", "propulsion and repulsion" (NP, 136). Because interactions with matter *matter* to the organism - they are literally a matter of life and death - *living* is a fundamentally *normative activity* (NP, 123, 126, 228). Canguilhem is adamant that the term 'normative activity' doesn't stand solely for the human faculty of producing normative judgements, i.e., judgements that evaluate facts in light of norms, but rather refers to something much more fundamental, namely "that which establishes norms" (NP, 127). The organism's engagements with its surrounding take place against the backdrop of "a permanent and essential vital need", the *vital need for self-maintenance/-perseverance*, as expressed in "reactions of hedonic value or self-healing or self-restoring behaviour" (NP, 127).

From the point of view of mechanical science, anchored in statistical and descriptive analysis, there can be no difference between anomaly and abnormality (NP, 155): why are some statistical divergences (anomalies) considered normal, while others (pathologies) are considered ab-normal? According to Canguilhem, this question can be adequately addressed only if we admit of a *new point of view*:

> That point of view is that of vital *normativity*. [...] It is because the anomaly has become pathological that it stimulates scientific study. The scientist, from his objective point of view, wants to see the anomaly as a mere statistical divergence, ignoring that the biologist's scientific interest was stimulated by the normative divergence. (NP, 136)

Objective science, like Minerva's owl, is always too late: it embarks on its investigative journey with the falling of the dusk, when the vital norms have already been established. Hence, although statistical constants investigated by the physiologist have epistemic merit, they do so only if set against the backdrop of the organism's engagements with its milieu: "Taken separately, the living being and his [milieu] are not normal: it is their relationship that makes them such." (NP, 143). For it is only *when life has already asserted itself*, i.e., when the organism has established a suitable milieu for itself, that it is possible to undertake the study of physiological constants which are the *expression*, and *not the bearer*, of vital normativity (NP, 165, 171, 175).

We can see, then, that Merleau-Ponty and Canguilhem, although traversing different paths, arrive at a similar conclusion: in order to properly understand vital phenomena - behaviour and health, respectively - the notion of vital normativity requires that we consider the *whole mode of life of the organism*[8] as expressed in-and-through *the history of its engagements with its milieu*. In the words of Paola Marrati and Todd Meyers: "[L]iving beings are not, and cannot, be indifferent to the conditions of their life, both the internal conditions of the organism [...] and the external conditions by the natural and social milieu in which they interact" (2008, ix–x). Just as it is inadequate to think of organismal behaviour as a mechanical process governed from the outside, so it is inadequate to think of organismal well-being (health) as being determinable solely from the inside; instead, the organism and its milieu constitute two aspects of an integrated whole.

4   Knowing Life: The Vital In-Between

Understanding life as normative (polarized) activity has far-reaching consequences for the knowledge of life, construed both as the "knowledge we have of life when we take it as an object" (*objective* dimension) and as the "knowledge that life itself produces" (*transcendental* dimension) (ibid., ix). The truth of immodest vitalism, as we will see shortly, is thus reflected not only in *how life is cognized* but also in *how life cognizes (itself)*. In this segment, we will focus on life as the object of knowledge, leaving the question of life as the transcendental condition of knowledge for the next section.

The critique outlined above has underscored a profound ambiguity surrounding the notions of 'organism' and 'milieu', an ambiguity which has to be dispelled if life sciences are to acquire a more solid epistemological footing. Let us look at each of the two notions in turn, starting with 'organism'. In the mechanical-objectivist picture, the term 'organism' stands for "a segment of matter", "a sum of physical and chemical actions" (SB, 151). However, this conception is problematic for at least two reasons. Firstly, since all organismal processes are couched in physico-chemical terms, the normal/pathological distinction becomes meaningless (ibid.). For

---

[8] Note that this is exactly how Canguilhem and Merleau-Ponty, again following Goldstein, understand the term 'behaviour'. See for instance: "We have spoken on several occasions of the *modes of life*, preferring this expression in certain cases to the term *behavior* in order to emphasize better the fact that life is dynamic polarity." (NP, 205; my emphasis)

instance, the build-up of waste in the organism, which, if not prevented, leads to deterioration and ultimately death, still proceeds in accordance with physico-chemical laws (NP, 128–9). Physico-chemical regularities are oblivious to biological realities, as vividly expressed by Canguilhem's music analogy: "[T]he states of an organism are like those found in music: the laws of acoustics are not broken in cacophony - this does not mean that all combinations of sounds are agreeable" (NP, 56).

Secondly, and relatedly, the mechanical point of view threatens to eliminate the very object it sets out to elucidate. In Merleau-Ponty's words:

> A total molecular analysis would dissolve the structure of the functions of the organism into the undivided mass of banal physical and chemical reactions. [...] In order to make a living organism reappear, starting from these reactions, one must trace lines of cleavage in them, choose points of view from which certain ensembles receive a common signification and appear, for example, as phenomena of 'assimilation' or as components of a 'function of reproduction'; one must choose points of view from which certain sequences of events, until then submerged in a continuous becoming, are distinguished for the observer as 'phases' - growth, adulthood - of organic development. (SB, 152)

Or as phrased more colourfully by Canguilhem:

> The laws of physics and chemistry do not vary according to health or disease. But to fail to admit that from a biological point of view, life differentiates between its states means condemning oneself to be even unable to distinguish food from excrement. Certainly a living being's excrement can be food for another living being but not for him. What distinguishes food from excrement is not a physicochemical reality but a biological value. (NP, 220)

All partitive analysis implicitly presupposes another understanding of the organism, one that doesn't conceive of it as a bundle of anatomico-physiological apparatuses, but as a "general attitude towards the world" (SB, 148) and a "center of actions which radiate over a 'milieu'" (SB, 157). The organism, on this reading, is not a thing, but a *center of normative (polarized) activity* (NP, 131), and as such, engenders a *form of (rudimentary) intentionality*, a directedness-towards-the-milieu. This intentionality, however, is not of the order of intellection, but of affect and action; it is not related to notion, but to (e)motion.

Now, concerning the concept of 'milieu', it can be said that the 'world' inhabited by the organism is not the "system of mechanical, physical and chemical constants, made of invariants" (NP, 197); instead, it is its *Umwelt* - "a being-for-the-animal," i.e., "a certain milieu characteristic of species" (SB, 125). In words of Canguilhem:

> The living creature does not live among laws but among creatures and events which vary these laws. What holds up the bird is the branch and not the laws of elasticity. If we reduce the branch to the laws of elasticity, we must no longer speak of a bird, but of colloidal solutions. (NP, 197-8)

The milieu, then, is "a world of qualified objects" (NP, 198), a domain of what Merleau-Ponty calls *sens*. The word is particularly fitting, for it designates both 'meaning' and 'direction' (cf. PP, 229). Thus, while structures in the milieu have *significance* for the organism, this is not an intellectual-conceptual but a *motor-affective* significance: they excite and orient the organism's interest, they function as solicitations and constraints, as demands and prohibitions. A given significative structure is thus neither (*pace* mechanicist) "physicochemical reality" devoid of value (NP, 220) nor (*pace* intellectualist) "logical signification" (PP, 7) with an "encyclopedic value" (NP, 98), but a biological reality imbued with "affective", "expressive" and "vital value" (NP, 136; PP, 7).

With these new conceptions of 'organism' and 'milieu' in place, the *relation* between them needs to be recast as well. According to Merleau-Ponty, causal explanations - even of the bidirectional type - are ill-suited for the task at hand, for they fail to account for the *relations of significance* that are characteristic for the organism-milieu interactions. For instance, a cat which, in the first trial, is taught to pull the string attached to a piece of food with its paw, is able to pull it in the second trial with its teeth with no additional training (SB, 96). The two movements, while *physiologically different*, i.e., involving different neuromuscular pathways, share the *same significance*, i.e., involve performances that have the same meaning for the organism. In other words, behaviour does not conjoin effective reactions and individual stimuli, but expresses a dialectic (internal, meaningful) relationship between an *aptitude* and a *typical situation*, "which are like two poles of behavior and participate in the same structure" (SB, 161).

Reverting once again to the comparison of the physical and the living systems, we can now better appreciate the fact that, in the latter but not in the former, the aptitude and the situation, the organism and the milieu, *constitute a new integrated whole*:

> In a physical form, dialectical relations between 'parts' stand out against a background (the environment) which is not part of the [...] form. In vital forms, both the organism and its milieu are 'parts' of one form. We perceive organisms as bound up with and engaged in their milieu as a place of relevance. (Sheredos 2017, 212)

The appropriate understanding of a given organism is predicated on our being acquainted with its vital norms as expressed in-and-through its interactions with the virtual milieu it lives in. This *qualitative (normative) aspect* of biological phenomena is ineradicable: "Quantity is quality denied, but not quality suppressed" (NP, 110). Thus, when a given state is described as 'biological' or 'physiological' by a scientist, it is, at least implicitly, evaluated as *positively qualified* by the organism in question (NP, 110–1). Quantitative inquiries are of course both useful and valid - both Merleau-Ponty (SB, 132) and Canguilhem (NP, 122) are adamant about that -, but they are so within the parameters implicitly set by the life form under scrutiny and intelligible to the life form performing the scrutiny.

5   Living Knowledge: From Technognosis to Metanormativity

Conceiving of life as polarised activity - as a dialectical *in-between* of the organism and its milieu - has another, and even more radical, corollary, for it casts fresh light on the fact that life appears in life sciences not only as an object (objective dimension) but, being an enterprise undertaken by scientists who themselves are living beings, also as a subject (transcendental dimension). Here, we touch upon an issue that has caused quite a stir in (at least French and German) philosophy, namely the issue of a (seeming) conflict between knowledge and life. This issue is usually depicted as allowing for (again, seemingly) only two resolutions: either *eradicating life through knowledge* and reverting to "a crystalline (i.e., transparent and inert) intellectualism," or *deriding knowledge through life* and plunging into "a foggy (at once active and muddled) mysticism" (TL, xvii). However, construing life as normative activity enables us to carve a *middle ground* between the two extremes, for it shifts our attention from the (seeming-but-non-existent) conflict between life and knowledge to the (difficult-but-tractable) tension between human being and its milieu (ibid.).

5.1 In the Beginning Was the Deed: On Technognosis

Can life be eradicated through knowledge? Let us start with an almost trivial observation that a scientist, being a living being himself, is, like all other living beings, actively involved in establishing and/or maintaining a dynamic equilibrium with his environment. As such, all his activities, scientific or otherwise, need to be situated against, and understood from, this general background. For instance, a physiologist examining the physiological constants and classifying them as (ab)normal,

> does more - not less - than the strict work of science. He no longer considers life merely as a reality identical to itself but as polarized movement. Without knowing it, the physiologist no longer considers life with an indifferent eye, with the eye of a physicist studying matter; he considers life in his capacity as a living being through whom life, in a certain sense, also passes. (NP, 222)

This has at least two significant implications. Firstly, and crucially, it follows that *life grounds, and calls forth, knowledge*. Thus, it is "life's setbacks" (ibid.) - the *dis*equilibria between the human being and his environment - that pave the way to cognition in general and scientific inquiry in particular. Canguilhem, for instance, underlines the epistemic import of *disease* for (life) sciences:

> Disease is the source of the speculative attention which life attaches to life by means of man. If health is life in the silence of the organs, then, strictly speaking, there is no science of health. Health is organic innocence. It must be lost, like all innocence, so that knowledge may be possible. Physiology is like all science, which, as Aristotle says, proceeds from wonder. But the truly *vital wonder* is the anguish caused by disease. (NP, 100-1; my emphasis)

This notion can be generalized. It is when the vital in-between is disrupted, that reflective inquiry, begotten from vital wonder, comes to the fore. Thus, if considered from the genealogical perspective, objective knowledge cannot be value-free; instead, it is *poli-valent (meta-normative)* in the sense that it allows transitions between different normative frameworks (more on this below).

The second implication is that the relation between technology and science becomes much more nuanced than the standard view would have it. Usually, technology is construed as a practical application of science: we need to know in order to be able to act (NP, 99, 104–5). However, such a view errs in at least two respects. To begin with, it ignores the constitutive role of technology in the epistemic enterprise. Technology is not simply "a docile servant carrying out intangible orders," but rather an "advisor and animator": it confronts us with concrete problems and urges us to tackle them without paying heed to the theoretical solutions which will eventually spring forth out of these encounters (NP, 101; cf. 105).

Secondly, by putting the cart (science) before the horse (technology), the classical approach severs life from knowledge, transposing it, in theory if not in practice, into the crystalline realm of intellectualism, in which disembodied gaze hovers over the mechanical world. However, it is possible to bridge this noetic chasm if technology is considered not as a product of science, but as a complexification and elaboration of

> vital impulses at whose service it tries to place systematic knowledge. […] All human technique, including that of life, is set within life, that is, within an activity of information and assimilation of material. It is not because human technique is normative that vital technique is judged such by comparison. Because life is activity of information and assimilation it is the root of all technical activity. (ibid.)

If knowledge is to be situated into the framework of life and action, more emphasis needs to be given to its *implicit, practical*, and *technical aspects*. That is to say, the ideal of *episteme* needs to be set in close(r) relation to the facticity of "praktognosia" (PP, 162), the *logos* furnished by the behavioural complexes of our biological bodies and their numerous culturo-technological extensions (habits, tools, institutions, etc.). If we overlook this constitutive relevance of *technognosis*, we are likely to misapprehend the domain of mind and thought.

5.2   Minded Life: On Hypervirtuality and Metanormativity

However, does this mean that science can be simply reduced to vital polarity, that mind is ultimately nothing but an extension of life? And if so, haven't we been impaled on the other horn of dichotomy, namely that of 'foggy mysticism'? It is true that Canguilhem and Merleau-Ponty both argue that (human) knowledge has to be understood against the backdrop of (vital) normativity, which makes it, first and foremost, a *new mode of behaviour*, "a new equilibrium with the world, a new form of organization of [human] life" (LT, xvii). However, this does not preclude a *qualitative* difference between life and mind:

> An organism's behavior can be in continuity with previous behaviors and still be another behavior. The progressiveness of an advent does not exclude the originality of an event. (NP, 87)?

To get a better handle on this delicate issue, let us look at Merleau-Ponty's account of the relation between what he calls the "vital order" (life) and the "human order" (mind). His contention is that, on the superficial level, the human order may look simply as *a variation* of the vital order: just as, say, animals live in their unique *organic* milieus, so human beings live in their admittedly more diversified, but structurally similar, *culturo-social* milieus:

> If life is the manifestation of an 'interior' in the 'exterior,' [human] consciousness is noth-
> ing at first but the projection onto the world of a new 'milieu' - irreducible to the preceding
> ones, it is true - and humanity nothing but a new species of animal. (SB, 162)

However, and crucially:

> [T]his lived consciousness does not exhaust the human dialectic. What defines man is not
> the capacity to create a second nature - economic, social or cultural - beyond biological
> nature; it is rather the capacity of going beyond created structures in order to create others.
> (SB, 175)

Thus, the difference between the two orders lies not so much in the greater complexity of human *Umwelten* but rather in the distinctly human ability to *transcend and modify* these selfsame *Umwelten*. According to Merleau-Ponty, what is characteristic of human behaviour and seems to be lacking in the comportment of non-human animals, is a certain *multiplicity and mobility of perspective*, which he expounds in his commentaries on Wolfgang Köhler's famous experiments on chimpanzees.[9]

In one set of experiments, chimpanzees, when confronted with a puzzle, had grave difficulties in alternating their viewpoint. For instance, having learned how to use a rod to obtain a piece of fruit, a chimpanzee left alone with a dried bush was unable to use its branches for the same purpose: "The tree branch as a stimulus is not even the equivalent of a rod, and the spatial and mechanical properties which permit it to assume this function are not immediately accessible to animal behavior" (SB, 113–4). Similarly, the animal did not use a branch as a tool - even if it had learned to do so in the preceding trials - as long as another monkey was sitting on it: "It leans against it; thus it cannot be said that it has not seen it; but it remains for him a means of support or rest; it cannot become an instrument" (SB, 114).

What are we to make of this? In Merleau-Ponty's view, the rod is not given to the chimpanzee in the same way as it is to the human observer, that is, as *a discrete thing* that can be seen from different perspectives and manipulated in different ways; instead, it is "invested with a 'functional value' which depends on the effective composition of the field" (SB, 116–7). More specifically, the situation in which the animal finds itself is given to it as a structured whole, in which the significance of each aspect is co-specified with the significance of all other aspects. Hence, if one segment of the situation changes, so do all the others. This is why, for the chimp, the branch-as-seat and the branch-as-instrument are not two aspects of the same thing, but literally *two different things* (SB, 145). Further, the motor-affective significance of the situation is predicated on the motor-affective aptitudes of the animal; it "is not

---

[9] For reasons of space, I provide but a brief overview of these experiments; for a more thorough and interpretatively brilliant account see Moss Brender 2017.

an object of knowledge for the chimp [...] Rather, the chimp lives its situation as an immediate call to action, an imperative or compulsion that cannot be denied" (Moss Brender 2017, 145).

In human beings, on the other hand, this motor-affective dynamism undergoes an important transformation, brought about by what Merleau-Ponty calls "multiplicity of perspective" (SB, 122). In words of Moss Brender:

> Like the chimpanzee, we experience our situation as Gestalt, an organized and oriented whole. But unlike the chimp, we are able to reorganize or reorient this whole, to 'Gestalt-shift' more or less at will between different possible configurations of the situation. (Moss Brender 2017, 146).

There is, then, a certain *distance* or *remove* from the motor-affective significance of a situation, which allows us to resist its imminent demands, to vary our perspective, and finally, to thematize it (more on this below).

The second example is even more telling. In one experiment, Köhler placed a piece of fruit behind a grill and taught the chimpanzee to use a stick to pull it within its reach. In another trial, he placed a three-sided frame around the fruit, whose open side was facing away from the animal. So, to get to the fruit, the chimp had to first push the fruit away from itself, then move it around the frame, and finally pull it towards itself. However, Köhler observed that the chimp had great difficulty in completing the task and was stubbornly trying to pull the fruit towards itself. This may strike us as odd for, while the chimp had enormous difficulties in moving the fruit along the required path, it is clear that it could have easily retraced the same route with its own body if the conditions allowed it to do so (SB, 117).

What to us look like two identical paths - from A to B and from B to A - are apparently not identical for the chimpanzee. Again, what are we to make of this? Merleau-Ponty's contention is that space for the chimp is not *geometric* but *corporeal*, i.e., space that has a "null-point", namely the chimp's own body (SB, 117). For this reason, the chimpanzee "does not experience itself as an object moving through a fixed landscape; rather, it is the landscape that shifts around the animal in response to animal's movements" (Moss Brender 2017, 148). In other words, movement for the chimp does not mean a *change of position* (a change *in* space) but rather a *change of situation* (a change *of* space).

To get a better understanding as to how this differs from human perception, let us consider what it actually takes to successfully solve the puzzle in question. To begin with, it requires that one dissociates oneself from one's embodied perspective and "take(s) up the object's point of view" (ibid., 149). Further, it requires that one sees oneself - one's own body - as yet another object, which stands in multiple relations with other objects. And finally, it requires the ability to translate a motor sequence that would be carried from the *object's* point of view into a visual sequence that would be seen from *our current* point of view, i.e., to "transcribe a kinetic melody into a visual [sequence] [...] establishing relations of reciprocal correspondence and mutual expression between them" (SB, 118). In short, the solution to the problem requires "mobility of perspective" (Moss Brender 2017, 149), which allows one to not only alternate perspectives from the same point of view (i.e., multiplicity of perspective) but to also *alternate the point of view itself* and imagine the action-to-be-taken *"from a perspective outside of the movement itself"* (ibid., 151; emphasis in the original). Note that this is not to say that human beings no longer have a center, but that, in our case, the center is *mobile*, and that we are able "to take up a virtual point of view without actually moving our body to that location" (ibid., 149).

What is of particular importance is that these two abilities - the multiplicity and mobility of perspective - allow for what Merleau-Ponty calls *symbolic behaviour*, i.e., a mode of behaviour in which behaviour itself becomes "the proper theme of activity" (SB, 103). The task of performing a detour in the example above requires that we "trace by our very gesture the symbol of the movement which we would have to make if we were in its place" (SB, 118). Put differently, to execute a detour requires establishing correspondences between various patterns of behaviour executed from various centers: the (actual) gesture I perform with my hand when I move the fruit along the relevant path *expresses the same significance* as the (virtual) movement of my body traversing the selfsame path. Symbolic behaviour thus stands for the ability to take up a *common significance* of various - *actual* or *virtual* - behavioural structures, and express it in (other) gestures, pictures, or words; these then stand as *symbols* or "structure(s) of structures," i.e., as structures expressing structural correspondences between various sensorimotor patterns (cf. Moss Brender 2017, 152).

The capacity of establishing "relations between relations" (SB, 118) is at the core of what we call *reason(ing)*. But note, firstly, that while not identical with vital behaviour, reason(oning) still is, essentially, a *form of behaviour*, a *new mode of being*; and secondly, that once constituted, it is *not* simply a contingent addition to former modes of behaviour, but rather a *novel structuration*, a transformation of the vital organisation *in its entirety*:

> Mind is not a specific difference which would be added to vital or psychological being in order to constitute a man. Man is not a rational animal. The appearance of reason and mind does not leave intact a sphere of self-enclosed instincts in man. [...] But if the alleged instincts of man do not exist apart from the mental dialectic, correlatively, this dialectic is not conceivable outside of the concrete situations in which it is embodied. (SB, 181)

Non-human living beings are embedded in their virtual domains of significance, but these domains are fixed by each organism's organization - the "a priori of the species" (SB, 122) - and are (almost) always given from a single perspective: "The animal *lives* in the meaning of its immediate situation without *perceiving* this meaning as such." (Moss Brender 2017, 146; emphasis in the original) In the case of the human mind, this vital dynamism is sublimated into a new structure: significance is not only *enacted* but also *thematized*. More specifically, human bodily attitudes are open to regular modification, whereby motor-affective significances in the human milieu also become modified. This is why, instead of "aptitudes," Merleau-Ponty uses the Hegelian term "work," which denotes "the ensemble of activities by which man transforms physical and living nature" (SB, 162). Work opens up not only the possibility that some*thing* can have more than one meaning, but also a realization that this shift in meaning corresponds to the alteration of the (bodily) attitude taken by some*one*: "It is only at the level of symbolic conduct [...] that, instead of seeking to insinuate his stubborn norms, the subject of behavior 'de-realizes himself' and becomes a genuine alter ego." (SB, 126).

Let us now, by way of summary, underline three main points that emerge out of these reflections. Firstly, a non-human living being is capable of *experiencing* in the sense of establishing meaningful relationships between physico-chemical stimuli and shaping them into structured wholes, its *Um-welt*; a human being, on the other hand, is capable of *reasoning* in the sense of establishing "relation(s) between relations" (SB, 118), and thus recognizing, beyond its current milieu, a *Welt* (SB, 176), "universe" (SB, 176) or "world of things" (SB, 175) - a virtual framework of constancies underpinning all possible body-situation alterations. Secondly, life is intrinsically *normative*, i.e., it is a polarized (non-neutral) activity whose dynamic equilibrium pertains not to physico-chemical conditions but "to *virtual* conditions which the system itself brings into existence" (SB, 145; our emphasis); mind, on the other hand, can be said to be intrinsically *metanormative*, i.e., it is a polarized (non-neutral) activity that can thematize and, to a certain degree, modify its virtual conditions of possibility. Finally, while non-human beings live in a domain of virtuality, human beings can be said to live in a *domain of hypervirtuality*: they live among meaningful structures of the second order, among *symbols* and *things*. Human beings can create, destroy, and re-create different virtual domains - the worlds of myth, art, science, etc.; put differently, they can enact multifarious virtual habitats, the outermost horizon of which - the virtual domain in which perspectives, constitutive of other virtual domains, coalesce - is what is called the 'objective world'.

We can now better understand what was said at the beginning of this section, namely that the position defended by Merleau-Ponty and Canguilhem transforms the question of life *vs.* knowledge into that of organism *vs.* milieu:

> Thought is nothing but a disentangling of man from the world that permits us to retreat from, to interrogate, and to doubt (to think is to weigh, etc.) in the face of obstacles that arise. [...] It is not true that knowledge destroys life. Rather, knowledge undoes the experience of life, seeking to analyze its failures so as to abstract from it both a rationale for prudence (sapience, science, etc.) and, eventually, laws for success, in order to help man remake what life has made without him, in him, or outside of him. (TL, xvii)

The relation of mind to life is the same as that of life to matter: it is a *novel structuration*, where the dialectical pair 'aptitude-situation' (normativity-virtuality) is restructured into the dialectical pair 'work-world' (metanormativity-hypervirtuality). What is crucial, however, is that, in this view, the originality of life (*a propos* matter) and mind (*a propos* life) is not one of *addition* but one of *transformation*, one of "a [wholesale] retaking and a 'new' structuration" of the preceding structural dynamics (SB, 184). Mind, for instance, is neither a non-material add-on to, nor a mere quantitative complexification of, life; instead, it is a new dynamic equilibrium, a qualitatively novel integrated whole. According to Merleau-Ponty, when it comes to such structural shifts (from 'matter' to 'life' and 'life' to 'mind'), "[i]t is not a question of two *de facto* orders external to each other, but of two types of relations, the second of which integrates the first" (SB, 181). This sits well with Canguilhem's claim that *quantitative continuity* doesn't preclude *qualitative discontinuity*: "The progressiveness of an advent does not exclude the originality of an event" (NP, 87).

It is not so much that, in living beings, physical phenomena are supplanted with novel vital phenomena, but rather that, after achieving a certain level of complexity, the behavioural pattern characteristic of physical things undergoes a wholesale transformation, whereby all constitutive processes acquire a new significance. And this, as we saw earlier, includes both continuity and discontinuity: the higher order "eliminates the preceding one as an isolated moment", yet it also "uses and sublimates it", "conserves and integrates it". There is, in short, a "double aspect" to the said relation, for it "both liberate[s] the higher from the lower and found[s] the former on the latter" (SB, 207, 208).

6 Conclusion: Life's Fecundity and Ouroboric Thought

Let us now return to the question of vitalism. We have seen that both Canguilhem and Merleau-Ponty agree that vitalism, when cast in the scientific mould, is false, even blatantly so; yet it is still a motivated error, one that "rest[s] on an authentic phenomenon" (SB, 216). There is more to vitalism than meets the scientific eye, which is why, as Canguilhem points out, it cannot be placed on the same footing or be refuted in the same way as, say, phlogiston or geocentrism (AV, 60–1). But what is this *more* that separates vitalism from the aberrant theories of old? What is the authentic phenomenon which vitalism so immodestly expresses?

The answer I have tried to sketch points in the direction of vital normativity. The latter, as we have seen, manifests itself in (at least) two aspects that are central to vitalism. The first aspect is the appreciation for *the generative and spontaneous dimension of life*, for the unexpected leaps, shifts, and transformations that characterize organic dynamism:

> It is certain that, for vitalists, the fundamental biological phenomenon [...] is the phenomenon of generation. A vitalist [...] is a man who is led to meditate on the problems of life more by the contemplation of an egg than by handling of a winch or iron bellows. (AV, 64)

The normativity of life manifests itself as *life's fecundity* (AV, 66–7), as life's capacity of both establishing *and transcending* new norms and thereby new life-forms, in relation to which (objective) knowledge always alights too late:

> Life is the formation of forms; knowledge is the analysis of in-formed matter. It is normal that an analysis could never explain a formation and that one loses sight of the originality of forms when one sees them only as results whose causes or components are to be determined. Because they are totalities whose sense resides in their tendency to realize themselves as such in the course of their confrontation with their milieu, living forms can be grasped in a vision, never by a division. (TL, xix)

The second, and in my opinion more important aspect, is best approached through Canguilhem's claim that vitalism "is first and foremost a *demand* [... a] need to keep the question of the sense of the relation between life and science open" (TL:

ix; my emphasis). There is, as we have seen, an ineradicable recursivity in all reflection on life: it is the impetus of vital normativity that calls forth knowledge, which, through its techniques and practices, recursively alters the dynamics of vital normativity. This idea, if taken seriously, has important implications on how we understand our epistemic practices and urges us to make conceptual space for a more dynamic type of reflection, which I would like to call *ouroboric thought*. We find references to ouroboric thought, under different labels, in both thinkers. Canguilhem speaks of *reasonable rationalism*, a rationalism that is able to "incorporate the conditions of its practice" by recognizing "the originality of life," and therefore acknowledge that "the thought of the living" ultimately takes "from living the idea of the living" (TL, xx). Similarly, Merleau-Ponty speaks of *radical reflection* (PP, 61–3) or *hyper-reflection* (VI, 38, 46), i.e., reflection that "knows itself as as reflection-on-an-unreflective experience" (PP, 72), and that "elucidate[s] the unreflective view which it supersedes and show[s] the possibility of this latter, in order to comprehend itself as a beginning" (PP, 73).

The crucial point is the same: Since we, as cognizers of life, at the same time live the lives of cognizers, our reflection, like the mythical snake of Ouroboros, must encompass not only its own embeddedness in the prereflective dynamism of life, but also how, in and through its operations, it modifies that selfsame dynamism in which it is embedded, thus altering both its conditions of possibility and its subsequent operations - and so on, indefinitely. It is here, I think, that the truth of immodest vitalism lies. Put simply, it is the *truth of vital wonder* (NP, 101), the truth of an existential blindspot of all our epistemic endeavours; it is the recognition that "one must know how to cede a place to the irrational, even and especially when one wants to defend rationalism" (MO, 95), that one must make room for "the darkness needed in the theatre to show up the performance" (PP, 115). Hence its epistemic immodesty, hence its existential impudence.

Bibliography

Abbreviations

TL Canguilhem, Georges. 2008. Thought and the Living. In *Knowledge of Life*, xvii–xx. New York: Fordham University Press.

AV Canguilhem, Georges. 2008. *Aspects of Vitalism*, 59–74. New York: Fordham University Press.

MO Canguilhem, Georges. 2008. *Machine and Organism*, 75–97. New York: Fordham University Press.

NP Canguilhem, Georges. 1991. *The Normal and the Pathological*. New York: Zone Books.

O Goldstein, Kurt. 2000. *The Organism*. New York: Zone Books.

SB Merleau-Ponty, Maurice. 1963. *The Structure of Behavior*. Pittsburgh: Duquesne University Press.

PP Merleau-Ponty, Maurice. 2002. *Phenomenology of Perception*. London\New York: Routledge Classics.

Secondary Sources

Marrati, Paola, and Todd Meyers. 2008. Foreword: Life as Such. In *Knowledge of Life*, ed. Georges Canguilhem, vii–xii. New York: Fordham University Press.

Moss Brender, Noah. 2017. On the Nature of Space: Getting from Motricity to Reflection and Back. In *Perception and Its Development in Merleau-Ponty's Phenomenology*, ed. Kirsten Jacobson and Russon John. Toronto: Toronto University Press.

———. 2013. Sense-Making and Symmetry-Breaking: Merleau-Ponty, Cognitive Science, and Dynamic Systems Theory. *Symposium* 17 (2): 246–270.

Peña-Guzmán, David M. 2013. Pathic Normativity: Merleau-Ponty and Canguilhem's Theory of Norms. *Chiasmi International* 15: 361–384.

Pollard, Christopher. 2016. Merleau-Ponty's Conception of Dialectics in Phenomenology of Perception. *Critical Horizons* 17 (3-5): 358–375.

Sheredos, Benjamin. 2017. Merleau-Ponty's Immanent Critique of Gestalt Theory. *Human Studies* 40: 191–217.

Smyth, Bryan. 2017. The Primacy Question in Merleau-Ponty's Existential Phenomenology. *Continental Philosophy Review* 50: 127–149.

Wolfe, Charles T. 2011. From Substantival to Functional Vitalism and Beyond: Animas, Organisms and Attitudes. *Eidos* 14: 212–235.

———. 2015. Was Canguilhem a Biochauvinist? Goldstein, Canguilhem and the Project of Biophilosophy. In *Medicine and Society, New Perspectives in Continental Philosophy*, ed. Meacham Darian, 197–212. Dordrecht: Springer.

Wolfe, Charles T., and Andy Wong. 2014. The Return of Vitalism: Canguilhem, Bergson and the Project of Biophilosophy. In *The Care of Life: Transdisciplinary Perspectives in Bioethics and Biopolitics*, ed. G. Bianco, M. de Beistegui, and M. Gracieuse, 63–75. London: Rowman & Littlefield International.

A 'Fourth Wave' of Vitalism in the Mid-20th Century?

Erik L. Peterson

Abstract In his 1966 John Danz lectures, Francis H. C. Crick decried vitalism in the life sciences. Why did he do this three decades after most historians and philosophers of science regarded vitalism as dead? This essay argues that, by advocating the reduction of biology to physics and chemistry Crick was: (a) attempting to imbue the life sciences with greater prestige, (b) paving the way for bioengineering and the reduction of consciousness to molecules, and (c) trying to root out religious sentiment in the life sciences. In service of these goals, Crick deployed vitalism as a straw man enemy. His wave of so-called vitalists in the middle of the twentieth century in fact raised legitimate questions regarding the relationship of organisms to their DNA molecules that Crick was ill-equipped to answer. Moreover, most were not vitalists at all but advocates for what I term *bioexceptionalism*—an argument for the methodological utility of keeping biological pursuits within their own domains, distinct from physics and chemistry, regardless of the ontological status of living things. Nevertheless, Crick's status as a "cross-worlds influencer" entrenched a philosophically-enervated reductionism in the life sciences for decades.

1 Introduction

Francis H. C. Crick (1916–2004) rose to the podium in the Roosevelt Senior High School auditorium in northern Seattle, Washington, USA, on the night of 24 February, 1966, to pose a simple question: "Is Vitalism Dead?" The entire event was odd. The location was odd: why at a high school and not at a university or scientific institute? The question was odd: historians and philosophers of science believed the mechanism-vitalism debate concluded three decades earlier, with the deaths of the major proponents of vitalism's third wave, Hans Driesch and Henri Bergson (Allen 2005; Beckner 1959). Why was Crick raising the dead? And Crick's answer to the

E. L. Peterson (✉)
Department of History, The University of Alabama, Tuscaloosa, AL, USA
e-mail: elpeterson@ua.edu

© The Author(s) 2023
C. Donohue, C. T. Wolfe (eds.), *Vitalism and Its Legacy in Twentieth Century Life Sciences and Philosophy*, History, Philosophy and Theory of the Life Sciences 29, https://doi.org/10.1007/978-3-031-12604-8_10

173

question was odd: vitalism thrived, he insisted. Three decades after the debate supposedly ended, a Nobel laureate took to a high school stage to inveigh against a wave of support for vitalism he sensed floating like some airborne virus through the sciences, infecting even the nonspecialist public.

Immediately after the lectures, University of Washington Press asked Crick to expand them into a book. The talks were already a mutation of a lecture Crick delivered to the Cambridge Humanists Society earlier, so it wasn't difficult for him to repackage them as *Of Molecules & Men* (publishers believed a title hinting of John Steinbeck would sell more copies in America). In the book, Crick claimed that prominent scientists still called on a phantasmal *élan vital* that eluded mechanistic explanations woven into biology. Nevertheless, vitalism's days were numbered, he assured his audience. Molecular biology—Crick's own research, in other words—was just then in the process of finally winding down that old debate. He issued a warning (again, quite odd considering the commonly held history of vitalism): "And so, to those of you who may be vitalists, I would make this prophecy: what everyone believed yesterday, and you believe today, only cranks will believe tomorrow" (Crick 1966: 99).

The purpose of this essay is to peel back the layers of Crick's lecture and subsequent book in order to discover what lay underneath this strange event. Who were the 1950s and 1960s vitalists against whom Crick made these pronouncements? How could vitalists exist in the professional life sciences at all, given that the term "vitalism" was more or less professionally hazardous by the 1960s (Peterson & Hall 2020)? Why did Crick, by then an internationally known pioneer of molecular biology, deliver his speeches in an American high school? What does this event have to tell us about the status of the life sciences mid-century? And, more to the point, what function did Crick's appellation "vitalism" play in that era, and what significance does it hold in our own?

Though it seems simple on its face, the story here consists of a complex tangle of arguments between Crick and his peers. It was a strange situation wherein physicists and chemists argued for what I term *bioexceptionalism*, while Crick argued for the reverse. By bioexceptionalism, I mean only an argument about the *methodological utility of preserving biological pursuits within their own domains, distinct from physics and chemistry*. Whether there are ontological reasons to hold biology as ultimately reducible to physics or not, there are disciplinary reasons to *act as if* life is special. Put a different way, this attitude—holding in abeyance ontological questions in favor of an anti-reductionist methodological stance—is what I'm deeming bioexceptionalism. It is no means clear that what motivated the advocates for bioexceptionalism in the mid-twentieth century was an ontic commitment to a vital force, entity, or fluid—something *essentially* unique about the living.

Why do Crick's moves here regarding vitalism matter? First, because of his stature as a Nobel laureate and champion of the Young Turks sweeping through molecular biology, they were rhetorically persuasive. Crick's aspersions deflected the critiques of his own reductionism and muted those who advocated for bioexceptionalism. Secondly, by insinuating these individuals were the "cranks" he mocked, Crick indirectly turned the lens of historians and philosophers of science away from investigating these figures and their influences on the life sciences. Thirdly, the existence of Crick's feared 'fourth wave of vitalism' in the mid-twentieth century a half-century after the third wave suggests the reductionistic solutions offered by earlier physiologists (e.g., Jacques Loeb) and geneticists (e.g., H. J. Muller) did not settle the issue earlier. Why, then, should we believe Crick's solutions settled them mid-century?

Perhaps the most significant reason to focus on Crick's *Of Molecules and Men* and the purported 'fourth wave of vitalism' is for the window it opens into longer-term trends within academic disciplines. These events reveal in miniature a growing split within the life sciences and between the sciences and academic disciplines not considered sciences. The "two cultures" divide lamented by C. P. Snow in the 1950s accelerated in the next decade. And Crick, in particular, saw in this trend cause for celebration, not regret. The future would replace biological explanations with exclusively physicochemical ones—meaning also that biological functions, including cognition, would be replaced by engineered, synthetic cells and computers. That was, Crick thought, the goal of the sciences and the conclusion (in both senses) of other knowledge traditions, such as what is traditionally known as the arts, literature, humanistic pursuits more generally, and, of course, religion. The vitalist "cranks," then, stood not only in the way of science but also his vision of a technologically mediated destiny—humanity without the humanities or the arts, and absent much of the social and even the life sciences.

2   Of Molecules and Crick

"It is notoriously difficult to define the word 'living'," began Crick in *Of Molecules and Men* (Crick 1966: 3). Viruses, especially, create a conundrum. They can certainly reproduce—as we in the SARS-CoV-2 era have become only too aware—which should mean they're living. But they're not especially multifaceted, not organism-like. What classifies viruses living things has to do with the fact that structurally, viruses are both sufficiently complex and sufficiently ordered at a chemical level. From Crick's perspective, ordered complexity at the tiniest level acted as the single marker between the living and non-living (Crick 1966: 5). Viruses demonstrate ordered complexity; rocks do not. Paradoxically, Crick also deployed viruses as a reminder that the gap between living and non-living was itself insignificant—crossable, for all intents and purposes. Natural selection excepted, there appeared no principle or law or operation unique to the domain of living things. The ordinary concepts of chemistry and physics could explain viruses and, by extension, all life (Crick 1966: 10). No vital forces necessary.

For Crick, this view comported with the ultimate goal of the field of biology under the new, confident leadership of molecular biologists, like himself: "to explain *all* biology in terms of physics and chemistry" (Crick 1966: 10, emphasis in original). And if this was the quest, Crick saw his role in it as two-fold: (1) to reorient the practice of the life sciences around this notion that all living things can be explained through physics and chemistry without reference to uniquely-biological forces, entities, or processes; (2) to undertake this reorientation because, to date, no interesting biological phenomena had yet appeared to discredit that view. The history of science would validate his role in the quest. Ultimately, Crick asserted, this would open up the world of bioengineering.

Yet some enemies remained to vanquish, if he hoped to complete his quest. These enemies belonged to a tribe of vitalists. For Crick, that meant they adhered to a particular creed:

> [T]here is some special force directing the growth or the behavior of living systems which cannot be understood by our ordinary notions of physics and chemistry. ...there must be something else in a biological system which cannot be included under the heading of physics or chemistry. There must be some sort of force, or some directing spirit—this is the sort of idea that appeals to nonscientists. Scientists on the other hand often prefer to think that there will be extra laws in biological systems which are not included in physics and chemistry (Crick 1966: 16–17).

These beliefs would not stand, according to Crick. History showed all earlier barriers to explaining biological systems in terms of chemical ones had been broken through. And thankfully, the one feature of the living world that didn't seem amenable to strictly chemical explanations, namely evolution via natural selection, now seemed at least partially reducible to chemistry. Mutations, Crick reminded his audience, were rare-but-consistently-occurring chemical events. Over the eons, very many of these rare-but-consistently-occurring events enabled a set of organisms to fit new environmental conditions. Like one of a nearly infinite set of chemical keys placed at random into a lock, there was always a chance the door could be opened. This had happened so often, in fact, that we could explain all of the living world merely through this process of chemical trial and error. The first gigantic mistake vitalists made, Crick insisted, was to make the living realm seem more complicated than this.

At present, there were but three areas still holding out against physics and chemistry gobbling up the life sciences for good: (a) the borderline between living and non-living; (b) the origins of life; and (c) "consciousness" (Crick 1966: 17). Crick's work on DNA, RNA—and the Central Dogma linking them in proper order—meant that molecular biology was close to solving (a). He spent much of the second section of his three-part John Danz Lecture, in fact, simply explaining to his audience the coded DNA to protein sequence. This level of understanding, he predicted, would allow bioengineers to assemble living organisms molecule by molecule by the turn of the twenty-first century at the latest. Simply completing this feat—engineering a living cell from non-living parts—would demonstrate two important facts. First, that there was no life/non-life boundary. Secondly, that the conditions of abiogenesis at the origins of life (hold out (b)) were not only plausible but uncomplicated. Consciousness presented a more intimidating problem, Crick admitted. Here he roamed far outside his expertise. Nevertheless, he confidently proclaimed consciousness also just a molecular problem to be solved by computing and, indeed, centered much of the rest of his scientific career around proving this point.[1] True, Crick conceded, the logical conclusion of these concepts would eventually render humans obsolete. On the way into the twilight, however, humans would get some great entertainment. We would be able to watch two computerized beings interact. How amusing, thought Crick, to pair an android programed for seduction with one programed to be a psychiatrist! "Explosively funny situations" such as these would more than make up for the sudden redundancy of humans and the concomitant realization that most of what humans valued about themselves turned out not to be important after all (Crick 1966: 83).

3 Decoding Crick's Depiction of Vitalism

University of Edinburgh developmental biologist C. H. Waddington took it upon himself to review *Of Molecules and Men* for *Nature*. That Waddington of all people would write the most prominent review makes sense: Crick had been a long acquaintance of Waddington's. More than that, Waddington pushed to write the review when he saw Crick veering beyond scientific claims, dabbling in the sorts of philosophy of biology that Waddington made a professional pursuit. "No Vitalism for Crick," Waddington's piece, came off less harshly than most of the reviews of *Of Molecules and Men*. Manhattan Project physicist Eugene Wigner, for instance, praised Crick for his lucid description of the DNA to RNA process but found the rest of the text dismissible at best (Wigner 2001). Waddington, by contrast, put Crick's claims into serious conversation with the longer-term discussion of the history of vitalism. Was there a wave of vitalism in the 1960s? Waddington thought Crick's formulation too sloppy to tell.

For him to attack vitalism, charged Waddington, Crick would have to be clearer on the implications of his definition of physics and the relationship of that physics to biology. Crick supposed physics was relatively solid. Waddington noted the strange behavior of particles and energies at the quantum level—something not predicted or even predictable in the physics of the nineteenth century, for instance— and doubted Crick's supposition. "If biologists should find it necessary to postulate an entity as odd as a quark, would that be vitalistic or not?" queried Waddington. If there was vitalism, Crick conflated two different varieties of it, Waddington warned. "Objective vitalists" claimed that there is something mysterious about living things, such that their behavior could not be fully explicated by any purely physical system. Waddington doubted if many of these sorts of vitalists existed in science anymore. "Subjective vitalists," however, focused on the gap between any description of consciousness or awareness or mind and the *experience* of it. Descriptions of neurons and their interconnections existed in a "different logical realm" from the subjective

---

[1] Even years later, after Crick admitted the existence of life itself seemed miraculous, he doggedly held to the notion that consciousness must be biochemistry (Nagel 2012: 124).

human experience; that much was "irrefutable," thought Waddington. Crick merely set up a straw argument when equating this realization of subjectivity with older notions of the soul quickening the human fetus.

"The only differences between us are matters of emphasis," Crick insisted to Waddington in private correspondence.[2] Though, he also admitted that his lack of philosophical training in or appreciation for philosophy meant that he had "never really thought about [these issues] in detail"—apparently even whilst delivering lectures and writing books about them. A month had transpired since Waddington's "very friendly review" appeared in *Nature*, and Crick had decided to sit down and more carefully consider his own claims. Vitalists, Crick decided, could really be divided into three camps. The "obvious sort" believed in souls irreducible to physics and chemistry. Crick really disliked this group, but belatedly admitted they were different than those who he attacked in the book. It was the second group, the "bio-tonic laws" group, that troubled Crick the most. Physicist Walter Elsasser fit into this category, though Crick admitted that he hadn't really read the recent works by Elsasser that Waddington referenced in his review. Waddington's treatment of Elsasser certainly made Elsasser's position into something stronger than Crick's strawman version of him, Crick admitted. The third group of vitalists, according to Crick, centered around Max Delbruck—a discovery that Crick had just made and that astonished him. Imagine, Delbruck a vitalist! These sorts of vitalists, if they deserved that appellation at all, thought biological investigations would reveal new laws in physics and chemistry. Crick seriously doubted those sorts of claims, too. Did he have evidence for it? Not really. But he doubted it, nonetheless. What would Waddington say to that?

The real problem with *Of Molecules and Men*, responded Waddington, had to do with the way Crick and others formulated the vitalism/mechanism dichotomy.[3] As formulated, the question being asked was: *Starting with a foundation of modern physics and chemistry, would we be able to account for biology?* Those who said 'yes' were said to be mechanists; those who said 'no' were dubbed vitalists. But that was the wrong formulation of the problem, according to Waddington. Instead, one should start by asking whether, beginning with whole systems in biology, we would find anything that physics and chemistry might not eventually accommodate—recognizing that physics and chemistry were evolving disciplines, just like biology. There were two features of biology that physics and chemistry could not deal with at present, Waddington suspected. The first, as Crick also noted, was natural selection. But the second, not of much interest to Crick nor, Waddington suspected, to other molecular biologists, was the process of development. This is the field of inquiry once situated in the old field of morphology. To Waddington—who specialized in paleontology and embryology alongside of philosophy before moving into genetics—development was still the most vexing problem. Concepts like natural

---

[2] Crick to Waddington, 9 November 1967, box 102, folder PP/CRI/I/2/6/5, Francis H. C. Crick papers, Wellcome Institute, London, UK.

[3] Waddington to Crick, 27 December 1967, box 102, folder PP/CRI/I/2/6/5, Francis H. C. Crick papers, Wellcome Institute, London, UK.

selection and development remained far removed from what bare-bones physics and chemistry presently had the power to explain, asserted Waddington. Of course, that didn't mean that physics and chemistry would *never* accommodate these foundational features of biology, just that they could not now. This was precisely Elsasser's point, Waddington reminded Crick. Years later, Waddington would clarify the problem with Crick's reductionism:

> The real snag of reductionism, it seems to me, arises if you suppose that we really know all there is to be known about the physical entities and laws. ... I was taught that molecules were made up of little groups of atoms which stick together with valency bonds, like little hooks sticking out of them .... This left absolutely no possibility of the very large scale protein molecules with their tertiary structure and allosteric behavior.... That concept didn't occur forty years ago ... it's been added on since. You must, I think, always realise that we don't know all about the basic physical entities ... (Waddington 1972: 30).

Crick didn't bother to answer Waddington's challenge. Perhaps by then, he'd moved on to discussions of astrobiology, among other things. Waddington invited him to contribute to the International Union of Biological Sciences (IUBS) meetings held in Lake Como, Italy, and Crick attended—he loved the Italian vistas. But reports from those philosophically robust meetings show Crick didn't contribute much to the discussions (Waddington 1970). We can discern his feelings, however, from his behavior immediately after. Among the more influential ideas to emerge from that year's IUBS theoretical biology meeting was Lewis Wolpert's "tricolor flag" model of embryonic development across gradients. Crick went after this model in 1970, attempting to discredit it (Crick 1970).[4] Despite much talking about big concepts, Crick never felt quite comfortable really diving into analytical philosophy of biology. Perhaps he felt, given his list of enemies to vanquish, he never needed to.

4 Identifying Crick's Vitalist Enemy

Crick dismissed as "the obvious sort" of vitalist folks like Pierre Teilhard de Chardin (1881–1955). Teilhard was an astute, geologically trained Jesuit priest who was on the 1929 team that discovered "Peking Man" (*H. erectus pekinensis*). But he had also been influenced by the work of vitalist Henri Bergson (Aczel 2007). Teilhard regarded evolution as fully teleological, a slow unfolding toward an Omega Point with humans as a key component of the creation shaping God's progressive vision: "The day will come when, after harnessing space, the winds, the tides, and gravitation, we shall harness for God the energies of love. And on that day, for the second time in the history of the world, we shall have discovered fire" (Teilhard de Chardin 2002: 87). Teilhard's vitalism wasn't so much an additive force or an *entelechy* as an overall condition of existence. "A directing spirit," as Crick dubbed it (1966:16).

---

[4] In this article, Crick picked up on the explanation already on offer by Turing (1952).

On the one hand, "obvious" vitalism like Teilhard's was not worth taking seriously. On the other, Crick suspected more than a whiff of Christianity—however heterodox Teilhard's faith appeared to a religious insider—wafted through all versions of "obvious" vitalism. And it was railing against Christianity that both energized Crick and made it difficult for him to find common ground with religious thinkers or institutions. As he wrote to the editor of Cambridge University's *Varsity* magazine, "In the past, religion answered these [enduring] questions, often in considerable detail. Now we know that almost all these answers are highly likely to be nonsense...."[5] The biology of Darwin and Mendel had already scrubbed this sort of thinking out of real science, Crick asserted.

But what about the other two kinds of vitalism? As a representative of the third type of vitalist, Crick could only identify Max Delbruck. As Delbruck was an insider, a founding figure of molecular biology, in fact, and as Delbruck claimed he found no in principle barrier between physics and biology, Crick turned away from attacking him.

The most bothersome version of vitalism, from Crick's point of view, was the second type—the "biotonic laws" version. Waddington affirmed that Walter Elsasser acted as a reasonable figurehead for this group, though he didn't dismiss this group as Crick did. On the surface, Crick's complaint was that Elsasser didn't understand biology. Underneath the bravado, however, Crick must have seen that Elsasser represented something very disconcerting. If Crick's major goal was to reduce biology to chemistry and physics, what did it mean when a physicist denied that physics had the power to accommodate the reduction?

Walter Elsasser (1904–1991) migrated from particle physics to geophysics to biology over the course of his career, achieving notoriety in the first two fields as a faculty member of prestigious universities from coast to coast in the United States. But aside from determining the physics at work in the earth's generation of a magnetic field, he felt his most important achievements were in theoretical biology. Elsasser dedicated the last four decades of his life attempting to hammer out the very relationship between physics and biology that Crick proclaimed to be his own ultimate quest. And, though, Elsasser definitely believed he was *not* advocating for vitalism, his concept of biotonic laws definitely rubbed practicing biologists the wrong way.

Biotonic laws, as advocated by Elsasser, was merely the idea that governing principles in biology are more expansive than the existing laws known to physics. He began promoting this concept in *The Physical Foundation of Biology: An Analytical Study*, the work that raised Crick's hackles (Elsasser 1958). The concept highlighted two bones of contention. The first one was obvious: Elsasser did not regard the 'ordered heterogeneity' of organisms as reducible to any physical process currently understood. The number of biochemicals arranged by cells was already large, and the ways in which they could be arranged to make a biological object like

---

[5] Crick to P. Medlicott, editor of *Varsity*, 3 October 1966, includes typescript draft (with handwritten additions) of Crick, "Why I am a humanist." Box 85, PPCRI/H/4/4, Francis H. C. Crick papers, Wellcome Institute, London, UK.

a cell was incomprehensible—a number larger than the number of stars. For example, something like 10^{80} particles are thought to exist in the observable universe. Around one thousand genes regulate the production of penicillin in *Aspergillus*, however. Imagining only two states of those genes—normal or mutated—the number of possible genetic combinations leaps to roughly 10^{300}—a clear indication of the possible heterogeneity of living things at the biochemical level (Rubin 1998: xiv). The question now becomes how this *possible* heterogeneity can be organized in precise ways to produce biological products that perform a range of *specific* functions when in the context of an organism. This is what R. D. Hotchkiss called the problem of ordered heterogeneity; Elsasser (1998: 39) adopted that terminology.

Crick admitted this, too, with his term "ordered complexity"—what viruses have that rocks do not. Elsasser ran afoul of Crick because of what Elsasser thought that ordered heterogeneity meant. The fact that this heterogeneity at the chemical level could be ordered precisely at the cellular level, and then that ordered heterogeneity passed on from generation to generation almost unaltered did not represent a principle that could be easily excavated from those already discovered in physics or chemistry. (In his elucidation of this point, Elsasser repeated the postulates of Niels Bohr (1933), for instance, who advocated complementarity in biology as well as in the wave-particle duality of light.) What's more, the inability of biology to cleanly reduce to physics was not merely due to our lack of present knowledge, but a problem of human intellect for as far out as Elsasser could envision. Life, for all intents and purposes, is just too complex to understand in the causal chain from proton to organism. This point, Elsasser claimed, was *not* calling forth vitalism because he invoked no supernatural intervention, no *élan vital*, and no *entelechy*—just really large numbers. Numbers too large, in fact, for any human to elucidate; even a computer-assisted one. Bringing biology down to the lowest physical level was epistemically ridiculous. Crick's reductionist quest to fit biology to physics, in other words, was not going to work.

More than that, it was unproductive. Elsasser's second bone of contention with Crick's reductionism was over what to do in the face of such immense complexity. He maintained that level of complexity meant that the life sciences would continue to draw on a methodology separate from that of physics and chemistry. Life wasn't limited to the study of molecules. Biology needed to acknowledge the order at the cellular level and above. This kind of holism reflected the only appropriate methodology in the life sciences. In effect, Elsasser argued for the permanent autonomy of biology from the physical and chemical sciences. Not because life was special in the old vitalistic sense, but because humans didn't have the capability of tracing all of the causal chains from the quantum realm, though the atomic, molecular, and to the organismic level. In this face of this great unknowing, biologists and those interested in living things should continue to employ traditional techniques already developed in the discipline without fear that they will one day be collapsed to the methodologies of chemistry and physics by molecular biologists like Crick. To put my terminology in Elsasser's mouth, the attitude of biologists should be one of *bioexceptionalism*. Why? Because when referencing only their basic physical and chemical structure, organisms are "highly indeterminate," while at the organismic level, living beings are highly constrained. Elsasser thought that gap could only be bridged by understanding "some sort of *morphological criteria*" not yet known (Elsasser 1998: 52).

From Crick's ill-defined perspective, this was just as much vitalism as Teilhard de Chardin's version. It might be hard to see exactly why this should be, since Elsasser offered this challenge from physics, not from Teilhard's religious motivation. Why did Crick conflate Elsasser's laws with religious considerations?

To an extent, it was simply an accident of timing. Crick's promotion of humanism and reduction of religious claims to "nonsense" in the Cambridge *Varsity* occurred at the same time that he attacked vitalism in the John Danz lectures that became *Of Molecules and Men*. As Waddington forced him to confess privately, despite his ire, Crick lacked familiarity with the decades of historical and philosophical examination of the claims of vitalism and had not carefully thought through his own claims before publishing.

Yet, it's worth mentioning that Elsasser's philosophy-rich rejection of reductionism rendered it strange (if not suspect) to Crick and the growing cadre of molecular biologists *just because it felt philosophical*, independent of its philosophical content. Apparently, many biologists wanted to avoid even the appearance of being too into philosophy. "Most working biologists…do not like to *think* and look with a very fishy eye on anything which savors of philosophy," wrote the editor of the *Quarterly Review of Biology*, Raymond Pearl, when pressed to publish pieces in theoretical biology (quoted in Peterson 2016: 74). It's at least conceivable, in other words, that merely the philosophical *feel* of Elsasser's work made it seem foreign and something to be avoided. The reductionism promoted by Crick in *Of Molecules and Men*, that organismic behavior could be reduced to molecules of DNA, seemed more conducive to the typical way of doing science by mid-century. As Waddington himself later explained: "[R]eductionism is a recipe for action…if you are confronted with a complex situation, for instance a living system, your best bet to get some sort of pay-off or other is to look for the physical or chemical factors which can influence the phenomenon in question." Of course, it made for "lousy philosophy." But when what is called for is making a "quick (scientific) buck by discovering some useful practical information," it inhibited the slow, complex search for multiple levels and a plurality of inputs (Waddington 1977: 23–25).

Perhaps by "vitalist," then, Crick meant that biotonic law "vitalism" stood in the way of his belief in reducing biology to physics and chemistry that facilitated the ultimate goal of bioengineering life (and presumably artificial intelligence) that such a reduction promised. Elsasser's biotonic laws placed a limit to Crick's reductionist quest and the progressive bioengineered utopia that was supposed to derive from it. One common objection to religious explanations of national phenomena is that they are "science stoppers"—religious advocates address scientific mysteries with a "because God made the world that way," and that's it; there's nothing more to say. Even if this wasn't what biotonic laws were intended to do, Crick sensed that they were still anti-progressive, just like religious objections were anti-progressive. "Vitalists" were anyone who thought Crick was wrong about the ease of reduction or who disagreed with the goal of engineering life. It didn't much matter from what quarter their objections had their origin.

5 Demystifying Mid-Century Vitalism

If we accept Crick's definition that "vitalists" were anyone who thought he was wrong about the ease of reduction or who disagreed with the goal of engineering life, then *indeed there was a fourth wave of vitalism in the middle of the twentieth century*. But I do not think we should accept Crick's definition. For one, it's not even his own.

Despite the fact that he claimed not to appreciate science fiction,[6] Crick's attack on "vitalism" parroted a line of attack by popular science fiction author Isaac Asimov. "Modern science has all but wiped out the borderline between life and non-life," proclaimed Asimov six years earlier in his very widely read *Intelligent Man's Guide to Science*. "[I]t is to biochemistry ('life chemistry') that biologists today are looking for basic answers to the secrets of reproduction, heredity, evolution, birth, growth, disease, aging, and death." As a biochemist himself, Crick seems to have taken this sentiment on board. "Once we get down to the nucleic-acid molecules, we are as close to the basis of life as we can get," Asimov continued. "Here, surely, is the prime substance of life itself. ... All the substances of living matter…depend in the last analysis on DNA" (Asimov 1960: 389). Biochemists and biophysicists, Asimov promised, would in the very near future offer all relevant information about DNA and, it followed from Asimov's perspective, all of secrets of life.

Just as Waddington challenged Crick's philosophy of biology in the late-1960s, physiologist Barry Commoner confronted Asimov's in the early-1960s. It is Commoner's position against Asimov that we can identify as the beginning of a wave of dissent against reductionism and the bioengineering impulse that flowed from reductionism, which both Asimov and Crick shared.

Commoner had professional standing (as did Waddington). As chair of a Committee on Molecular Biology and chair of the American Association for the Advancement of Science (AAAS), Commoner ranked neither as a disciplinary out-sider nor as a member of the fusty "old guard" defending his legacy. And from that position, Commoner encouraged biologists—even those who agreed with the "increasingly important place in *all* areas of biological research" now occupied by biochemistry and biophysics—to view statements like Asimov's not as a corrective against vitalism but as direct attacks on their fields (Commoner 1961: 1746, empha-sis in original).

---

[6] The "necromancy of science," Crick called science fiction. Crick to P. Medlicott, 3 October 1966, includes typescript draft (with handwritten additions) of Crick, "Why I am a humanist." Box 85, PPCRI/H/4/4, Francis H. C. Crick papers, Wellcome Institute, London, UK.

> Since biology is the science of life, any successful obliteration of the distinction between living things and other forms of matter ends forever the usefulness of biology as a separate science. If the foregoing [statement of Asimov's] is even remotely correct, biology is not only under attack; it has been annihilated (Commoner 1961: 1746).

And, in fact it did appear to Commoner and others that the discipline of biology had dramatically lurched toward annihilation in the years since the Second World War. Instead of becoming botanists, embryologists, or even geneticists, new ambitious students migrated to biochemistry and biophysics. Instead of plants or even *Drosophila*, students learned on model organisms such as bacteria and viruses. The very meaning of "organism" had changed as a result. That trend, feared Commoner (1961: 1747), signaled darker days ahead.

Earlier generations of biologists had also noted how these sorts of philosophical divisions around the theory of life could end up altering the practice of biology as discipline. T. H. Huxley, for instance, long ago noted distinctions between experimental and taxonomic biologists, and the drift in interest and expertise toward the former (Desmond 1997: 628–30). But this time was different, Commoner believed. Organismal biologists felt real apprehension. Headwinds of interest and attitude and, perhaps most importantly, grant money was blowing biology away from the study of systems and organisms and toward a complex but not very exciting molecule—DNA (Schaffner 1969).

The "crisis in biology" was larger than this, however; its root cause, philosophical. Reductionists like Crick presumed "the unique capabilities of living things" had their ultimate explanation in "separable chemical reactions." The competing anti-reductionist view found those unique capabilities to be "a property of the whole cell" and, beyond that, the whole organism (Commoner 1966: 44). Recent research in physics and chemistry led to the conclusion that cellular organization could not be explained with reference only to chemical and physical constituents. In other words, life was not reducible to non-life; the cell represented the least complex system that counted as living (Commoner 1961: 1747).[7]

That root philosophical cause not only lured practitioners toward studying molecules, Commoner asserted that it also compromised the future welfare of humankind. For too long, physicists and chemists had limited solutions to larger human problems to merely technical fixes. For instance, when applying their scientific knowledge to social problems like food procurement, many scientists exhibited an "increasing tendency to ignore the facts of life" and defaulted to the use of "insecticides, herbicides, fungicides, nematocides, pesticides, and other assorted agents" while ignoring other causes and consequences. By promoting the view of that there was no barrier between living and non-living, these scientists created, with naïvely clear consciences, materials that fatally restricted the "adaptive latitude of the ecological environment, which is so vital to the success of plant, beast, and man" (Commoner 1961: 1748). What's more, by altering the present adaptive landscape,

---

[7] For this point of his argument, Commoner relied upon the testimony of physicists like Niels Bohr and Cyril Hinshelwood, and, of course, Walter Elsasser.

chemists and physicists also impacted the future in unpredictable ways. Overconfidence in the physico-chemical sciences led to "blindness," then pollution by an underregulated chemical-industrial complex. If this happened in a large-scale way, like it had in the nuclear-military-industrial complex, Commoner feared not simply for his occupation or his discipline, but for the viability of many kinds of life on Earth, including human. Only "sickness and death" would follow (Commoner 1966: 46). He longed for a more perspicacious alliance between physics, chemistry, and biology in which, organismic biologists would take the reins rather than bio-chemists like Crick, focused merely on "mating strands of DNA" (Commoner 1961: 1746).

This is the crucial point. For Commoner, the debate over reductionism—so often regarded as an abstraction wrangled over only by professional philosophers of science—took on grave ethical weight.

Other powerful voices joined Commoner's. Several emphasized the philosophical shortcomings of reductionism. Physical chemist Michael Polanyi's might have been the loudest; his protégé Marjorie Grene's the sharpest. They both argued that biology requires a different conceptual scheme.

> For whether an organism operates more as a machine or more by a process of equipotential integration, our knowledge of its achievements must rely on a comprehensive appreciation of it which cannot be specified in terms of more impersonal facts, and the logical gap between our comprehension and the specification of our comprehension goes on deepening as we ascend the evolutionary ladder. ... [W]hat we observe about the capacities of living beings must be consonant with our reliance on the same kind of capacities for observing it. Biology is life reflecting upon itself... (Polanyi 1962: 347).

Biological processes, according to Polanyi, were those able to maintain stable and open systems. This claim wasn't new or interesting, of course. Others had compared living thing to flames in slow motion, Polanyi admitted—consuming material, leaving chemically altered by-products, moving, spreading. What differentiates living things from flames, from Polyani's perspective, is the ability to capture material perturbations large and small and use them to perpetuate themselves in formations more suited to the environment. They develop in new configurations, in other words, and then they pass on those newly developed configurations. These were the new rules of the biological level that emerged from the purely physicochemical. A sentient level rose out of the purely biological, as well. Evolution had to explain these emergent levels, which rendered the theory of evolution into a biotonic law (Polanyi 1962: 344–64).

Marjorie Grene emphasized that the divisions between biology on the one hand and physics and chemistry on the other aren't denials of the continuity of nature—though it isn't clear that "emergence" does violate that continuity in the first place, as Crick might believe. Causes and explanations just look different in biology than they do in physics or chemistry. In the physical sciences, phenomena follow logically from theory. Given the kinetic theory of gas, for instance, we know that if pressure is thus-and-so and temperature is thus-and-so, then volume is logically predictable. Molecular biologists might say that they're doing something similar when predicting phenotypic changes after genotypic manipulation—what happens phenotypically when we knockout genes X, Y, and Z. But biological explanations presuppose functioning organisms. And though machines exist for some end beyond themselves—they are made to produce a result—organisms are ends in and of themselves. They possess their own histories. So, causes and explanations in biology specify causal connections that biologists *induce* from concrete, living, operating individuals. In the physical sciences, identifying causation requires *deduction* of particulars from abstract logical systems (Grene 1962). Thus biology simply is exceptional: it requires a different logical scheme than physics or chemistry using a system (an organism) that has ends in itself. Geneticists might deny that *E. coli* or *D. melanogaster* has ends at all. But, according to Grene, it turns out to be crucially different to be a bacterium than to be a benzene ring and even more crucially different to be a creature that *studies* both bacteria and benzene.

By the late-1960s, a genuine "wave" of opposition to reductionism overflowed the life sciences and attracted followers in the social sciences as well. Several of these individuals collected at prominent conferences in Europe and the United States well into the 1970s, including the Theoretical Biology symposia at the International Union of Biological Sciences meetings in Italy from 1967 to 1970, the Alpach Symposium in Austria in 1968, the "Biology and the History of the Future" at Chichen Itza, Mexico in 1969, and the "Biology and the Future of Man" conference at Paris in 1974. This is as close to a "fourth wave of vitalism" as the life sciences came.

The question remains: was this "fourth wave" what Crick derided in *Of Molecules and Men*? The first two waves of vitalism in the seventeenth through nineteenth centuries transitioned the theory of life from "frank spookism" to "diluted animism," but held fundamentally to "non-materialistic points of view" (Beck 1957: 133). Hans Driesch, Henri Bergson, William MacDougall, Charles Otis Whitman, and other "third wave" vitalists early in the twentieth century continued to emphasize an ontological difference between the living from the non-living. Compared to these first three eras of vitalism, the "fourth wave," from Commoner to Elsasser to Polanyi to Grene to attendees of the 1960s–70s symposia appears dramatically different in several respects. First of all, a larger number and broader range scientists and scholars were involved mid-century. Secondly, the "fourth wave" groups and individuals explicitly *disavowed* vitalism to a person. Even the most vociferous opponents of reductionism, Arthur Koestler (1969), for instance, did not claim to be vitalists. By these measurements, Crick's confrontational "...what everyone believed yesterday, and you believe today, only cranks will believe tomorrow" attacked something that was not there.

6   Not Vitalism; Bioexceptionalism

Still, what Crick surely sensed in the mid-1960s, even if he misidentified what it was and what it wasn't (i.e., religiously motivated) was an attack on his own simplistic reductionism. And, perhaps to defang that attack, Crick lumped under the heading of 'vitalist': (1) those who were skeptical regarding his claims about the *eminent ability* to reduce the life sciences to the physical sciences and (2) those who were skeptical regarding the *desirability* of doing so. There were, by my count, at least two dozen prominent life scientists who, following Commoner, openly opposed these two points (Peterson 2016; Peterson & Hall 2020). Counting the numerous physical scientists and philosophers who also opposed these two points, that indeed could constitute a wave. And such a trend seems to have continued. In fact, some philosophers would go so far as to say the twenty-first century has produced a kind of *anti-reductionistic consensus*, at least among their peers in the humanities (Waters 1990). This is perhaps another cause of the disciplinary canyon dividing the humanities and the life sciences (Callebaut 2010; Grene 1983).

To be clear, this mid-century swell was not the fourth wave of *vitalism*, as Crick styled it. *Organicism* is an appropriate term for this twin rejection of vitalism and of reductionism simultaneously. This term has its own history as part of a longer set of anti-reductionistic explanations, though less attention has been paid to it than to the earlier fights over vitalism (Peterson 2016).

Here, though, I want to adopt "bioexceptionalism" as an even more capacious term. This term captures what were loosely defined beliefs held by a very broad, multidisciplinary group including physicists and psychologists as well as philosophers and biologists. They shared doubts about reductionism for ontological reasons (e.g., Niels Bohr, Arthur Koestler, who organized the Alpbach Symposium as "Beyond Reductionism" and espoused the concept of the "holon") and epistemological ones (e.g., Elsasser, Polanyi, Grene, Waddington, other organicists). Bioexceptionalism also captures the attitudes of those like Commoner, who—while he may have rejected reductionism for the other reasons—feared the ecological and even professional consequences of a reductionist worldview and explicitly rejected reductionism on those grounds. Whether biology could be reduced to physics and chemistry or not, there were good ecological reasons to act *as if* it could not. The so-called "fourth wave," then, was a broad collection of those who rejected the reductionistic quest Crick had embarked upon.

I still feel uncomfortable with the metaphor of a "wave," however. A few dozen individuals pushing back against Crick might appear large. But Crick was a "cross-worlds influencer" both well positioned professionally and highly motivated to discredit those who challenged his reductionism (Aicardi 2016). As Crick's son, Michael, revealed at his father's 2004 memorial service:

> My thesis here today is that Francis' driving quest was to knock the final nails in the coffin of vitalism. My father wanted to put these ideas to bed, first of all with the structure of DNA .... Francis was a man who was trying his whole life to win an argument. I never understood who he was arguing against, but he had this total conviction that he had to win this argument (quoted in Aicardi 2016: 87).

In lectures and, especially, personal relationships with younger scholars, Crick leveraged his prestige to marshal supporters across disciplines (e.g., Jacob 1970) to fiercely root out vitalism. Moreover, Crick tied his anti-vitalism to devout anti-religious beliefs. "Francis Crick was an evangelical atheist," proclaimed two of his memorializers from the neurosciences. "He had a consistent and completely rational world view without a need to invoke vitalism, or any non-material force" (Siegel & Callaway 2004: 2029). As Crick moved from the chemistry of the DNA molecule, to its transcription and translation, to "molecular psychology"—exemplifying his role as a "cross-worlds influencer"—he maintained the conviction expressed in *Of Molecules and Men* to squeeze religious sentiment out of all science (Aicardi 2016). That zealous leadership allowed Crick to build an "army" of followers even outside his original field of molecular biology (Siegel & Callaway 2004: 2030).

Even leaving aside the question of how much of a wave it was, how could an anti-reductionist alternative rise at all in the middle of Crick's crusade against it? Bioexceptionalism of the kind advocated by Elsasser, Commoner and others rose because, counterintuitively, the dismissive attitude of reductionists intensified, fueled by Crick and those molecular biologists who sympathized with his condemnation of "vitalism," his suspicion of philosophy, his anti-religious fervor.

Their rejection of reductionism reaction is understandable. To have one's subject matter be reduced to something simpler feels insulting somehow—like the reducer is missing the crucial *something* about one's knowledge area. One well-known biochemist objected to that dismissive attitude by those committed to physico-chemical reduction this way:

> While Occam's Razor is a useful tool in the physical sciences, it can be a very dangerous implement in biology. It is thus very rash to use simplicity and elegance as a guide in biological research …. [T]his may make it very difficult for physicists to adapt to most biological research. Physicists are all too apt to look for the wrong sort of generalizations, to concoct theoretical models that are too neat, too powerful, and too clean.

Ironically, this quote belongs to Francis Crick himself (1989: 138–139). Physical scientists, he complained, came into biology with the wrong sort of oversimplifying attitude. Instead of waiting to learn the methods, the way of seeing the world—acquiring personal knowledge, to use Polanyi's term—they shoved their way through using an inappropriate reductionist methodology. Perhaps as Crick aged, he came to appreciate the insistence of Commoner and others that biology really does require a method distinct from the physical sciences and equally valuable. Reductionism, as even Crick admitted alongside his detractors, is a kind of *devaluing*. And it takes a kind of hubris to do that devaluing. According to philosopher Mary Midgley (1994: 16), "When we say that any actual thing in the world … is quite simple and needs only one sort of explanation, we are, almost unavoidably, saying that it is something fairly trivial."

7   Conclusion

If there was a bioexceptionalist wave mid-century, why does Crick's reductionism seem more persuasive and closer to the heart of modern biology even now? There are at least two reasons: (1) training, and (2) utilitarianism bordering on commodification.

What Elsasser, Polanyi, Grene, and others did not fully appreciate at the time, is that nuanced appeals to the separateness of biology from physics and chemistry made use of philosophical distinctions—between theory and technique, epistemology and metaphysics, *is* versus *as if* statements. This is not customarily part of training in biology. According to biologists themselves, biology is not a discipline focused on contemplation so much as on action. For example, when, in the 1970s, the scion of biomedicine Peter Medawar was asked whether the philosophical "attitude with which a man approached biology might help him to focus on certain problems that otherwise might be ignored," Medawar responded with "a flat negative…it made no difference at all. A man was a good scientist, or he was not" (Goodfield 1974: 87). Thus, as an unintended consequence, advocates for bioexceptionalism asked for philosophical reflection, a skill quite alien to training in the life sciences. Crick's reductionism was easier, more familiar (de Chadarevian 2002).

Secondly, and perhaps more saliently, Crick expressed hope in bioengineering, by which Crick—though he expressed it very flippantly—seems to have meant three things: (1) the creation of new products generated by modified or completely synthetic organisms that would enhance the human experience; (2) the augmentation of human cognition by computers; (3) the eventual replacement of our present-day biologics, perhaps even humans, by synthetic "organisms." We certainly do not have to take his futurism seriously. But the history of the life sciences in the second half of the twentieth century and the first quarter of the twenty-first partly comports with the attitude Crick exemplified and that frightened Commoner. Waddington warned that reductionism was good for a "quick (scientific) buck" (Waddington 1977). One wonders whether to what degree today's followers of Crick still seek the parenthetical.

8   Postscript: Why Did Crick Speak at a High School?[8]

"I feel I ought to say a word or two about delivering the lectures in the University Presbyterian Church," wrote Crick to Frank Thomas Watkins, Dean of the Graduate School, University of Washington, and the man who was to make arrangements with Crick to deliver the John Danz lectures. "Although I have not finally decided on the precise title of my lectures," he continued, "my tentative title is: 'Is vitalism dead?' The lectures will be concerned with the impact of biological ideas, both present and future, on our concept of the world."

It had been a rough few years for Dean Watkins. He came to UW in the early-1960s after a distinguished career as Vice Admiral and head of the Pacific Submarine Fleet for the US Navy. Though he spent a lot of time underwater during World War II, he had no experience as an academic, least of all with graduate

---

[8] From the Crick to Vice Admiral Frank T. Watkins exchange, 14 December 1965, PPCRI/E/1/14/5. Francis H. C. Crick papers, Wellcome Institute, London, UK.

students and faculty at what was rapidly becoming the premier research institution in the American Northwest. The university was growing fast and, even with his experience in the military, change was difficult. Worse, on April 29, 1965, a 6.7-magnitude earthquake rattled the whole Olympia-Tacoma-Seattle area, killing several people and damaging thousands of buildings. The University of Washington lost key structures including Meany Hall, the largest lecture space on campus (Buhain 1999; Lange 2000). Crick's talk, originally scheduled to be delivered in Meany Hall, had to be postponed almost a year. Trustees and administrators wrangled over repairing or replacing the building, creating further delays. The University Presbyterian Church in Seattle stood near campus, seated nearly as many as old Meany, and the church was known for its progressive social vision (Staniunas 2016). It would be a fine substitute. Dean Vice Admiral Watkins made the arrangements.

But the English co-discoverer of the molecular structure of DNA was being difficult.

Crick, later described as an "evangelical atheist," promised that his talks on the structure of DNA and vitalism would not be "militantly anti-Christian." Still, he thought an argument against vitalism might offend "Church Authorities," given the degree to which he was convinced that vitalism concealed religious beliefs. He had backed out of commitments before merely because they overlapped with events *sponsored* by churches, let alone actually taking place *inside* one. Maybe the Seattle Presbyterians wouldn't like "Is vitalism dead?" Maybe they would regard his talk as "anti-Christian propaganda." It was a telling admission. Rather than engage with the bioexceptionalist challenges articulated by Elsasser, Commoner, Polanyi, Grene, and others through the 1960s against the loose espousal of reductionism like his, Crick waved his hands at the whole thing. They were merely religious objections (Aicardi 2016). Though, he admitted to Dean Vice Admiral Watkins, "Whether this is widely true I must confess, I do not know."

I can imagine Dean Vice Admiral Watkins clenching his jaw as he picks up the phone to begin again the hunt for a new venue. Eventually, the UW dean moved the John Danz lectures to the student's auditorium at Roosevelt High School, where, for 3 days in the winter of 1966, a Nobel laureate preached to his audience that vitalism was out there—alive.

References

Aczel, Amir D. 2007. *The Jesuit and the Skull: Teilhard de Chardin, Evolution, and the Search for Peking Man*. New York: Riverhead Books.
Aicardi, Christine. 2016. Francis Crick, Cross-Worlds Influencer: A Narrative Model to Historicize Big Bioscience. *Studies in History and Philosophy of Science Part C: Studies in History and Philosophy of Biological and Biomedical Sciences* 55: 83–95. https://doi.org/10.1016/j.shpsc.2015.08.003.
Allen, Garland. 2005. Mechanism, Vitalism and Organicism in Late Nineteenth and Twentieth-Century Biology: The Importance of Historical Context. *Studies in History and Philosophy of Biological and Biomedical Sciences* 36: 261–283. https://doi.org/10.1016/j.shpsc.2005.03.003.
Asimov, Isaac. 1960. *The Intelligent Man's Guide to Science*. Vol. 2. New York: Basic Books.

Beck, William S. 1957. *Modern Science and the Nature of Life*. New York: Harcourt, Brace & Co.

Beckner, Morton. 1959. *The Biological Way of Thought*. New York: Columbia University Press.

Bohr, Niels. 1933. Light and Life. *Nature* 131: 421–423. https://doi.org/10.1038/131421a0.

Buhain, Venice. 1999. *The Life and Death of Old Meany Hall*. The Daily of the University of Washington (April 13). http://www.dailyuw.com/news/article_543d7526-607a-5950-8b0b-feccc0761263.html

Callebaut, Werner. 2010. The Dialectics of Dis/unity in the Evolutionary Synthesis and Its Extensions. In *Evolution the Extended Synthesis*, ed. M. Pigliucci and G.B. Müller, 443–481. Cambridge, MA: MIT Press.

Commoner, Barry. 1961. In Defense of Biology: The Integrity of Biology Must be Maintained If Physics and Chemistry are to be Properly Applied to the Problems of Life. *Science* 133 (3466): 1745–1748.

———. 1966. *Science and Survival*. New York: The Viking Press.

Crick, Francis H.C. 1966. *Of Molecules and Men: The John Danz Lectures, 1966*. Seattle, WA: University of Washington Press.

———. 1970. Diffusion in Embryogenesis. *Nature* 225: 420–422.

———. 1989. *What Mad Pursuit?* New York: Penguin Books.

de Chadarevian, Soraya. 2002. *Designs for Life: Molecular Biology After World War II*. Cambridge: Cambridge University Press.

de Teilhard Chardin, Pierre. 2002. *Toward the future*, trans. R. Hague. New York: Harcourt,

Desmond, Adrian. 1997. *Huxley: From Devil's Disciple to Evolution's High Priest*. Reading, MA: Addison-Wesley.

Elsasser, Walter M. 1958. *The Physical Foundation of Biology: An Analytical Study*. London: Pergamon Press.

———. 1998. *Reflections on a THEORY of organisms: Holism in Biology*. Baltimore, MD: Johns Hopkins University Press.

Goodfield, June. 1974. Postscript in the Light of Discussions at Serbelloni. In *Studies in the Philosophy of Biology: Reduction and Related Problems*, ed. F.J. Ayala and T. Dobzhansky, 85–88. Berkeley, CA: University of California Press.

Grene, Marjorie. 1962. The Logic of Biology. In *The Logic of Personal Knowledge: Essays Presented to Michael Polanyi on His Seventieth Birthday, 11th March 1961*, 191–205. London: Routledge & Paul.

———. 1983. Introduction. In *Dimensions of Darwinism: Themes & counterthemes in twentieth-century evolutionary theory*, ed. M. Grene, 1–15. Cambridge: Cambridge University Press.

Jacob, Francois. 1970. *La logique du vivant. Une histoire de l'hérédité*. Paris: Gallimard.

Koestler, Arthur. 1969. Opening Remarks. In *Beyond Reductionism*, ed. A. Koestler and J.R. Smythies, 1–2. New York: Macmillan.

Lange, Greg. 2000. *Earthquake Rattles Western Washington on April 29, 1965*. HistoryLink.org (March 2). https://www.historylink.org/File/1986.

Midgley, Mary. 1994. *The Ethical Primate: Humans, Freedom, and Morality*. New York: Routledge.

Nagel, Thomas. 2012. *Mind and Cosmos: Why the Materialist Neo-Darwinian Conception of Nature is Almost Certainly False*. New York: Oxford University Press.

Peterson, Erik L. 2016. *The Life Organic: The Theoretical Biology Club and the Roots of Epigenetics*. Pittsburgh, PA: University of Pittsburgh Press.

Peterson, Erik L., and Crystal Hall. 2020. "What is Dead May Not Die": Locating Marginalized Concepts Among Ordinary Biologists. *Journal of the History of Biology*. https://doi.org/10.1007/s10739-020-09618-1.

Polanyi, Michael. 1962. *Personal Knowledge: Toward a Post-Critical Philosophy, Corrected edition*. Chicago: University of Chicago Press.

Rubin, Harry. 1998. *Introduction to Walter M. Elsasser, Reflections on a Theory of Organisms: Holism in Biology*. Baltimore, MD: Johns Hopkins University Press.

Schaffner, Kenneth F. 1969. The Watson-Crick Model and Reductionism. *The British Journal for the Philosophy of Science* 20 (4): 325–348.

Siegel, Ralph M., and Edward M. Callaway. 2004. Francis Crick's Legacy for Neuroscience: Between the α and the Ω. *PLoS Biology* 2 (12): 2029–2032. https://doi.org/10.1371/journal.pbio.0020419.

Staniunas, David. 2016. *Presbyterian America: Seattle*. Presbyterian Historical Society (July 20). https://www.history.pcusa.org/blog/2016/07/presbyterian-america-seattle.

Turing, Alan. 1952. The chemical basis of morphogenesis. *Philosophical Transactions of the Royal Society of London B, Biological Sciences* 237: 37–72.

Waddington, C.H. 1967. No vitalism for Crick. *Nature* 216 (Oct. 14): 202–203.

———., ed. 1970. *Towards a theoretical biology: An IUBS symposium*. Vol. 3. Edinburgh: University of Edinburgh Press.

———. 1972. Comment to H. C. Longuet-Higgins, "The Failure of Reductionism". In *The Nature of Mind*, ed. M. Swann. Edinburgh: University of Edinburgh Press.

———. 1977. *Tools for Thought: How to Understand and Apply the Latest Scientific Techniques of Problem Solving*. New York: Basic Books.

Waters, C.K. 1990. Why the Antireductionist Consensus Won't Survive the Case of Classical Mendelian Genetics. In *Proceedings of the Biennial Meeting of the Philosophy of Science Association, vol. 1*, ed. A. Fine, M. Forbes, and L. Wessels, 125–139. East Lansing, MI: Philosophy of Science Association.

Wigner, E.P. 2001. Review of *Of Molecules and Men* by F. Crick. In *Historical and Biographical Reflections and Syntheses. Historical, Philosophical, and Socio-Political Papers, vol B /7*, ed. J. Mehra, 501–502. Berlin: Springer.

Metabolism in Crisis? A New Interplay Between Physiology and Ecology

Cécilia Bognon-Küss

Abstract This chapter investigates the hybrid relationships between metabolism, broadly and a-historically understood as the set of processes through which alien matter is made homogeneous to that of the organism, and forms of vitalism from the eighteenth century on. While metabolic processes have long been modeled in a reductionist fashion as a straightforward function of repair and expansion of a given structure (either chemically, or mechanistically), a challenging vitalist view has characterized metabolism as a creative, organizing, vital faculty. I suggest that this tension was overcome in Claude Bernard's works on "indirect nutrition", in which nutrition, rightly conceived as a general vital phenomenon common to plants and animals, was both characterized as an instance of the general physico-chemical determinism of all phenomena and as the sign and condition of the "freedom and independence" of the organism with respect to the environment. I propose that Bernard's theory of indirect nutrition was central in the elaboration of his general physiology and has, at the same time, underpinned a self-centered view of biological identity in which the organism creates itself continuously at the detriment of its external *milieu*. I further argue that this conception of biological individuality as metabolically constructed has since, and paradoxically, supported a view in which the organism appears as an autonomous and self-creating entity. I then contrast this classical view of the metabolic autonomy of the organism with the challenges raised by microbiome studies and suggest that these emerging fields contribute to sketch an ecological conception of the organism and its metabolism through the reconceptualization of its relationship with the environment. The recent focus on a "microbiota – host metabolism" axis contributes to shift the focus away from the classical concept of organism, somehow externalizing vitalism out of the autonomous individual in favor of an ecological, collaborative, and interactionist view of the living.

C. Bognon-Küss (✉)
Université Paris Cité, Labex "Who am I", Laboratoire Épigénétique et Destin Cellulaire, Paris, France

© The Author(s) 2023
C. Donohue, C. T. Wolfe (eds.), *Vitalism and Its Legacy in Twentieth Century Life Sciences and Philosophy*, History, Philosophy and Theory of the Life Sciences 29, https://doi.org/10.1007/978-3-031-12604-8_11

1   Introduction: The Metabolic Roots of Living Beings

A century-long medical-philosophical tradition has associated nutrition, the operation by which organized bodies grow, preserve, and maintain themselves, with life.[1]
In contrast to inorganic bodies, whose preservation seems to be merely the persistence of a previous material state, living organisms preserve themselves by undergoing continuous change: it is by the renewal of their substance in a double movement
of composition and decomposition that they succeed in maintaining themselves.
Organized beings therefore seem to escape the alternative of identity and otherness
since, *via* the process of nutrition, it is certainly foreign constituents that penetrate
them and renew their matter, but according to specific modalities by which they
appropriate an exteriority to constitute it as their own identity. Thus, for a living
body, feeding itself does not only mean compensating for the losses suffered by
ingesting foreign matter, or circulating flows of matter and energy in its interior, but
also and above all transforming this matter into its own. Now, beyond this equation
between life and nutrition, how can we picture the relationships between a metabolic-
centered approach to life and vitalism – if the latter isn't to be understood as a mere
heuristic initiating inquiries into the nature of the living and its distinctiveness but
also in the complexity of its historical manifestations and the diversity of its philosophical commitments[2]? In other words, if nutrition-metabolism is pictured as an
essential organizing force, as the manifestation of life itself or at least as its main
distinctive feature, what kind of vitalist claim does this tradition endorse? Conversely,
what philosophical effects can we expect from this confrontation between metabolism and forms of vitalism with respect to contemporary developments in biology?

While metabolic processes have long been modeled in a mechanistic fashion as
a straightforward function of repair and expansion of a given structure (either chemically, or physically[3]), a challenging vitalist view, developed in the eighteenth century, that could be labelled "structural-functional",[4] has characterized nutrition as a
creative and organizing vital faculty.[5] Nutrition, far from assembling preformed elements in a preexisting structure, could come to designate processes of alterations
and syntheses that were both chemical and vital. If life partly consists in the continuous destruction of one's own material constituents, then it has to be, simultaneously, a continual regeneration of one's own bodily matter. Nutrition could then
refer to a continuous production of the living being by itself (production of

---

[1] See Aristotle, *De Anima*, II, 4415b–25.

[2] For precise characterizations of varieties of vitalism both synchronically and diachronically, see
Cimino and Duchesneau (1997), Wolfe (2010a), (2017a), (2019), (2021), Normandin and
Wolfe (2013).

[3] For an analysis of mechanistic conceptions of nutrition, see Bognon-Küss Forthcoming.

[4] See Duchesneau (2012) [1982], Wolfe (2017b). This "structural–functional" conception involves
a theoretical effort to understand the circular causality (vs. linear) between the parts and the whole
and the emergentist nature of the organism's organization.

[5] See de Bordeu (1752), Kant (2000), C.F. Wolff (1789), Bognon-Küss (2019).

specialized parts – flesh, muscle, tendon, etc. – and production of new qualities – irritability and/or sensitivity) by virtue of a certain relationship to its environment (the continuous exchanges of materials and reciprocal transformations). What gradually emerged in the eighteenth century, is the possibility of thinking of life as an unstable rather than a static equilibrium, in which decomposition is less a price to pay for life than a condition for its realization. In this respect, the instrumental and technical framework in which the organism had been thought of up to the eighteenth century had to be broken down and replaced by a model in which functional integration, solidarity and self-production of the parts refer to a continuous vital dynamic. In other words, the living is defined as a specific productivity: it organizes itself by organizing foreign matter. When it assimilates, an organism produces its own substance, a new composition of its own which is not reducible to the sum of the components that enter into it. This shift in the way nutrition was understood, as a dynamic and creative process of organization, supported the elaboration of a conception of organisms as self-organized and autonomous entities, capable of producing their organization and maintaining their form beyond their continuous material renewal, due to their exchanges with their environment.

My hypothesis is that this tension between mechanistic and vitalist views was overcome in Claude Bernard's works on "indirect nutrition", in which nutrition, rightly conceived as a general vital phenomenon common to plants and animals, was both characterized as an instance of the general physico-chemical determinism of all phenomena and as the very manifestation of "life in its state of nudity", *i.e.* as the unification of organic destruction and creation. I suggest that Bernard's theory of indirect nutrition was central in the elaboration of his general physiology and has, at the same time, underpinned a self-centered view of biological identity in which the organism creates itself continuously at the detriment of its external *milieu*, to the extent that nutrition becomes, in the case of Bernard, the sign and condition of the "freedom and independence" of the organism with respect to its environment (Sect. 2). Nutrition obviously refers to the trivial fact that all living beings depend on other cells and on an environment for their survival, that living beings are therefore relational – to the point that this circulation of the outside into the body could indicate a possible dissolution of the organism into the world ("Man is what he eats" – "Der Mensch ist was er isst" as in Feuerbach's famous pun[6]). However, this intuition has been replaced by a focus on the autonomy and independence of organisms – a vision that "metabolism" as a concept for thinking about the dynamic maintenance of organisms has initially reinforced: nutrition and then metabolism rather indicated the way by which organisms dialectically construct their own boundaries and their own internal *milieu*.

But what happens to this "metabolic self" when it is confronted with recent findings in microbiology, metagenomics, or evolutionary ecology? The progressive recognition of the diversity, ubiquity, and functional capacity of microorganisms[7] and

---

[6] See "The Mystery of Sacrifice, or Man is what He Eats" in Feuerbach (1990) 26–52.
[7] Wu et al. (2009).

the correlative exponential development of studies on the participation of microbes in the vital functions of complex multicellular organisms[8] have had deep impacts on our understanding of the biology of plants and animals. They have profoundly challenged our traditional views of the physiology of nutrition, the performance of metabolic functions, and biological individuality itself, as they shifted the focus to the myriad of bacteria that enable organisms to perform these functions essential to their maintenance and survival. Therefore, what microbiome studies[9] force us to think about is a progressive decentering of a biology focused on autonomous organisms towards a biology of relations,[10] integrating not only microorganisms as independent entities, but also "their theatre of activity", *i.e.* the host and the environment (biotic and abiotic) understood as ecological components of these relations.[11] The last section aims at examining how these research programs have led to profoundly challenge the classical view of the "metabolic self" – this closed, autonomous, individual organism, capable of constituting its identity alone and at the expense of its food – and how they contribute to sketch an ecological conception of the living and its metabolism (Sect. 3).

2 Metabolism, "Freedom and Independence" or the Self-Production of the Organism

Metabolism is classically understood as the continuous transformation of matter and energy captured from the environment in order to, for any living entity, realize and maintain itself as a system. Ruiz-Mirazo and Moreno (2013) for example describe metabolism as a "persistent far from equilibrium dynamics of self-construction and self-repair" and argue that metabolism is constitutive of biological autonomy in a "dialectical way", to the extent that the autonomy, distinctiveness, and identity of the organism are precisely achieved through its metabolic openness: organisms reconstruct sets of components of their own (proteins, DNA, RNA, membranes…) that are not present in their environment, through processes of interaction, transformation, and synthesis. Through metabolism, living systems fabricate themselves continuously and "this is precisely what it means to be autonomous." This equation between autonomy, identity and self-production through metabolic processes is the main philosophical problem I wish to address. I will question the kind of philosophical claim that can be associated with this metabolic-centered

---

[8] Jones (2013).

[9] On the term "microbiome" and its history, see in particular Eisen (2015), Prescott (2017), Berg et al. (2020).

[10] See in particular Gilbert et al. (2012), McFall-Ngai et al. (2013), Sélosse (2012), Tipton et al. (2019).

[11] Whipps et al. (1988).

view of life, and to what extent this latter grounds a self-centered conception of biological individuality as being autonomous and self-creating.[12]

2.1   Direct Assimilation and the Alienation of Organic Life

The concept of "metabolism" has gradually stabilized as a set of mechanisms linking the vital specificity of organisms to their chemical conditions of existence, and as a scheme through which self-production and the maintenance of biological identity could be apprehended in a naturalistic perspective. What is at stake in the constitution of the concept of metabolism in the second half of the nineteenth century is the establishment of a certain autonomy, independence or freedom of the organism in relation to its environment through the identification of the mechanisms that allow the organism to carry out syntheses. While post-Lavoisieran chemistry thought of animal nutrition as the absorption of preformed immediate chemical principles from digestion, any histochemical element of the animal body having its origin in the food ingested (fat gives fat, muscle substance gives muscle), in a theory of nutrition that can be called "direct assimilation", metabolism conceives of the organism's independence from food through the elucidation of its capacity to manufacture complex organic substances (without necessarily finding them preformed in the food). The theory of "direct assimilation" indeed relies on the general hypothesis that physiological processes are caused by a change in the proportion of the elements contained in the substances of the organism, and by a continuous exchange between the interior and the exterior.[13] From this point on, physiological functions can be conceived as the maintenance of an equilibrium between the proportions of the elements that make up the animal substance: albumin, fibrin, muscle, and gelatin. This leads to the conclusion of the chemical distinction between the two kingdoms of animals and plants: plants synthesize chemical substances (nitrogen) that will after be consumed by animals.

The chemical operations carried out by animals and plants therefore appear as reversed processes: while plants are a laboratory of organic synthesis, animals carry out the analysis of their products, by extracting the nitrogen contained in the plants they consume. Animal nutrition is therefore no more than the absorption of a matter already chemically identical to that of the organism that assimilates it. A conception of nutrition is then established, certainly chemical, but which can be apprehended with two pairs of concepts: on the one hand, static *vs.* dynamic, on the other hand, preformation *vs.* epigenesis.

"Chemical statics" is the name Dumas and Boussingault gave to their approach to organized beings in 1842, centered on the preeminent role of combustion in

---

[12] For an in-depth analysis of the ontological implications of the concept of organism, see Wolfe (2010b).

[13] See for example Berthollet (1784, 120–125); Cuvier (1810, 99); Holmes (1974).

animal physiology. It is noteworthy that this view of combustion in animals, and thus of the animal-plant complementarity, is not based on direct evidence of the processes, but proceeds from a general understanding of the elementary compositions and chemical properties of the three basic categories of organic substances: carbohydrates, fats, and nitrogenous albuminoid bodies. In these lessons, they argue that if in animals "new organic matters can be born", these are nonetheless "always simpler matters, closer to the elementary state than those which they received" from the plants. Animals merely undo (assimilate, absorb, analyze) the organic materials formed by the plants,[14] hence this lapidary chemical qualification of the animal: "animals constitute from the chemical point of view real combustion apparatuses".[15]

Consequently, the histochemical elements that make up the animal substance must be a simple aggregate of ingested nutrients, whose immediate principles are selected and separated during digestion. In the end, what does the animal that feeds itself do? It separates these chemical principles according to a chemical reaction that is similar to the one theorized by Lavoisier when he studied combustion *in vitro*. The place of this reaction for these authors is the blood, rather than the tissues – simple receptacles of the principles extracted by digestion. The blood is then conceived as a liquid resulting from the dissolution of the elements that make up the nutrients and which then come to aggregate in the organism. As Claude Bernard summarizes it, in this theory "every histochemical element of the animal body had to have its origin in the food ingested": the organism is in the end, and in a very literal sense, only what it eats, that is, on the chemical level, an aggregate of "immediate preformed principles of food or digestion."[16] Ultimately, animal nutrition, chemically conceived, turns out to be "preformationist": the forms are no longer individual organic forms, but the chemical elements that the animal receives ready-made, and that it cannot produce. One thus understands the theoretical solidarity between the chemical *statics*, the renewed *preformationist* conception of the nutrition and finally, the *complementarity* between chemical passivity of the animals and chemical activity of the plants: "(...) it is in the vegetable kingdom that the great laboratory of organic life resides; it is there that vegetable and animal matter are formed (...), from the plants, these matters pass all formed into the herbivorous animals".[17]

A strong contradiction thus emerges between the philosophical role given to nutrition at the turn of the nineteenth century and this theory of nutrition as direct assimilation. This contradiction has the spectacular effect of alienating organic life from itself, of dissociating its operations on the grounds of a chemical division of nature: the chemical complementarity of analysis and synthesis, of destruction and creation, of decomposition and composition, no longer refers to the phases of a

---

[14] Dumas et Boussingault (1842, 46).

[15] *Ibid.*, 4.

[16] Bernard (1878, t. II, 382).

[17] Dumas et Boussingault (1842, 6).

single process taking place within the living being, but to operations distributed in nature to agents that a wise economy will have cunningly arranged. This massive discrepancy with the physiological elaboration which, at the same time, tries to think the coexistence of composition and decomposition processes in organic life, seemed to be a major obstacle for the elaboration of a general physiology.[18] If post-Lavoisierian chemistry thinks of the animal as the site of a continuous self-combustion, this no longer refers to an inner conflict or to the necessary complementarity of contradictory processes, but to a harmonious distribution of chemical work in nature. The chemical conflict is externalized as organic life dissociates from itself and as animal organization resolves itself in that of the vegetable.

2.2   Indirect Nutrition as a Means for the "Freedom and Independence" of the Organism

We can therefore measure the distance that separates such a model from that of metabolism, the prodromes of which can be found in the development of a theory of indirect nutrition in Claude Bernard's general physiology. What Bernard points out, with the complementarity of chemical destruction and creation in physiology, is the solidarity of what will later be called anabole and catabole, *i.e.* the times of a cyclic processuality that will be named "metabolism" as the articulation of these two moments. The problem solved by this double chemical movement is, in a way, as old as the recognition of the originality of living beings: the unity of order and continuity with constant change.

Nutrition is no longer the simple ingestion and analysis of elements already formed, so that we would find at the end by analysis what we put in at the beginning: it is indirect in the sense that the organism's own activity interposes itself between ingestion and the final product to create a material specifically appropriate to the conditions and needs in which the organism finds itself. The experiment with the washed liver (Bernard 1855) allowed Bernard to show that an intermediate material (glycogen), secreted and stored in the liver, could produce glucose under certain circumstances. Thus nutrition, insofar as it is indirect, is a sign of this internal environment, the *milieu intérieur* that the organism secretes itself, for itself. It is true that the internal environment is a set of liquids, blood, lymph, etc., in which the cells are immersed and which in a way buffer the variations of the environment so that the cells remain within low ranges of value of their environmental parameters. The relationship of the interior of the organism with the environment is therefore indirect, and this clearly becomes thinkable from the moment when a physiological function so obviously oriented towards the exterior, *i.e.* nutrition, turns out to be in

---

[18] Bichat for example thought of the characteristics of organic life, common to both plants and animals, as the duality of a process continually engaging opposing forces (1800, §1).

reality indirect. More profoundly, this means that the internal constitution of the organism is not so much given by the ingestion of external elements then distributed in the organism, as by the creation, from external simulations such as food, of stocks of tissue or liquid substances from which the organism then draws the elements it needs.

The very notion of glycogen illustrates this idea that the organism is no longer dependent on the vagaries and fluctuations of glucose supply – the organism is not "at the mercy of the slightest whims and narrow necessities of food. The truth is that it is independent of it to a very large extent, and that the living machine still possesses here a kind of chemical elasticity that is its safeguard".[19] On the contrary, it creates a continuous supply and a stable source of glucose from its own internal processes. Consequently, one must conceive the inner environment not only as a set of liquids but even before that as the effect of a process of continuous constitution of this environment via chemical processes that somehow decouple the organismic inner chemical dynamics, at all levels, from the dynamics of external environment/ organism relationship. If the organism includes an interior environment which buffers it with regard to the external variations, it is because it has in it processes likely to create this interior environment on the basis of the external circumstances, and the indirect nutrition is primarily one of these mechanisms.

The corollary of this conception of the internal environment directly affects the teleological conception of nature that was presupposed by the advocates of direct nutrition. To put it in a few words, if there is a teleology of life – in the deterministic framework that Claude Bernard subscribes to – it must be folded inside the organism itself, in terms that are reminiscent of Kantian teleology and its insistence on "internal purpose". The organism is not in view of anything other than itself, it is, in a way, totally in view of itself, and it is the existence of the internal environment that proves it: "the living organism is made for itself; it has its own intrinsic laws. It works for itself and not for others."[20]

In this conception of nutrition decoupled from digestion, food disappears ("In a word, one does not live on one's actual food, but on that which one has eaten previously, modified and, as it were, created by assimilation"[21]) to the benefit of the vital operations that allow the organism to produce its own substance, its own *milieu*, and finally its independence with regard to the external environment: "(...) the perpetual changes of the cosmic environment do not affect it; it is not chained to them, it is free and independent".[22] Nutrition is therefore an authentic organic creation, "and in this respect, everything is created in the living organism, and nothing comes to it from outside".[23] The organotrophic or nutritive phenomena are therefore the physico-chemical conditions by which the "nutritive center" (*i.e.,* the nucleus of the

---

[19] Bernard (1878, II, 382).

[20] Bernard (1878, I, 148).

[21] Bernard (1878, I, 122).

[22] Bernard (1878, I, 113).

[23] Bernard (1867, 92).

cell) "creates the organism" under appropriate conditions.[24] Indirect nutrition thus points towards two constants: a series of conversions in which the identity of the organism is dialectically constructed at the end of chemical elaborations (composition and decomposition, assimilation and de-assimilation); the persistence of this identity through time despite the constant molecular renewal to which the organism is subjected – the organism remains itself by virtue of the constitution of this internal environment through the mechanisms of its internal secretions. In a note of the *Cahiers* edited by Mirko Grmek, Bernard castigated the chemical statics of living beings, the Stoffwechsel of the Germans, because for him it was "nothing other than metempsychosis. This is not correct (…) Individualism is the master of physiology. It is found even in digestion, and everywhere".[25] In this somehow enigmatic note, Bernard probably means that the chemical statics, which we have seen serves as the chemical basis for a theory of direct assimilation, *i.e.*, the unimpaired passage of food into another body, is no more than a palingenesis of souls. On the contrary, nutrition, as a creative and organizing force is at the same time the cellular basis of organic individuality and the means through which the autonomy and individuality of the total organism is built.

Independence and freedom thus become the pivots around which a metabolic conception of the organism is developed, and this conception refers to a double dialectic: (1) a dialectical relation to food, since assimilation is conceived as a negation of the otherness it represents; (2) a dialectical relation to the environment, since the constitution of the organism's interiority and autonomy is achieved by this metabolic openness to the environment. What is at stake in this opposition between theories of direct and indirect nutrition, is the elaboration of a concept and a process supporting the vision of a living body that is autonomous and identical to itself over time. The constitution of this biological problem thus deploys, within the concept of organism, a new dimension of autonomy: it is by virtue of its constant exchanges with its environment that the living body can acquire a certain autonomy and individuality with respect to it.

In "The whole and the part in biological thought" Canguilhem wrote that "it would be wrong to say that Claude Bernard ignored the romantic prestige of the concept of organism, at the very moment when he was developing experimental techniques and clarifying the ideas that allowed him to break, in the field of biology, the logical circle of the whole and the part."[26] Among these ideas, I would like to add, was first and foremost the concept of metabolism: the very concept of metabolism by which organic life was reconciled with itself in the uniformity of its processes, and within which the biological individual could maintain its autonomy from cosmic conditions in spite of and by virtue of the exchanges it has with its environment, and in fact create himself. And although one should be aware of Bernard's distrust of realistic of substantial forms of vitalism, I have nevertheless

---

[24] *Ibid.,* n°218, 228.
[25] Bernard (1965).
[26] Canguilhem (1968, 327–328).

sought to show that, through the elaboration of the theory of indirect nutrition, Bernard's general physiology had at the same time sketched out the contours of a concept of organism.

2.3   The Dialectical Autonomy of the "Metabolic Self"

This preliminary conceptualization of metabolism provided a framework for thinking about the notion of biological identity, outlining a vision of organisms as self-organizing, autonomous entities by virtue of their relationship to the environment. This elaboration can in fact be understood philosophically within the framework of dialectics, and in particular the Hegelian dialectics, which is known to have taken a close interest in the mechanisms of assimilation in order to think about the self-determination and autonomy of living beings: "The essential point here is the process of nutrition. (…) The process of nutrition is nothing other than this transformation of the inorganic nature into that organic nature which belongs to the subject.", writes Hegel.[27]

By feeding itself the animal turns to the outside world and introduces an external substance into itself – this is the moment of externality. Then the organism must overcome, or digest, this externality. Hence, through assimilation the animal returns to itself, and produces its own substance: it transforms the inorganic into its own organic nature. Although food is already chemically organized, it nonetheless represents the inorganic substance for the animal organism that aims at overcoming this alterity. Interestingly, Hegel refers to Spallanzani's experiments on digestion (Spallanzani 1783) to support his theory: the dialectical nature of nutrition is established by physiological experiments that show that nutrition is not a direct and passive process, but a process in which the organism is engaged in a series of chemical transformations which express the realization of its own essence. Through digestion and assimilation, the organism aims at realizing its own identity. In nutrition the organism does not lose its distinctiveness, nor does it dissolve into the world, but it rather imposes its own determination on what it eats.

This philosophical model, by which metabolic identity is thought of as dialectical, has been quite robust. Hans Jonas, for example, has devoted this understanding of metabolism to the constitution of an interiority, autonomy, and self-identity of the organism. In his 1966 book, the *Phenomenon of Life*, Jonas holds that it is in the dialectical relationship that organisms establish with their environment through metabolism that they realize their identity: "The exchange of matter with the environment is not a peripheral activity carried out by a persistent nucleus: it is the mode

---

[27] Hegel (2004), addition to §365.

of total continuity (self-continuation) of the subject of life itself".[28] In this sense, the organism must be understood as the result of its constant metabolizing activity, and this activity – this capacity of building up its parts by the mediation of external matter – is precisely "its being self-centered individuality". Not only does metabolism emancipate the form of the organism from its immediate identity with matter but it also elaborates a genuine, *i.e.* organic, "mediate and functional kind of identity".

More recently, philosophers who, following Varela's work (Varela 1979; Maturana and Varela 1982), favor metabolism in the characterization of life as an auto-poietic system, similarly insist on the dialectical character of the autonomy of organisms: while metabolism refers to processes of continuous transformations of matter and energy taken from the environment, it is precisely in this metabolic openness to the environment that organisms realize their autonomy and identity as they construct from them components that are their own as well as the organizational closure without which the system could not subsist as such (Moreno and Mossio 2015). In the formal perspective of self-organization, living beings are autonomous biological systems whose identity is described as a "circular process of reflexive interconnection, whose main effect is its own production" (Varela 1997). The metabolic relationship to the environment has the effect of this organizational closure, which is understood as circular causality (since there is a feedback loop between the boundaries of the organism, or the cell membrane, and the metabolic circuit). In this model, the coherence and sustenance of the system are therefore the direct effects of the activity of the system as such, and not those of the action of the environment (in the sense that no internal change in the system can be induced by a direct action of the environment). The auto-poietic system is certainly not a unit isolated from its environment by a protective closure, since the condition for the realization of its autonomy lies precisely in the metabolic interactions it is likely to establish with its environment. But precisely, the environment seems to be characterized as a moment in the dialectical realization of an identity: this relationship of metabolic openness to the environment is a question of constructing the delimitation or the closure that ensures the autonomy and individuality of living organisms.

How can this conception of the "metabolic self," which I have described as self-centered, be reconciled with the reconceptualization of metabolism to which recent discoveries in biology on the role of heterospecific entities[29] oblige us?

---

[28] Jonas (1966, 76).

[29] Here I refer to cooperative associations of *heterospecific individuals*, namely collectives of biological entities of distinct species (*i.e.* with distinct genealogical, evolutionary histories) that live in close association with one another and are supposed to behave evolutionarily and physiologically as one individual, at one given timescale. For detailed discussions see for example Wilson and Sober (1989), Ereshefsky and Pedroso (2015), Pradeu (2016).

3   Metabolism, Identity, and Microbiome Studies: Challenges from the Ecological View of Life

Biological identity is often held to be circumscribed by three major sets of properties, namely an adaptive immune system discriminating between self and non-self, a genome conditioning unique phenotypic traits, and, among animals that have a central nervous system, brain functions that support personality and cognition.[30] One hypothesis I have defended so far is that the concept of metabolism has underpinned, in the historical complexity of its development, a strong conception of biological identity as self-determined and autonomous.[31]

Making sense of the concept of identity means two things: explaining how x is identical in the sense that it is x and not y and can therefore be consequently distinguished from any y; and explaining how x is capable of being identified as x in different contexts and time intervals - in particular, how x can last as x.[32] I will call the first question "distinction" (Id), and the second "persistence" (Ip). In relation to biological identity, the question of "distinction" has often been framed around the general concept of information and has focused in recent decades on narratives that foreground genes and genomes; whereas the question of "persistence" traditionally mobilizes the concept of metabolism, as it encompasses the processes by which individual organisms maintain their identity over time. The question of "persistence", *i.e.* the maintenance of identity over time, is often framed either in terms of multi-level selection processes (how a collective of cells is protected from selfish mutants, for example[33]), or in terms of self-organization, involving studies of metabolism at all levels.[34]

However, it can be argued that metabolism intimately links these two issues of identity, namely distinction (Id) and persistence (Ip). I underscored that the concept of metabolism emerged in the mid-nineteenth century in opposition to a conception of identity maintenance as direct assimilation. The emergence of this concept of metabolism has thus provided biology with a conception of organisms as autonomous individuals (Id), capable of self-organization and maintenance (Ip) in a wide variety of contexts.

---

[30] See for example Rees et al. (2018), and for a critique Park et al. (2018).

[31] Note that other metabolic views of physiological individuality have been defended from a a-historical perspective: see Dupré and O'Malley (2009), Gilbert et al. (2012), Godfrey-Smith (2013), Roughgarden et al. (2017).

[32] Snowdon (1995), Wiggins (2001).

[33] Michod (1999).

[34] Moreno and Mossio (2015).

3.1   Challenges

However, the metabolism around which this conception of identity has been reinforced is now contributing to redraw its boundaries. Indeed, if metabolism underpinned a conceptual figure where the identity of the living being could be grasped and expressed as emerging from the chemical world and the environment, new advances in biology in the field of microbiology and evolutionary biology of symbioses call into question this figure which seemed to equate identity, autonomy and closure.[35] For several decades now, we have been measuring the extent to which symbiotic bacteria and organisms are intertwined. Various functions in animals and plants are carried out by symbionts or mutualists of different species: termites use bacteria to digest cellulose,[36] healing processes in mammals are due to symbionts, the origin of the placenta in mammals is most probably due to a symbiosis with a virus[37], and in vertebrate physiology, lipid metabolism, xenobiotic detoxification, vitamin synthesis and intestinal permeability are all functions performed by bacteria.[38] And microbes themselves are now considered as heterogeneous entities: bacteria and archaea contain many viruses (bacteriophages), some of which seem to contribute to their functioning. As far as genetic identity is concerned, this means that the genome of an organism, although essential to its life, is not only the genome of the host. The notion of "metagenome" emerged in the early 1990s[39] from the development of computational methods to better understand the genetic composition, activities, and reciprocal interactions of these complex communities, therefore transcending the level of the individual organism. Although the role of the microbiota in physiology and development is widely recognized, there is still no consensus on the impact it should have on our conceptions of biological identity (Id and Ip).[40]

The recognition of the extent of the intertwining of symbiotic bacteria and hosts has thus led to theoretical and semantic innovations: some redefine the microbiota, in a move that contributes at the same time to redefine the very notions of inside and outside of organisms, as a genuine organ of the body, constituted by a collectivity of heterogeneous organisms[41]; others argue, under the concept of the holobiont,[42] that

---

[35] Here I focus on symbiotic associations understood as the cooperative associations of *heterospecific individuals*, see supra, n.33. In this restricted definition, symbiosis is characterized as a subclass of mutualisms, West et al. (2007), and not as a more general type of close and long-termed relation between heterospecific individuals, be it mutualistic, commensalitsic or parasitic.

[36] Breznak & Brune (1994).

[37] Mallet et al. (2004).

[38] Nicholson et al. (2012).

[39] Handelsman et al. (1998).

[40] Bordenstein and Theis (2015) *vs* Moran and Sloan (2015), Douglas and Werren (2016), Rees et al. (2018).

[41] O'Hara and Shanahan (2006).

[42] Zilber-Rosenberg and Rosenberg (2008).

the "true" biological individual is the host-symbiont unit, which is supposed to be a unit of evolutionary processes and physiological functioning, and not the classical concept of organism. Some philosophers have suggested that biological identity requires conceptual clarification in light of these developments[43] or that the concept of individual should be abandoned[44]. I propose that what is at stake in this tension between a metabolic conception of physiological identity and the reconfiguration of the relations between the inside and the outside, unity and plurality, identity and otherness is less the abandonment of the notions of identity or individuality than the way in which this contemporary crisis of metabolism forces us to deploy new strategies for thinking about these concepts. Indeed, it is more a certain strong conception of biological identity as self-determined and autonomous, a conception that has been solidified around the concept of metabolism, that is now in crisis.

Rearranging such a conception in the light of contemporary work does not mean having to give up defining biological identity in this context. If we are willing to admit that nutrition and metabolism allow us to think precisely through these paradoxes (identity in change, autonomy through openness, and these points of passage between the exterior and the interior, the inert and the animate, the other and the same) then they undoubtedly conceal conceptual resources that the history of the stabilization of the concept of metabolism – as a dialectical scheme for thinking about the autonomy of organisms – will have covered up. It is therefore less a question of undertaking a radical critique of the concept of metabolism than of redefining it in the light of these new developments in microbiology, a redefinition that could also be conceived as a reactivation of abandoned or forgotten scientific avenues, and which should in turn contribute to drawing up a renewed conception of biological identity.

3.2   Conceptual Issues: Biology of Organisms and Metaphysics of Identity

These new challenges contribute to questioning the boundaries and the very relevance of the concept of organism and its traditional equation with the idea of biological individuality: first by making the organism the unification of a plurality of heterogeneous entities; second by focusing attention on the networks of interactions between species rather than on discrete and isolated entities. The crisis of the concept of metabolism thus becomes a crisis of the concept of biological identity. The upheavals that I try to think about are therefore both internal mutations in contemporary biology and the impact of these mutations on metaphysical conceptions of biological identity and individuality.

---

[43] Hutter et al. (2015).

[44] Gilbert et al. (2012).

In terms of biology, I assume that the shift from an evolutionary biology of competition between organisms to an evolutionary biology of symbiosis[45] finds its metabolic translation here in the gradual shift from an egocentric conception of appropriation to a vision centered on cooperation and the persistence of otherness. It is clear that this twofold evolutionary and metabolic shift towards plurality and interaction networks in turn disrupts the way in which we compartmentalize the entities that make up the living world. While biology has historically been built around the notion of organism and this traditionally refers to the idea of an autonomous individual with clearly discernible boundaries, it has become clear that the isolated organism is incapable of functioning properly independently. Recent work in plant physiology has shown, for example, that the metabolic, immune, and developmental functions of plants can only be achieved through their membership of a vast network of mycorrhizal fungi.[46] Although nutrition initially referred to the obvious fact that all living organisms depend on other cells and an environment for their survival, and that living organisms are therefore relational, this intuition was replaced by a focus on the autonomy and independence of organisms – a view that metabolism as a concept for thinking about the dynamic maintenance of organisms initially reinforced. What these different works force us to think about is, on the contrary, this progressive decentering of a biology of organisms towards a biology of networks: in a biological world made up of interactions, can we still identify discrete and autonomous units? Can a multiplicity of bacterial species united around a metabolic function (*e.g.* nitrogen fixation with functional relays between different bacterial species) be characterized as a biological individual? Finally, it is our capacity to identify collective entities as functional and evolutionary individuals that is being questioned here (Bouchard and Huneman 2013).

On the metaphysical level, such a complexification of the picture of life clashes with the concepts with which we traditionally describe the objects that make up the living world. Thus, the concept of individual envelops three dimensions, epistemic and ontological: it is both that which is indivisible, *i.e.* that which remains at the end of processes of logical division (classically in Aristotle), referring correlatively to the properties that make such an individual distinct from any other,[47] and that which is re-identifiable as the same individual at different moments in time, *i.e.* that which persists or is identical to itself over time.[48] While biology seems to provide paradigmatic cases of individuality - as Aristotle does in the *Categories* (2a 10–13) in relation to the definition of substance Aristotle 1963 - since "such a man, such a horse" persist in time, can be re-identified as the same man or the same horse at different intervals of time, and are indivisible in the sense that their division would not result in two distinct men or horses, it is at the same time obvious that this intuition is

---

[45] Doolittle and Bapteste (2007), Sapp (2009), Sélosse (2012), Tripp et al. (2017), Bapteste and Huneman (2018).

[46] Sélosse and al. (2006), Roy and Sélosse (2015).

[47] Lowe (2001).

[48] Strawson (1959).

contradicted by many biological facts, in both aspects of spatial indivisibility and temporal continuity. For example, the discovery of the phenomena of organic rein-tegration in the eighteenth century, such as the regeneration of freshwater polyps after a longitudinal cut (Trembley), shows that two functional individuals can emerge from the division of an initial individual. Similarly, the idea of temporal continuity clashes with both the constant molecular renewal to which living beings are subject and the existence of holometabolous organisms, *i.e.*, organisms whose development is characterized by a complete metamorphosis of states, such as lepi-doptera, beetles, etc. Finally, this common preconception of the indivisibility of biological individuals classically encounters the question of compositionality and levels of hierarchy within living organisms.

This question, which is not new, can be considered as constitutive of the elabora-tion of the concept of organism, which classically integrates a reflection on the articulation between structures (organs, apparatuses, systems) and functions, with a reflection on hierarchically integrated levels of organization (genes, cells, organs). However, such a question has a new dimension today, since the upheavals I have mentioned - in the field of the evolutionary biology of symbioses, microbiology, or metabolic physiology - no longer only concern the mereological relations between the whole and its parts, unity, and plurality. These works shift the attention of biolo-gists to the nature of dynamic and emergent interactions between a host and hetero-specific entities, without which homeostasis, development or immunity are not possible. To confront these questions is therefore to question the relevance of the metaphysical concept of the individual for contemporary biology, or to elaborate the conditions that would be required for such a concept to be consistent.

These questions about the relevance of a concept of individual for biology are based, as one can see, on a metaphysics of identity, in such a way that it seems dif-ficult to disentangle what, for the living, is identity and what is individuality. Indeed, as has been pointed out, the criteria for attributing individuality are indexed to both distinction (synchronic identity) and persistence (diachronic identity) and therefore imply agreement on a metaphysical concept of biological identity. However, in both respects, the focus on metabolism and the undermining of the biological topics described above pose a challenge to classical metaphysical conceptions of identity. Whether they are metaphysics of substance (what is identical is what remains under change, *i.e.*, substance being understood as an ontologically independent discrete entity, and persistent subject of changes in accidental properties[49]) or of resem-blance (x is the same if x resembles itself at two intervals of time), these ontologies are centered on what does not change, change being always derivative, secondary.

The consideration of metabolic processes, on the other hand, obliges us to think about the centrality of change for the living being at all levels, and therefore to elaborate an ontology that allows us to give an adequate account of biological iden-tity, *i.e.*, in a way that is consistent with what science describes. One of the theoreti-cal challenges for today's philosophy is to contribute to the development of a

---

[49] Robinson (2004).

concept of identity in which change and relationships are central, without being subsumed by an autonomous assimilative center.

Symmetrically, the way in which we classically conceive of biological identity along the two axes of distinction and persistence is strongly undermined here. While the distinction of an individual is supposed to be ensured by the form, it is generally held that persistence is achieved by the continuous renewal of the matter that composes it. This dichotomy between matter and form can be called "hylomorphism", in a very general sense that goes beyond the content that Aristotle gave to this concept. The way in which this conception was inscribed in twentieth century biology and which refers to the hierarchical distinction between reproduction and metabolism - that is to say, a genome conferring on living beings their distinction, *i.e.*, their form, and a metabolism ensuring the continuous material production of this form - contributed to reinforcing a vision of living beings that is both 'genomecentric' and metabolically egocentric. From this point of view, the classical question of the priority of the genetic or the metabolic, of form or matter, of distinction or persistence, in the scenarios of appearance and history of living forms is in a way already inscribed in the structure of the concepts since it presupposes their strict separation. This question ceases to be relevant as soon as the dichotomy between matter and form is blurred by the new ways of defining the concepts of organism and biological identity that this crisis of the concept of metabolism opens up.

3.3   Ecologicizing Biology

The investigation of theses metaphysical questions needs to rely on a paradigm shift in physiology, *i.e.* the ecologicization of its focal concepts. In light of the indispensability of the microbiota for the development, functioning and maintenance of organisms, should the microbiota be considered as an organ[50]? If so, should the very notion of organ be redefined to include the contribution of heterospecific and adaptive elements during the life of the organism in its functional dimension? Can the organism itself, as a community of species, then become the object of an ecological type of analysis[51]?

Is the concept of holobiont relevant for thinking about this integration between the host and the set of micro-organisms it harbors? and if so, how should it be defined? The holobiont concept, *i.e.*, the set of hosts and microbes designated as an interdependent and co-evolving unit, has been proposed as an alternative to traditional conceptions of the organism as an isolated and autonomous biological individuality.[52] But such a conception of the living organism is not without questions: does the whole constituted by the host and the micro-organisms constitute a

[50] Baquero and Nombela (2012).

[51] Costello et al. (2012).

[52] Zilber-Rosenberg I., Rosenberg E. (2008).

functioning unit for physiology? Whether the holobiont is a unit of selection or not, as evolutionary theorists debate, one must ask what kind of integration between these different organisms would be required for the holobiont to be characterized as a functioning unit. The focus on the metabolic processes carried out by the interaction patterns between host and microbes leads to question the robustness of a conception of identity built on the collaboration between discrete entities. Therefore, some even propose to go beyond the holobiont concept in that it conceptualizes interactions between species from autonomous and separate entities and enshrines boundaries to the detriment of considering networks.[53]

From this perspective, it is the processes themselves, rather than the individuals who carry them out, that seem robust.[54] Indeed, although there is great diversity of species and strains in the bacterial taxonomic groups present in the gut microbiota, the essential functions performed by these organisms remain remarkably stable.[55] Bacteria of different species participate in metabolic cycles, converting nutrients into metabolites that are used by other bacteria to produce other metabolites, which in turn are used by the host. These functional steps can be carried out by a multiplicity of strains present in the gut, so that these cycles can continue regardless of the identity of the organisms carrying them out. This contrast between the stability of metabolic processes and the diversity of the actors that carry them out in some way doubles the complexity of organisms. The compositionality of living organisms cannot be thought of in the mode of a swarm of bees since, in such a model, already appreciated by eighteenth-century vitalists, the discrete entities that make up the whole are identical to each other, just as the multiple bees are genetically equal. Conversely, the holobiont model requires us to think not only of the heterogeneity of the parts, but also of their lack of stability in the execution of metabolic processes. As Falkowski et al. express it: "in essence microbes can be viewed as vessels that ferry metabolic machines through strong environmental perturbations into vast stretches of relatively mundane geological landscapes. The individual taxonomic units evolve and go extinct, yet the core machines survive surprisingly unperturbed."

To acknowledge the functional inclusion of heterospecific entities in any organism, to conceive their irreducible relevance for its physiological functioning, to integrate ecological concepts of interaction to analyze the very dynamics of the organism and to redefine its health: if these research directions prove robust and consistent, they prompt the substitution of the metaphor of the bee swarm classically used to represent the organism (the parts all serving the whole, the division of labor between food and reproduction instantiated by castes...) with a metaphor of the organism as an ecosystem.[56] Yet, to what extent is this simply a metaphor? Is there an ontological significance to the assimilation of the organism to an ecosystem of individuals of

---

[53] Simon et al. (2019).

[54] Doolittle and Zhaxybayeva (2010), Falkowski et al. (2008).

[55] Doolittle and Booth (2017).

[56] Van Baalen and Huneman (2014).

different species in mutualistic, predatory and competitive relationships? If the trend in classical community ecology initiated by Clements (1916) and often labeled organicism, has been criticized for its overly metaphorical assimilation of ecosystems or ecological communities to organisms with metabolism, do not contemporary approaches to the organism evoked in this project allow us, on the contrary, to think of organisms as ecosystems, seeing this "like" less as an analogy than as an ontological identity? If there is to be a scientific metaphysics of the organism and its identity, it cannot do without questioning the metaphorical character of the organism/ecosystem analogy that we see unfolding today in physiology, evolutionary biology or even cellular biology (Scadden 2014 speaks of the "cell niche").

4 Conclusion

If we were to map the stakes of a philosophical reflection on biological identity informed by a genealogical study of the constitution of the concept of metabolism in which this problem crystallized for general physiology and therefore post-Bernardian biology, we would have to distinguish what we would call the traditional self-centered vision of identity, where identity is based on metabolism as a process justifying a constitutive homogeneity of the organism while integrating foreign substances. There are, of course, theoretically varieties of this egocentric vision, depending on whether they focus on thermodynamics, organic chemistry or kinetic chemistry as frameworks for formulating the requirements required by any metabolism, and whether they focus on integration into the body or cell self-maintenance, as key objectives of the metabolic process.

To this self-centered view, we would oppose an alternative, relational view in which identity is the result of the cooperation and conflict of organisms of different species that compose an organism – what the concept of holobiont tries to grasp. Such a vision would integrate the theoretical contribution of the evolutionary biology of symbioses or the metagenomics of cellular activity, but I argue it would struggle to accurately reflect the limits of identity, namely the distinctions between all those microbes, viruses that contribute to the identity of a given living organism. It contradicts the traditional equation between metabolism and biological identity, but at the same time requires a renewed understanding of metabolism.

To what extent does each aspect of the self-centered view prevent the development of processes in which heterogeneous elements – organisms of other species, such as bacteria or fungi – perform or contribute to a vital function? In particular, which aspects of the complex meanings of the concept of metabolism are most problematic in integrating the role played by heterogeneous elements into an inclusive account of biological identity? These are the questions that would probably emerge from taking into account this new frontier of the metabolic concept that is being built today by the renewed knowledge of heterospecific contributions to the biological construction of identity, or by the ecological turn in physiology.

What then becomes of the question of the relations between metabolism and vitalism with which I opened this chapter? In other words, what does this movement of the concept of metabolism, from the elaboration of the self-centered individual to the consideration of networks of interaction, do to vitalism as such? While vitalism traditionally served as a foil for both biology and philosophy, being attached to metaphysical and nondeterministic claims, *i.e.* the hypothesis of nonmaterial forces governing the generation, development and vital functions in living beings, I recalled that vitalism should be understood in the diversity of its historical manifestations and philosophical presuppositions. Minimally, vitalism can be depicted as the commitment to the existence of irreducible vital properties or dispositions, a commitment that seems to have constantly been challenged – and threatened – by the possibility to reduce biological processes to a set of physico-chemical reactions. To Foucault[57] vitalism plays "an essential role as an indicator in the history of biology": an indicator of problems to be solved (what constitutes the originality of life without constituting an independent empire in nature?), an indicator of reductions to be avoided (emphasizing the pervasiveness and necessity of concepts of preservation, regulation, adaptation). Interestingly, in the space of these complex relations between the biological and the inorganic, it appeared that the understanding of nutrition as an indirect process of self-organization had, at the same time, contributed to sketch the contours of the concept of organism. With this crisis of metabolism that I have briefly outlined, however, it is not so much the physico-chemical reductions that seem to threaten the integrity of the self-centered individual, than its openness, the dissolution of its boundaries, or its decentering. What is being questioned then in this interplay between vitalism and metabolism is no more the dualism between the organic and the inorganic, but rather the reassuring dualism between the organism and the environment, the very existence of an autonomous and self-creating "milieu intérieur" that delimited the boundaries and independence of the organism from its environment. To this extent, this progressive decentering and complexification of metabolism is at the same time a call to overcome a dialectical conception of the interactions between organisms and their environment and an appeal to externalize vitalism.

References

Aristotle. 1963. *Categories and De Interpretatione*, translation and notes, J. Ackrill, Oxford: Clarendon Press.

Bapteste, Eric, and Philippe Huneman. 2018. Towards a Dynamic Interaction Network of Life to Unify and Expand the Evolutionary Theory. *BMC Biology* 16 (1): 56.

Baquero, F., and C. Nombela. 2012. The Microbiome as a Human Organ. *Clinical Microbiology and Infection* 18 (juillet): 2–4.

---

[57] "Introduction" to *The Normal and the Pathological*, Canguilhem (1991, 18).

Berg, Gabriele, Daria Rybakova, Doreen Fischer, Tomislav Cernava, Marie-Christine Champomier Vergès, Trevor Charles, Xiaoyulong Chen, et al. 2020. Microbiome Definition Re-Visited: Old Concepts and New Challenges. *Microbiome* 8 (1): 103.

Bernard, C. 1855. Sur le mécanisme de la formation du sucre dans le foie (Comptes rendus par l'Acad. des sciences, 24 septembre 1855).

———. 1867. *Rapport sur les progrès et la marche de la physiologie générale en France. Recueil de rapports sur les progrès des lettres et des sciences en France.* Paris: à l'Imprimerie Impériale.

———. 1878. *Leçons sur les phénomènes de la vie communs aux animaux et aux végétaux.* Paris: J.-B. Baillière et fils.

———. 1965. Cahier de Notes 1850–1860. ed. Mirko Dražen Grmek and Robert Courrier. Paris: Gallimard.

Berthollet, Claude-Louis. 1784. Recherches sur la nature des substances animales et sur leurs rapports avec les substances végétales. *Mémoires de l'Académie Royale Des Sciences*: 120–125.

Bichat, Xavier. 1800. *Recherches physiologiques sur la vie et la mort.* Paris: Béchet Jeune Gabon.

Bognon-Küss, Cécilia. 2019. Between biology and chemistry in the Enlightenment: How Nutrition Shapes Vital Organization. Buffon, Bonnet, C.F. Wolff. *History and Philosophy of the Life Sciences* 41 (1): 11.

———. Forthcoming. Nutrition, Vital Mechanisms, and the Ontology of Life. In *Mechanisms, Life and Mind in the Modern Era*, ed. A. Clericuzio, P. Pecere, and C.T. Wolfe. International Archives of the History of Ideas, Springer.

Bordenstein, S.R., and K.R. Theis. 2015. Host Biology in Light of the Microbiome: Ten Principles of Holobionts and Hologenomes. *PLoS Biology* 13 (8).

Bouchard, Frédéric, and Philippe Huneman, eds. 2013. *From Groups to Individuals: Evolution and Emerging Individuality.* Cambridge, MA: The MIT Press.

Breznak, J.A., and A. Brune. 1994. Role of Microorganisms in the Digestion of Lignocellulose by Termites. *Annual Review of Entomology* 39: 453–487.

Canguilhem, Georges. 1968. *Études d'histoire et de philosophie des sciences.* Paris: Librairie philosophique Vrin.

———. 1991. *The Normal and the Pathological.* New York: Zone Books.

Cimino, Guido, and François Duchesneau, eds. 1997. *Vitalisms from Haller to the cell theory. Proceedings of the Zaragoza Symposium* XIXth International Congress of History of Science, 22–29 August 1993. Biblioteca di physis 5. Firenze: L. S. Olschki.

Clements, F. 1916. *Plant Succession: An Analysis of the Development of Vegetation.* Washington DC: Carnegi Institution.

Costello, E.K., K. Stagaman, L. Dethlefsen, B.J.M. Bohannan, and D.A. Relman. 2012. The Application of Ecological Theory Toward an Understanding of the Human Microbiome. *Science* 336 (6086): 1255–1262.

Cuvier, Georges. 1810. *Rapport historique sur les progrès des sciences naturelles depuis 1789 et sur leur état actuel.* Paris: De l'Imprimerie impériale.

Bordeu, Théophile de. 1752. *An omnes organicae corporis partes digestioni opitulentur ?* Paris: Typis viduae Quillau.

Doolittle, W. Ford, and Eric Bapteste. 2007. Pattern Pluralism and the Tree of Life Hypothesis. *Proceedings of the National Academy of Sciences* 104 (7): 2043.

Doolittle, W. Ford, and Austin Booth. 2017. It's the Song, Not the Singer: An Exploration of Holobiosis and Evolutionary Theory. *Biology and Philosophy* 32 (1): 5–24.

Doolittle, W. Ford, and Olga Zhaxybayeva. 2010. Metagenomics and the Units of Biological Organization. *Bioscience* 60 (2): 102–112.

Douglas, Angela E., and John H. Werren. 2016. Holes in the Hologenome: Why Host-Microbe Symbioses Are Not Holobionts'. Edited by Margaret J. McFall-Ngai and R. John Collier. *MBio* 7 (2).

Duchesneau, François. 2012. [1982] *La physiologie des Lumières empirisme, modèles et théories.* Paris: Classiques Garnier.

Dumas, Jean-Baptiste, and Jean-Baptiste Boussingault. 1842. *Essai de statique chimique des êtres organisés*. Paris: Fortin, Masson et Cie, libraires.

Dupré J., and O'Malley M. 2009. Varieties of Living Things: Life at the Intersection of Lineage and Metabolism. *Philosophy and Theory in Biology* 1 (3): 1–25.

Eisen, JA. 2015. *What Does the Term Microbiome Mean? And Where Did It Come From? A bit of a surprise*. Microbiol Built Environ Netw.

Ereshefsky, Marc, and Makmiller Pedroso. 2015. Rethinking Evolutionary Individuality. *Proceedings of the National Academy of Sciences* 112 (33): 10126–10132.

Falkowski, P.G., T. Fenchel, and E.F. Delong. 2008. The Microbial Engines That DRIVE Earth's Biogeochemical Cycles. *Science* 320 (5879): 1034–1039.

Feuerbach, L. 1990. *Gesammelte Werke*. Berlin: Akademie Verlag.

Gilbert, Scott F., Jan Sapp, and Alfred I. Tauber. 2012. A Symbiotic View of Life: We Have Never Been Individuals. *The Quarterly Review of Biology* 87 (4): 325–341.

Godfrey-Smith, P. 2013. Darwinian Individuals. In *From Groups to Individuals: Perspectives on Biological Associations and Emerging Individuality*, ed. F. Bouchard and P. Huneman, 17–36. Cambridge MA: MIT Press.

Handelsman J, Rondon MR, Brady SF, et al. 1998. Molecular biological access to the chemistry of unknown soil microbes: A new frontier for natural products. *Chem Biol* 5: R245–R249.

Hegel, Georg Wilhelm Friedrich. 2004. *Encyclopedia of the Philosophical Sciences, Part II. Philosophy of Nature*. Oxford\New York: Clarendon Press, Oxford University Press.

Holmes, Frederic Lawrence. 1974. *Claude Bernard and Animal Chemistry: The Emergence of a Scientist*. Cambridge: Harvard University Press.

Hutter, T., et al. 2015. Being Human Is a Gut Feeling. *Microbiome* 3 (1): 9.

Jonas, Hans. 1966. *The Phenomenon of Life. Toward a Philosophical Biology*. New York: Harper & Row.

Jones, S. 2013. Trends in microbiome research. *Nature Biotechnology* 31 (4): 277–277.

Kant, Immanuel. 2000. *Critique of the Power of Judgment*. Cambridge: Cambridge University Press.

Lowe, Ernest J. 2001. *The Possibility of Metaphysics: Substance, Identity, and Time*. Oxford: Clarendon Press.

Mallet, F., et al. 2004. The Endogenous Retroviral Locus ERVWE1 is a Bona Fide Gene Involved in Hominoid Placental Physiology. *PNAS* 101 (6): 1731–1736.

Maturana, Humberto R., and Francisco J. Varela. 1982. *Autopoiesis and Cognition: The Realization of the Living*. Dordrecht: Kluwer.

Mcfall-Ngai, Margaret, Michael G. Hadfield, Thomas C.G. Bosch, Hannah V. Carey, Tomislav Domazet-Lošo, Angela E. Douglas, Nicole Dubilier, et al. 2013. Animals in a Bacterial World, a New Imperative for the Life Sciences. *Proceedings of the National Academy of Sciences* 110 (9): 3229–3236.

Michod, R. 1999. *Darwinian Dynamics*. Oxford University Press.

Moran, N., and D. Sloan. 2015. The Hologenome Concept: Helpful or Hollow? *PLOSBio*. 13 (12).

Moreno, Alvaro, and Matteo Mossio. 2015. *Biological Autonomy. A Philosophical and Theoretical Enquiry*. Dordrecht: Springer.

Nicholson, et al. 2012. Host-gut Microbiota Metabolic Interactions. *Science* 336 (6086): 1262–1267.

Normandin, Sebastian, and Charles T. Wolfe. 2013. History, Philosophy and Theory of the Life Sciences. In *Vitalism and the Scientific Image in Post-Enlightenment Life Science*, vol. 2, 1800–2010. Dordrecht\Heidelberg\New York: Springer.

O'hara, Ann M., and Fergus Shanahan. 2006. The Gut Flora as a Forgotten Organ. *EMBO Reports* 7 (7): 688–693.

Parke, Emily C., Brett Calcott, and Maureen A. O'Malley. 2018. A Cautionary Note for Claims about the Microbiome's Impact on the "Self". *PLoS Biology* 16 (9): e2006654.

Pradeu, Thomas. 2016. Organisms or Biological Individuals? Combining Physiological and Evolutionary Individuality. *Biology and Philosophy* 31 (6): 797–817.

Prescott, Susan L. 2017. History of Medicine: Origin of the Term Microbiome and Why It Matters. *Human Microbiome Journal* 4 (June): 24–25.

Rees, T., et al. 2018. How the Microbiome Challenges Our Concept of Self. *PLOS Bio* 16 (2).

Robinson, H. 2004. Substance. In *The Stanford Encyclopedia of Philosophy*. http://seop.illc.uva.nl/entries/substance/

Roughgarden, J., S.F. Gilbert, E. Rosenberg, I. Zilber-Rosenberg, and E.A. Lloyd. 2017. Holobionts as units of selection and a model of their population dynamics and evolution. *Biological Theory* 13 (1): 44–65.

Roy, M, and M.-A. Sélosse. 2015. "Les Réseaux Mycorhiziens, Des Réseaux Mutualistes Entre Champignons et Racines Des Plantes." In Analyse des réseaux sociaux appliquée à l'éthologie et l'écologie, edited by C. Sueur, Éditions Matériologiques, 291–328.

Ruiz-Mirazo, Kepa, and Alvaro Moreno. 2013. Synthetic Biology: Challenging Life in Order to Grasp, Use, or Extend It. *Biological Theory* 8 (4): 376–382.

Sapp, Jan. 2009. *The new foundations of evolution: On the Tree of Life*. Oxford\New York: Oxford University Press.

Scadden, David T. 2014. Nice Neighborhood: Emerging Concepts of the Stem Cell Niche. *Cell* 157 (1): 41–50.

Sélosse, Marc-André. 2012. Symbiose et mutualisme versus évolution: de la guerre à la paix? *Atala* 35 (35): 49.

Sélosse, Marc-André, Franck Richard, Xinhua He, and Suzanne W. Simard. 2006. Mycorrhizal Networks: Des Liaisons Dangereuses? *Trends in Ecology & Evolution* 21 (11): 621–628.

Simon, Jean-Christophe, Julian R. Marchesi, Christophe Mougel, and Marc-André Sélosse. 2019. Host-Microbiota Interactions: From Holobiont Theory to Analysis. *Microbiome* 7 (1): 5.

Spallanzani, L. 1783. *Expériences sur la digestion de l'homme et les différentes espèces d'animaux*. A Geneve: chez Barthelemi Chirol.

Snowdon, P. 1995. Persons, Animals, and Bodies. In *The Body and the Self*, ed. J.L. Bermúdez, A. Marcel, and N. Eilan, 71–86. Cambridge, MA: MIT Press.

Strawson, Peter Frederick. 1959. *Individuals: An Essay in Descriptive Metaphysics*. London: Methuen.

Tipton, Laura, John L. Darcy, and Nicole A. Hynson. 2019. A Developing Symbiosis: Enabling Cross-Talk Between Ecologists and Microbiome Scientists. *Frontiers in Microbiology* 10 (February): 292.

Tripp, Erin A., Ning Zhang, Harald Schneider, Ying Huang, Gregory M. Mueller, Hu Zhihong, Max Häggblom, and Debashish Bhattacharya. 2017. Reshaping Darwin's Tree: Impact of the Symbiome. *Trends in Ecology & Evolution* 32 (8): 552–555.

van Baalen, M., and P. Huneman. 2014. Organisms as Ecosystems/Ecosystems as Organisms. *Biological Theory* 2014 (9): 357–360.

Varela, Francisco J. 1979. *Principles of Biological Autonomy*. New York: Elsevier.

———. 1997. Patterns of Life: Intertwining Identity and Cognition. *Brain and Cognition* 34 (1): 72–87.

West, S.A., A.S. Griffin, and A. Gardner. 2007. Social Semantics: Altruism, Cooperation, Mutualism, Strong Reciprocity and Group Selection. *Journal of Evolutionary Biology* 20 (2): 415–432.

Whipps, J., K. Lewis, and R. Cooke. 1988. Mycoparasitism and plant disease control. In *Fungi Biol Control Syst*, ed. M. Burge, 161–187. Manchester University Press.

Wiggins, D. 2001. *Sameness and Substance Renewed*. Cambridge University Press.

Wilson, David Sloan, and Elliott Sober. 1989. Reviving the Superorganism. *Journal of Theoretical Biology* 136 (3): 337–356.

Wolfe, Charles T. 2010a. From Substantival to Functional Vitalism and Beyond. *Eidos: Revista de Filosofía de la Universidad Del Norte* 14: 212–235.

———. 2010b. Do Organisms Have An Ontological Status? *History and Philosophy of the Life Sciences* 32 (2/3): 195–231.

———. 2017a. Varieties of Vital Materialism. In *The New Politics of Materialism : History, Philosophy, Science*, ed. John Zammito and Sarah Ellenzweig, 44–65. London: Routledge.

———. 2017b. Models of Organic Organization in Montpellier Vitalism. *Early Science and Medicine* 22: 1–24.

———. 2019. *La Philosophie de la biologie avant la biologie: Une histoire du vitalisme*. Paris: Classiques Garnier.

———. 2021. Vitalism and the Metaphysics of Life: The Discreet charm of Eighteenth-Century Vitalism. In *Life and Death in Early Modern Philosophy*, ed. Susan James. Oxford: Oxford University Press USA - OSO.

Wolff, Caspar Friedrich, ed. 1789. *Zwei Abhandlungen über die Nutritionskraft welche von der Kayserlichen Academie der Wissenschaften in Saint Petersburg den Preis getheilt erhalten haben. Nebst einer fernen Erläuterung eben derselben Materie von Caspar Friedrich Wolff*. St. Petersburg: Kayserl. Akademie der Wißenschaften.

Wu, Dongying, Philip Hugenholtz, Konstantinos Mavromatis, Rüdiger Pukall, Eileen Dalin, Natalia N. Ivanova, Victor Kunin, et al. 2009. A Phylogeny-Driven Genomic Encyclopaedia of Bacteria and Archaea. *Nature* 462 (7276): 1056–1060.

Zilber-Rosenberg, I., and E. Rosenberg. 2008. Role of Microorganisms in the Evolution of Animals and Plants: The Hologenome Theory of Evolution. *FEMS Microbiology Reviews* 32: 723–735.

Vitalist Arguments in the Struggle for Human (Im)Perfection: The Debate Between Biologists and Theologians in the 1960s–1980s

Victoria Shmidt

Abstract In this chapter, I explore and offer critical reflections on the widespread practice of attributing negative value to "vital forces" in debates on health and disease, as the direct result of the extensive dissemination of genetics and its implications since the late 1960s. This historical reconstruction focuses on the most heated debates in popular science periodicals and editions, having the longest-lasting public "echo," which have shaped an intergenerational continuity in the reproduction of vitalist arguments in discursive practices regarding health, disease, and their genetic factors.

Mapping attacks on vital forces as various forms of negation addresses three different debates in the historically interrelated repertoire of potentially rival approaches to health, disease, and their genetic components: (1) the attribution of negative value to primal instinct as an obstacle to the progress of human civilization; (2) the normative vitalism mainly associated with French philosophers George Canguilhem, Michel Foucault, and Gilles Deleuze; and (3) the movement for the deinstitutionalization of health care within the negative theology presented by Ivan Illich.

The reproduction of vitalist arguments in the each of the three realms is seen as a historical continuity of the medical vitalism that appeared in the Enlightenment and that produced a less monolithic and more conceptually coherent continuum of the positions regarding health, diseases, and their causes. In line with the Lakatosian division into internalist and externalist histories of science, I focus on the multiple functions of vitalist arguments: as a main force in the contest among rival theories regarding health and disease (as a part of the internalist narrative); as a signifier of the boundary work delineating science and not-science, whether labeled as theology or as "bad" science aimed at legitimizing science (as a part of externalist history); and as an ideological platform for bridging science and its performance in policies concerning reproduction .

V. Shmidt (✉)
SOEGA, University of Graz, Graz, Austria
e-mail: victoria.shmidt@uni-graz.at

217

C. Donohue, C. T. Wolfe (eds.), *Vitalism and Its Legacy in Twentieth Century Life Sciences and Philosophy*, History, Philosophy and Theory of the Life Sciences 29, https://doi.org/10.1007/978-3-031-12604-8_12

Keywords Vitalism · Internal and external histories of biology · Reproduction ·
Georges Canguilhem · Ivan Illich

1 Introduction

In the late eighteenth century and throughout the nineteenth century, the classical
epistemological status of vitalist powers as "extra-causal agents powering the living
body" (Wolfe 2008, 461) began to be transformed through reinforcing epistemological skepticism as a driving force for reproducing vitalist arguments (Benton
1974, 18,20). Observing that physiological functions as "too intimately intertwined
to be quantified according to purely mechanical laws of force and motion" (Wolfe
and Terada 2008, 552), this launched the "systematic negation of strictly linear causality in explaining health and disease." (ibid) The shifting ontological status of
vitalist arguments determined the split in accepting the potential (and indisputable)
cognizance of vital phenomena and claiming the necessity of such argumentation as
a "covering law." (Osborne 2016) Vitalism began operating as "an intermediate,
pragmatic position located in between two metaphysically rigid extremes," (Wolfe
and Terada, 2008 543) namely, mechanism and animism.

The "irreducibility of the knowledge about vital forces to a nomothetic explanation" (Huneman 2008 619) became a shared assumption for the various movements
within medical vitalism, which fixed its role for the next generation of thinkers who
theorized health and pathology in the contexts of evolutionary biology and genetics.
Moreover, the animal economy, a defining concept of medical vitalism (Wolfe and
Terada 2008 548), introduced the active language of function and of usage, which
was opposed to the static, atemporal character of purely anatomical approaches
(ibid 549).

Along with focusing on the complexity of the connection between actions and
efficient causality, medical vitalism started differentiating irritability and sensitivity,
which not only assisted in avoiding reference to a soul in the debates about the
causes of mental diseases (Huneman 2008 617), but also precipitated a transition from a rationally devised vision on the developing organism to the investigation
of an irrationally developed entity (Benton 1974 22,35). In terms of agential thinking (Okasha 2018 18), this trend nurtured a powerful platform for attributing to
vitalist forces both negative and positive value to an organism.

The established interconnection between a skeptical vitalist view of explaining
disease and the ascendency of a "cautionary" view on vital forces as potentially
dangerous may be viewed as the primary tool for the survival of vitalist arguments
and their conceptual continuity (Shan 2020 138) in theorizing health and pathology
in the twentieth century. This text brings into focus the strong historical connection
between the arguments introduced by medical vitalism, and their applications, and
the arguments developed by biologists and theologians in the 1960s and 1980s.

In the late 1960s, ascribing destructive value to vitalist forces operated as a central assumption in different views on health and pathology and their role in human

evolution. Anne McLaren (1968), one of the most publicly recognized reproductive biologists in the late 1960s, underscored in her preliminary article for the special issue of *Contact*, a journal affiliated with the British and Irish Association for Practical Theology that :

> Even after millions of years of evolution, frogs return to the water to spawn; seals return to dry land to breed. Although most people seem reasonably content with the way in which we reproduce our species at present, recent advances in biological understanding make it certain that we shall soon be able to exercise much greater control over the processes of reproduction. (McLaren 1968 13)

Georges Canguilhem (1988), a philosopher and physician, asserted: "In distinguishing biological variety from negative vital value, we have, on the whole, delegated the responsibility for perceiving the onset of disease to the living being itself."[1] Ivan Illich (1974), a philosopher and priest, challenged such a view: "[We] adapt to the stress of the second industrial revolution and over-population ...this kind of survival with fear adaptability is also a heavy handicap: the most common causes of disease are exacting adaptive demands." (Illich 1974 83)

I explore and offer critical reflections on the widespread practice of attributing negative value to vital forces in debates on health and pathology, as a direct result of the extensive dissemination of genetics and molecular biology and its implications since the late 1960s. The migration of genetics into the practice of reproductive politics on national and global levels had played a central role in reinventing the well-known debates from the interwar heyday of eugenics in the twentieth century about the criteria for health and disease, normality and abnormality. These debates around the dichotomy between the individual and the social space put into question the future of humanism and consequently the future of religion (Deane-Drummond 2012, 29). The exchange between scientific and theological arguments regarding heredity became a leading forum for shaping the variety of views on eugenics and its measures, as in the early 1930s. Since the 1960s, along with the flourishing of human genetics and its recruitment to various policies, these debates began to involve the most prominent representatives of both camps and to be widely disseminated. While Wolfe and Terada have called for attention to the recurring features of vitalism as non-doctrinaire, and the coherence that occurred in the late eighteenth and nineteenth centuries (Wolfe and Terada 2008, 542), this text aims to provide a philosophically informed historical reconstruction of vitalist arguments since the 1960s, as a construction mobilized for solving multiple issues concerning genetics and its social and ethical implications. These debates between theologians and biologists reflect the most significant turns in the resurrection of vitalist arguments after the Second World War.

---

[1] Georges Canguilhem (1991) *The normal and the pathological* p. 155, cit. in Stuart Elden (2019) *Canguilhem* Cambridge Polity p.25.

2 Methodology

This text focuses on the most heated debates published in popular science periodicals, which have the longest-lasting public "echo," and which have shaped intergenerational continuity in the reproduction of vitalist arguments in discursive practices regarding health, disease and the contribution of genetics . It addresses three streams of debates at the international level that focused on vital forces as decisive risks for humanity and its collective health between the late 1960s and 1980s: (1) the denial of "quasi-animal" reproductive patterns of "backward" people and societies, which connected human genetics with the longue durée of developmental idealism, and reinforced the division between doctrinal (systematic) theology and various theological interpretations of genetic arguments and their implications; (2) the deconstruction of the binary opposition between the norms of health and pathology in the approach invented by Georges Canguilhem and his successors, Michel Foucault and Gilles Deleuze; and (3) the rejection of genetics as a part of modern medicine in the movement for deinstitutionalization spearheaded by Ivan Illich and his supporters. The historical reconstruction of these debates provides the methodological grounds for answering the question: "How did the argument stemming from ascribing negative value to vital forces operate and develop in different cultural and political contexts?" Such reconstruction is motivated by the mission to move from the general and almost non-historical concept of "knowledge circulation" to a philosophically informed historization of medical science.

I reconstruct the history of vitalist arguments through a Lakatosian division of the history of science into internalist ("repertoires of thought, action, and technique, that shape the organisation of scientific inquiry") (Knorr-Cetina 1999) and externalist (social, cultural, economic or political conditions make scientific progress possible) narratives (Lakatos 1970). The scientific "career" of vitalist arguments is embedded in the struggle of medical science for its legitimacy, often sought by rearranging of the dichotomy between holism (or organicism), and mechanistic models of explanation.

This struggle cannot be attributed wholly to either the internalist or externalist history of medical science. In the 1960s, the rapid rise of human genetics confronted an increasing contest between Mendelian and Lamarckian explanatory schemes to integrate the role of the environment in development and evolution (Koonin and Wolf 2009). This struggle included a whole range of possible arguments, including a long-lived theoretical and empirical rivalry between major research programs in human genetics (the internalist history, in Lakatosian terms), as well as various efforts aimed at rearranging the boundaries between genetics and eugenics in favor of legitimizing genetics (the wholly externalist history). Introducing vitalist arguments into this contest diversifies the models that explain the interconnection between nature and nurture. I connect the arguments for ascribing negative value to vital forces (internalist history) to their operation in the debates between geneticists and theologians that addressed the societal role of genetics (externalist history). The reproduction of vitalist arguments in these debates fixes the multifunctionality of such arguments.

The Bhaskarian taxonomy of negation into "real" (which creates a distance between animals and humans in their patterns of reproduction), "transformative" (which reconstructs the pre-existing binary opposition of health and disease into a scale for measuring the degree of normality and pathology), and "radical" (which welds biology with a well-disseminated, exceptionally positive view on human adaptability as a driving force of institutionalizing medical assistance in favor of overcoming such dependency) types, which maps *the internalist narrative* of ascribing negative value to vitalist forces. Mapping the attacks on vital forces as various types of negation joins different debates in the historically interrelated repertoire of potentially rival approaches to health, pathology and the genetic contribution to health and disease.

This continuum connects the operation of vitalist arguments with their transformation in the nineteenth century and the challenges produced by rooting genetics in various practices of demographic policy since the 1960s. Since Bhaskar accepts *real negation* as the necessary ground for developing transformative and radical negation, the continuum of these approaches is seen as a reflection of the dialectics of a skeptical vitalism, which "takes an opportunity to develop to a systematic methodology aimed at testing the new facts and implications by multiple criteria." (Benton 1974 18–20) Rather than being separate efforts in negating vital forces, their interconnection had become of the grounds for adaptation, for example, in the countries of Eastern Europe.

The imperative to revise the social and political contexts of genetics, or its *externalist history,* will be explained as a feature of the "control society," a social order recognized as predominant, and which has replaced the previous "disciplining society" since the 1960s. Technoscientific development, including the achievements of biomedicine, a core driving force of the transfer to a control society, had shifted from the previous practices of forming individuals through placing them into institutional spaces of confinement by "modulating" certain "individuals," to defining individuals as a part of "samples, data, markets or "banks'" (Patton 2018 205).

Playing a significant role in establishing the first biopolitical strategies of the eighteenth century, medical vitalism had populated the models that problematized the division between the "normal" and the "pathological" through various arguments targeted at the criteria and signifiers of health and disease (Rabinow and Rose 2006). The transfer from a disciplining society to a control society that called for rearranging the boundaries between genetics and eugenics began to be viewed as an agent of control society. The non-scientific production of knowledge provided an argument for or against a control society in which "tacit assumptions about the contents of science are forced to become explicit." (Gieryn 1999 24) Paul Rabinow and Nikolas Rose (2006 211) underscore that "contemporary biopolitics operates according to logics of vitality, not mortality: while it has its circuits of exclusion, letting die is not making die."

Thus, I trace the role of vitalist arguments in the boundary work engaged by biologists and physicians on the one side and theologians on the other. The goal is to learn "where they may not roam without transgressing the boundaries of legiti-macy" (Gieryn 1999 16) of their position regarding the control society. I explore the operation of vitalist arguments in various forms of boundary work[2] through decon-structing multiple analogies between organism and society or, more generally, through connecting vitalism and naturalism as signifiers for specifying the contexts of adapting vitalist arguments within societies that experienced various degrees of geopolitical dependence.

3   The Longue Durée of Developmental Idealism: In Search of the Best Path Toward Human Perfection

Since the end of the eighteenth century, recognizing reproduction as a "conservative pattern for all [that is] vital" (McLaren 1968 13), produced a particular view of human beings. It did so through the lenses of controlling reproduction and the abil-ity to engage in the self-control of reproduction as the next stage of evolution toward becoming more human: "If conception were to become a deliberate act in its own right, it would less often be undertaken irresponsibly. Perhaps conception might then be awarded the same moral status as birth and death receive at present." (ibid 15).

During the interwar period, the dissemination of eugenic ideas and politics aim-ing to distance people from "backward" reproductive patterns attained the function as a sustainable supra-national ideology for establishing a global agenda of repro-ductive politics. These ideas and practices received their "second wind" in the 1960s within "genetic demography," an interdisciplinary program for applying the achieve-ments of human genetics to reproductive policies.[3] Along with the revision of the genetic causes of disease, health, and fertility, genetic demographers struggled with the inevitable stigmatization of their explanatory schemes (and of themselves) due to the close connection of those schemes with eugenics.

Managing the stigma of eugenics utilized various methods for reestablishing the boundaries between "good" and "bad" science, from the blatantly ahistorical oppo-sition of "sober" eugenics and "perverted" *Rassenhygiene* among the geneticists who cooperated with German scholars, including the most prominent members of

---

[2] Vitalism had already taken on a primary role in boundary work in the end of the eighteenth cen-tury, when psychiatry appeared during the systematic rearrangement of the boundaries among medical, religious, and judiciary approaches to insanity. See more in Philippe Huneman (2008) Montpellier vitalism and the emergence of alienism in France (1750–1800): the case of passions. *Science in Context*, 21 (4), pp.615–647: p.616.

[3] The rise of scientific warnings regarding reproduction had started in the early 1960s and was debated within a few conferences: Man and his future (G. Wolstenholme, ed.) Ciba Foundation Symposium, 1962; Control of human heredity and evolution (T. M. Sonneborn, ed.) MacMillan, 1965; and Biological aspects of social problems (Meade and Parkes, eds.) Plenum Press, 1965, which document many other ideas and references to primary literature.

NSDP (Koch 2004 317), to attempts to reform eugenics as a useful tool for bringing together the social and biological realms of degeneration as processes important to understanding the vicissitudes of evolution (Ramsden 2006 7). This boundary work aimed at distinguishing genetics from eugenics as "the less reliable, less truthful, less relevant source of knowledge about natural reality." (Gieryn 1999 16) This interpretation of eugenics did not mean to label it "ersatz," but sought to extend its frontiers.

Such an *expansion of scientific boundaries* resonated with the evolution of the complicated history of eugenics as an agent of the disciplining society and its very palpable presence in the history of institutional violence against disabled people and minorities, to the more refined methods of control over reproductive behavior offered by genetic demography where, "Limiting population in the interests of national economic prosperity does not operate according to the biopolitical diagram of eugenics, and is not the same as purification of the race by elimination of degenerates." (Rabinow and Rose 2006 210) The duality of this evolution was reflected in the two paradigms of genetic counseling: "genetic cleansing" and "informed choice" (Clarke 1997), which were disseminated with the aim to regulate reproductive behavior among less socially responsible societies that needed to remain under surveillance and more advanced social groups that required intermediate forms of social control that avoided the centralization of government (Dean-Drummond 2012 106).

The growing controversy over nature vs. nurture added to the role of vitalist arguments in revising the connection between genetics and eugenics and their implications. In his publications for the public, Joshua Lederberg called for reintroducing vitalist argument in the name of saving progress and avoiding its risks. Lederberg recognized in the mechanistic understanding of heredity (though the inevitable output of the progress of molecular biology) the threat of a monolithic and sophistic rationalization of fundamental human policy and the decay of medicine wisely dedicated to the welfare of individual patients (Lederberg 1966 519). The solution offered by Lederberg and his prominent colleagues was "euphenics", an explanatory scheme that sought to rival both eugenics and Mendelian biology. Constructing disease as "any deficit relative to a desired norm" adherents of euphenics relied upon epigenetics to produce "the healthy genotype as the product of communicating between environment and humans." (ibid 521)

The mission of euphenics would be to maximize the efforts aimed at treating individuals affected by genetic diseases through all possible means for ameliorating genotypic maladjustment. Bringing forward the role of adaptability in coping with genetic diseases brought this question to the fore: "How to identify the most adaptable genotypes now living and what is the price, to the detriment in special skills of this adaptability?" (ibid)

Euphenics relied on the "match of pragmatic expectations of the milieu of the individual and his descendants." (ibid 524) Generally, euphenics was seen as a remedy for all the applications that totally failed within the framework of eugenics, such as genetic counseling. Even more, prioritizing euphenics should prevent the "grave danger that the minority view will lead to a confusion of the economic and social aims of rational population policy with genocide." (ibid) Remarkably, the main concern regarding the risk of abusing genetic implications was "a disastrous impediment to the adoption of family planning by just those groups whose economic and educational progress most urgently demands it." (ibid 530)

Despite the constant rivalry with eugenics, euphenics functioned as a kind of developmental idealism, a concept used as an umbrella notion for the complex of ideas, politics, and practices, including population genetics, which directly link the ideal of "modern industrial and urban society with high levels of education, wealth, and health" (Thornton et al. 2012) and proper demographic policy, "considered under four headings: decrease, increase, improvement and substitution." (McLaren 1968 15) The negation of primal instinct as insufficiently human for progress and its risks achieved all the qualities of real, basic, negation in terms of the Bhaskarian approach. It attacked the absence of consciousness regarding primatal instinct and launched the process of distancing from uncontrolled reproduction (Bhaskar 1993). Negating primal instinct as vital but destructive became one of the main responses to the duality of attributing to humans the immanent qualities of social life, especially culturation and biological evolution. Duality in understanding humans as both animals and God's children has shaped the Christian version of developmental idealism.

Developmental idealism, a secular Enlightenment paradigm arising from the European religious context (Nickolson 1990), has been accepted by the Roman Catholic Church, despite the rhetoric of the geneticists who often opposed scientific collective responsibility to the selfish individualism of Christian mercy regarding those individuals considered "useless." Well-known Scottish geneticist Patricia Ann Jacobs (1968) simplified the complicated history of the relation between genetic and theological arguments in favor of achieving efficient control over human reproduction:

> In discussions on suffering and the love of God, there is a traditional view that we need our weak brethren to develop our attitudes of compassion and thus become spiritually refined. To many scientists, such a view seems of monumental and repulsive selfishness… Many ethnic groups have exposed their defective babies after birth, but now genetic knowledge can predict and detect many such children in good time and take remedial action. (Jacobs 1968 20)

Along with the proliferation of eugenic movements since the end of the nineteenth century, those associated with Christian churches embraced a wide range of positions in relation to the debate over strict control over reproduction. While the attempts to resist eugenic movements from the side of Christian associations and clergy were fragmented and based upon ethical concerns,[4] the acceptance of eugenic ideas among theologians, priests, and, even more, elite clergy established an

---

[4] An entirely ethical argument against sterilization in the United States and euthanasia in Nazi Germany stemmed from the idea of "natural rights" and violation of such rights through applying negative eugenic measures; see more in Leon, S. (2004). "A Human Being, and Not a Mere Social Factor": Catholic Strategies for Dealing with Sterilization Statutes in the 1920s. *Church History, 73*(2), 383–411.

intersectionality between theological arguments and eugenic ideas that transformed Christian epistemology. Despite the historically controversial bifurcation of negative and positive eugenics between Protestants and Catholics (Gillette 2014), Christian epistemology tied together the various driving forces that determined the wide range of strategies to put together religious and genetic arguments in favor of humanity as an exceptional quality, to be protected from the barbarity of those who remained more animal than human.

Opposing humans and animals fit well with the racial hierarchies that often included the distance between a particular "race" and the ideal of humanity as an argument to justify negative eugenics: "The races are different and, measured against certain ideals, may be more or less capable of striving towards these ideals."(Muckermann 1934 29) Agostini Gemelli and Hermann Muckermann, two Catholic priests and prominent eugenicists, embraced mutually polar positions regarding forced sterilization (Dietrich 1992),[5] but both consistently developed an argument in favor of limited dynamic and vital racial "intermixtures," in contrast to widespread negative opinions concerning intermixture with "aliens," which were perceived as dangerous to the interest of "domestic raciality" (*Heimrassigkeit*) or Italian "purity" (Turda and Gillette 2014).

This motive dovetailed with the prioritization of improving the health of the nation. Muckermann explained the necessity to introduce forced sterilization, defined by him as "un-fertilization" (*Unfruchtbarmachung*), by referencing the vital interests of the German people: "Ethical concerns are therefore out of the question, because consideration for the common good requires this addition to the existing measures to maintain the health and work endurance of the people." (Muckermann 1934 124) The mission to prevent hereditary diseases (*schweren Erbleiden zu ver-hüten*) inclined Muckermann to weld together not only neo-Lamarckian and Mendelian explanatory schemes for pathologies, but also to also label hereditary factors as given by Mother Nature (ibid 135), all in favor of stressing the impossibility of preventing so-called malformation in ways other than forced sterilization.

Remarkably, he built this argument through the division of families into "unnatural large families" (*unnatürliche Großfamilie*), "unnatural dwarf-families" (*unnatürliche Zwergfamilie*) and "true-to-nature normal families" (*naturtreue Normalfamilie*) (ibid 136). Only the last category was acceptable, able to bring religion and eugenic arguments together in mutual accordance in favor of a sustainable practice of health (ibid 157).

Blurring the boundaries between genetic and religious arguments for promoting developmental idealism had become a long-term trend, along with the regular attempts of the Roman Catholic Church to reestablish Christian views on reproduction and the ways to intervene in it. Between 1930 and 1995, the Catholic Church issued three different Encyclicals aimed at limiting the dissemination of

---

[5] Donald J, Dietrich (1992) Catholic Eugenics in Germany, 1920–1945: Hermann Muckermann, S.J. and Joseph Mayer Journal of Church and State, 34 (3): 575–600.

eugenics and genetics-based politics.[6] Nevertheless, the preparation of these documents was accompanied by debates among Catholic experts that blurred the unity of the Catholic position regarding the regulation of reproductive policy. For the Catholic Church, the argument developed by the Methodists and accepted by Catholics (Ramsey 1970), shifted slightly from the non-acceptance of eugenics and genetic improvement as a dangerous practice of "playing God," to moderate acceptance of the idea which posits "ourselves as in some sense co-creators with God." (Deane-Drummond 2012 30) Seeing humans as "created co-creators" (Hefner 1993 37) reinforced the view on genetic medicine as a possible ally in the struggle by the Church for healthy "normal" families, putting the Roman Catholic Church in the position of "theologian interrogation of genetic science or conversation partner." (Deane-Drummond 2012 29). In his salutary exhortation to the participants of the Twentieth International Conference on "The Human Genome," Cardinal Javier Lozano Barragán claimed:

> [W]e will start from the consideration of the genome as the structural element that orga-nizes the human body in its individual and hereditary dimensions; it involves the set of genes, but it goes further to embrace all the other elements that together with genes consti-tute the original energy to develop through all existence and which signify the key mystery of human life. (Barragán and Lozano 2005)

This fundamental dualism, "which also positioned the Roman Catholic Church unconditionally on the side of the good," was reinforced in a UNESCO report on culture and development (1995), which claimed "religion appears to be a resurgent force in human affairs today." (Sjørup 1999 389)

One of the most visible outputs of this trend has been reestablishing the mission of Catholic Church (Deane-Drummond 2012 46) to balance the priority of the truth of scientific knowledge with the ideal of humanity as dependent on God. Since the late 1960s, in scientific and Catholic versions of developmental idealism, the uncon-trolled instinct toward reproduction began to be viewed as similar to the risks of nuclear war and other ecological hazards directly linked with population growth (Castel 1973; Ornstein and Lederberg 1967).

In 1968, the editor of *Contact* wrote in an introduction to a special issue aimed at shedding light on the contest between scientific progress and the threat of non-civilized reproductive patterns that, "If we fail to control population size we may expect not only increasing dangers of famine and damage to our natural environ-ment but a far-reaching erosion of privacy and self-respect." (Contact 1968 13) The eschatological theme stemming from this non-doctrinal theology was balanced by the new official Catholic rhetoric on the interpretation of the sexual act as "a perfor-mance of love and dignity, valued and irreplaceable part of harmonious parenting and family life,"[7] part of the systematic "transformation of natural law under the

---

[6] Encyclical Casti Connubii, 1930, by Pius XI, opposition to eugenics especially forced steriliza-tion; Humanae Vitae, 1968, by Paul VI, the regulation of birth, reproductive technologies; Evangelium Vitae, 1995 by John Paul II, sanctity of human life.

[7] See development of the argument in Evangelium Vitae, 1995.

new idea of love" (Deane-Drummond 2012 32) begun in the 1970s. This started the processof "develop(ing) theologically informed ethics which would be accepted by people." (ibid 35)[8]

This framework continues to underpin developmental discourse on religion, in contrast to a situated knowledge about religion "which is not blurred by Eurocentric universalist epistemologies." (Sjørup 1999 406) But what are the options for negating these epistemologies for those who implement them? George Canguilhem and his colleagues focus on the binary oppositions of health vs. disease and norms vs. pathology in order to transform the approaches to the human individual and space in favor of balancing the interests of society and individuals, and medicine as a mediator between them.

4   Normative Vitalism: Transformative Negation of Medically Proven Imperfection

In his critical reflections concerning the strict opposition between health and disease in Western medicine, Georges Canguilhem aims at deconstructing the accusations of medical professionals concerning the faults in theories and the institutional violence resulting from such faults: "Is this a question of biology or parasitism of biology?" (Canguilhem 2009 72)[9] This question stems from the personal experience of Canguilhem, who had participated in the French Resistance and who grasped the interrelation between European medical traditions and Nazi medicine. Canguilhem introduced the concept of "scientific ideology" in order to reconnect internalist and externalist histories of sciences, or more precisely the history of medicine, to clarify the interrogation between science and society. To him, scientific ideology operates as "aligned with an already instituted science whose prestige it recognizes and whose style it seeks to imitate" (Canguilhem 1988 44), but generates its own norms of "scientificity" (Elden 2019 132). Science remains connected to and infiltrated by one or another scientific ideology.

Through his historical inquiry, Canguilhem traces the interrogation between science and scientific ideology to elucidate the role of global societal changes in shaping the research agenda for biology. Canguilhem explains the acceptance of the Malthusian view on population politics by Darwin through the transformation from

---

[8] The dramatic struggle between the adherents of sexual pessimism as the only suitable position for the Roman Catholic Church and the promotion of authentic love as the only reason for sexual relationships remains beyond the aims of this chapter, but it is remarkable that sexual pessimism did not prevail in this rivalry due to the systematic rejection of biological miracles typical for the cults of Mary and other streams that shaped the agenda of sex as a performance of God's love. See more in Uta Ranke-Heinemann *Eunuchen für das Himmelreich Katholische Kirche und Sexualität* Hoffmann und Campe Verlag Hamburg 1988.

[9] Here and further all references to the writings by Canguilhem are derived from Stuart Elden (2019) *Canguilhem* Polity Press.

agrarian capitalism to industrial capitalism, as the main historical context for the scientific ideology of the "struggle for life," with its imperative of free competition (Canguilhem 1988 100). Other remarkable examples of the interconnection between biology and scientific ideology include a class-based explanation of division into neo-Lamarckism and Mendelism that Canguilhem had brought into focus. He underscores that Lysenkoism was correct to suggest that it was possible to change the hereditary constitution of the organism, but he does not accept the political arbitrariness with which the authorities dealt with geneticists who were labeled as the enemies of the Soviet people. In a quite provocative way, Canguilhem defines Mendel "as the head of retrograde, capitalist, and idealist biology." (Canguilhem 2009 149) Expressing such an extreme view on the division into neo-Lamarckism and Mendelism reflects not only Canguilhem's political affiliation, but also his intention to support an explanatory scheme that remained in the shadows for political reasons.

It is plausible to argue that Canguilhem accepted the call to emancipate biology from wide range of scientific ideologies as a way towards the *protection of autonomy*: a "kind of boundary-work results from efforts of outside powers, not to dislodge science from its place of epistemic authority, but to exploit that authority in ways that compromise the material and symbolic resources of scientists inside." (Gieryn 1999 17) The task to emancipate biology from scientific ideology is complicated, according to Canguilhem, because "biology has taken concepts from politics, and it may be that here politics is simply taking them back."(Canguilhem 2009 98)

Gieryn compares the efforts of scientists to prevent science from becoming "a hand-maiden to political or market ambitions" with establishing "interpretative walls to protect their professional autonomy over the selection of problems for research or standards used to judge candidate claims to knowledge." (Gieryn 1999 17) The bulk of Canguilhem's reflections can be interpreted as a kind of wall-building aimed at emancipating biology from one of the pillars of developmental idealism, the overarching contradiction between health as the norm and disease as pathology. The role of science in the appearance and disappearance of scientific ideologies remains a primary framework for critical writings by Canguilhem, with vitalism as a driving ontological assumption that would redraw boundaries between what medical scientists do and "consequences far downstream – the possible undesired or disastrous effects of scientific knowledge." (ibid) As an alternative to "the susceptibility of medicals to 'error', sickness and pathology," (Osborne 2016) Canguilhem redefines the processual dimension of human life because life itself invents and invests in norms.

Canguilhem may be aligned with the long and extended tradition of French naturalism, namely, providing analogies between physiology and medicine on one side and sociology and social science, more broadly speaking, on the other. Even more, Giuseppe Bianco (2013) explains the transfer of Canguilhem from rigid, rational, and mechanistic Cartesianism to a particular kind of vitalism through his involvement in the contest between two rival traditions informed by vitalism, the philosophy of concept and the phenomenological subjective analysis of experience. But which of these traditions prevailed in Canguilhem's thought? It is noteworthy that the answer varies depending into which philosophical and historical contexts Canguilhem is situated.

Bianco focuses on the role of Alain's philosophy in Canguilhem's devotion to vitalism to highlight his critical response to the tradition of French naturalism and specifically to the multiple analogies between the human body and the body of the state, since both practice a hierarchy of obedience to certain social laws, which, in turn, obey biological laws. Canguilhem's position is described as only partially accepting of such parallelism, operating "as an instrument devoid of an intrinsic teleology but with a teleology given to it by man (Bianco 2013 265)." Pasquinelli (2015), who claims that German *Lebensphilosophie* and the catastrophe of Nazi *Staatsbiologie* is the only context for understanding the intransigent position of Canguilhem against norms, recognizes that "Canguilhem is keen to deny any dominant function by a superior power and he recognizes the logical primacy of the abnormal over the normal." (ibid). Exploring the normative vitalism by Canguilhem as *transformative negation of destructive vital sources* can add nuance to this debate.

Canguilhem's rejection of violent intervention in biological laws by the means of social norms stems from the intention to "not change the powers, but rather render them rational" (ibid 260), and perfectly aligns with the core of transformative negation, namely, negating the source of the existential incompatibility between the elements of the totality, but not the totality itself (Bhaskar 1993 58). Since transformative negation often considers previously established stratifications as giving rise to internal contradictions, it seeks to transform not the whole subject of negation but that particular element in which change is decisive for overcoming the contradiction (Creaven 2002 86).

Canguilhem seeks to transform the opposition between health as a "normal" status and disease as a manifestation of pathology into a more refined scale of health/disease statuses, which he considers a demandable tool for the progress of medical knowledge and its implications. Canguilhem embedded this task in the mission to revise the impact of physiology producing knowledge of the body as a machine, as one of the primary grounds for institutionalizing medicine (Bianco 2013 256).

In this way, he follows the doctrine by Francois Broussais, whose systematic critique of the nosological viewpoint of Canguilhem, together with Michel Foucault, recognized the grounds for the systematic deconstruction of an essentialist view on health and pathology (Williams 1994). Canguilhem first developed his argument in 1943 in his MD thesis, *The normal and the pathological*, in the late 1960s. He revised his text as a critical reflection on the growing influence of developmentalist agenda of medical science. Apparently, Canguilhem used all the dialectical tools available to him to resolve the internal contradiction[10] between accepting all that is vital and attributing negative value to some of its manifestations, such as well-known diseases from Broussais' time (Williams 1994 169).

In line with the Broussainian holistic intention to view "human beings in the lack of absence of health" as a performance of variation, Canguilhem directly defines biological normativity "as a response on the absence of impossibility of [biological] indifference." (Canguilhem 1991 105) The idea of normativity serves the dynamic concept of health significant to Canguilhem, who rejects perfect health in favor of understanding health as organic well-being.

Instead, through two extremes, norm and pathology, Canguilhem divides anomalies into varieties (minor deviation), structural defects (deformations of a more significant kind), "heterotaxy" (complex structural shifts that nonetheless do not impede function), and monstrosities (both complex in appearance and serious in impairment) (ibid 109). This scale not only prevents thinking about anomaly in terms of general ideas on harmful or, even more, mortal deformities, but also separates extreme cases of monstrosity and places them in opposition to anomalies. While deviations are seen within normativity, the monstrous "holds to a different logic, a chaos of exception without laws, an imaginary, murky and vertiginous world." (Canguilhem 2009 184)

Ascribing monstrosity with rarity reverberates with Canguilhem's expectations that medical scientists and practitioners describe phenomenon in objective terms, as well as the impossibility to pass judgment based on a purely objective criterion: "only varieties or differences can be defined with positive or negative vital values." (Canguilhem 1991 200) Lorna Weir (1998) explores the strengths and weaknesses of this conceptualization for prenatal diagnosis and achieving more sensitivity toward the needs of mothers.

While, according to Weir, Canguilhem's approach "serves as a corrective to ideological criticism of the life sciences which fails to have a concept of cultural differentiation," it fails with regard to "the understanding of the interactions between scientific and non-scientific culture." (Weir 1998 256) Weir observes that biomedical rationality absorbs and subdues cultural contexts, instead of individualizing medical judgement in favor of the autonomy of patients, through the claims of physicians and scientists shaped by the concurrent medical ideology.

Claude Debru explains that there exists an antinomy between normativity and autonomy and the inevitable development of disease through which normativity and the dynamic concept of health have ceased to matter (Debru 2011 7). Ivan Illich had

---

[10] According to Bhaskar, "internal contradiction" refers to a double bind or self-constraint because of mutually excluded expectations of the systems in which subject tries to perform. (Bhaskar, Roy. Philosophy and scientific realism. In *Critical realism: Essential readings*. ed. M.Archer et all, 16–48. London: Routledge.) Due to his negative view on possible institutional violence within medicine, Canguilhem could not accept real negation of any vital force, even as it posed a threat to further human progress. Moreover, he remained unpersuaded regarding the totality and polarity of life, which excluded any opportunity to evaluate particular vital forces either negatively or positively.

offered a more fundamental explanation for the loss of autonomy by contemporary people: "[Modern] man is the animal who has lost his instinct and it desperately dependent upon such extra-genetic, outside-the-skin control mechanism without which human behaviour would be ungovernable." (Illich 1974 88)

5 Vital Human Imperfection: Unending Vulnerability vs. Lethal Adaptation

Ivan Illich, one of the most radical critics of the institutions of modern society, starts attacking the submission of medical science to the task of controlling people immediately after developing his campaign for "deschooling" people whom he saw as "the victims of an effective process of total instruction and manipulation once they are deprived of even the tenuous pretense of critical independence." (Illich 1970 11)

Illich recognizes medicine as another agent of "paralyzing life" - through impeding people "to search perfect health" (Illich 2017 287).[11] Historian Barbara Duden explains the intervention of Ivan Illich with the debates regarding human health and disease through a particular moment in history when one agency, namely medicine, achieved a monopoly over the social construction of bodily reality (Duden 2002 222). Illich caustically suggests that all Christians would complete their daily prayer by saying: "Lead us not to succumb to diagnosis, but deliver us from the evils of health." (Illich 2017 290)

His call for demedicalization may be recognized as an *expulsion*, boundary work that opposes "orthodox science against heterodox, mainstream against fringe, established against revolutionary."(Gieryn 1999 17) For Gieryn, expulsion does not aim to challenge or attenuate the epistemic authority of science itself, but rather to deny privileges to the space of science. Illich aims to deny the monopoly of medicine in ordinary life in favor of "autonomous confidence in biological vigour, the wisdom of traditional rules, and the compassion of neighbors." (Illich 1974 92) The initial argument against medicine, an inevitable transformation of "the human being who needs healthcare in a subsystem of the biosphere, an immune system that should be controlled, regulated and optimized, as 'a life,'" (ibid 107) moves far beyond the traditional opposition of mechanism vs. vitalism and rearranges the already mentioned positions regarding health, disease, and human perfection.

In contrast to mainstream Christian theology and aligned with the Kierkegaardian tradition of negative theology, Illich places the complicity of the human and animal nature of "man" as a product of civilization at the center of his argumentation. For Illich, the fundamental distinction between animal and human health is the positioning of pain, impairment, and death as a gift of human culture (Illich 1974 90). He explains the unique role of pain in the human experience by its function to give rise

---

[11] *The obsession with perfect health* was written in the early 1980s but translated into English and published recently.

"to a cultural program whereby individuals could deal with reality in those situations in which reality was experienced as inimical to the unfolding of their lives." (ibid 94) Besides the anxiety over the nature/culture boundary, personal attention and shared interpretation equip the individual to make pain tolerable, sickness understandable, and the life-long encounter with death meaningful. Illich focuses on the social and political functions of facing and experiencing pain to set limits on humanly constructed abuses when these became intolerable. However, "the modern cosmopolitan medical civilization denies the need for man's acceptance of pain, sickness and death." (ibid 90)

Illich explains the sustainable struggle of medicine against pain with the intention to gain power: "Pain only the part of human suffering over which the medical profession can claim competence or control."(ibid 96) By ceasing to conceive pain as a "natural" or "metaphysical" evil in modern society, medicine had become, for Illich, a main driving force of separation from self-regulation through emotions: "grief, guilt, sin, anguish, fear, hunger, impairment, and discomfort." (ibid) Replacing personal responsibility for pain by medical control triggers an atomization of the social order: "Compassion turns into an obsolete virtue. The person in pain is left with less and less social context to give meaning to the experience overwhelming him." (ibid 96) The efforts of medicine to present pain, impairment, and death as abnormal make people dependent on seeking medical treatment.

This pessimistic view on modern medicine is exemplified by another of Illich's key stances on becoming human, through discovering "a particular program ...to conduct themselves in their struggle with nature and neighbor." (ibid 105) This entirely ironic paraphrase of the Darwinian "struggle for life" is reinforced by the no less ironic elucidation of the stance that "people are more the product of their environment than of their genetic endowment." (ibid)

Indeed, the binary opposition of nature vs nurture, one of the main concerns for biology and genetics, is viewed by Illich as only meaningful for medicine itself but not for ordinary people. However, Illich abandons his irony regarding "the weapons and the rules and the style for the struggle...supplied by the culture in which [people] grew up" at the moment he starts to discuss the "main achievement of human evolution," namely, adaptation: "Mankind has so far shown an extraordinary capacity for adaptation. Man has survived with very high levels of sub-lethal breakdowns." (ibid 89)

While Illich agrees that "[M]an, unmodified by a particular place and companionship, simply does not exist," (ibid 82) breaking with the underlying, unchanging normative type of the human as the formal object of scientific enquiry. He proves his stance by opposing the "natural man" generated by Enlightenment anthropology and the "consensual man" introduced by classical anthropology. This division leads Illich to an internal contradiction between his negation of "given genetic make-up, given history, given geography," (ibid) the whole range of attributions impeded by progress, and the obvious acceptance of individual development through personal experience, described by Illich as a two-sided process. He concludes, "In determining their health, people create their physical being, just as, more generally, by determining their culture, they create themselves." (ibid 93) He tries to solve this contradiction by the systematic opposition of culture as system of meanings vs. civilization as a system of techniques.

Culture, a necessary "cocoon" for people's ability to cope, determines their competence to live with their own recollection of past injuries and their certainty of unending vulnerability. Illich returns to pain as a key signifier in differentiating culture and civilization: "Culture makes pain tolerable by integrating it into a meaningful system, cosmopolitan civilization detaches pain from any subjective or intersubjective context in order to annihilate it." (ibid) Using the attitude to pain as a signifier for differentiating culture and civilization can be explained by Illich's multiple affiliations with various theories regarding civilization, but probably one of the key connections is represented by his intention to reinvent the mission of socialist-Christian medicine to move beyond the individualist focus of human physiology to the level of the collective. (Williams 1994 218) Clearly, Illich opposes the authentic collective sympathy that turns pain "into a political issue which gives rise to a snowballing demand on the part of anaesthesia consumers for artificially induced insensibility, unawareness, and even unconsciousness." (Illich 1974 94)

Illich does not indicate a specific date for the transition from culture to a "cosmopolitan civilization," but introduces a criterion for determining whether medicine works in favor of culture or serves civilization: "Human survival rather than disease; the impact of stress on populations rather than the impact of specific agents on individuals; the relationship of the human niche in the cosmos to the species with which it has evolved rather than the relationship between the aims of people and their ability to achieve them." (ibid, 82) Civilization transforms "the concrete person into a resource fostering and legitimizing the proliferation of agents and services of high-tech biomanagement." (Duden 2002 227) Health within civilization "designates a cybernetic optimum. It is conceived as an equilibrium, between the socio-ecological macro-system and the population of its human-like subsystems. By submitting to the optimization process, the subject denies himself." (Illich 2017 288) This negation of adaptation as a vital force that started operating on the verge of mechanization follows the logic of negating primal instinct as compatible in its destructive potential to "technogenic" catastrophes in developmental idealism.

The aim to deconstruct the technogenic science-attributed and self-inflicted body, an output of civilization, would require *radical negation* of adaptation as a vital force coopted by civilization for alienating people from their subjectivity. Radical negation involves the auto-subversive overcoming of a previous vision and consciousness, and, for Bhaskar, performs the dialectic life of consciousness. (Bhaskar 1993 57) The demand of this negation for Illich is of highest priority because of his radical critical position regarding the role of the Catholic Church which, to Illich, had become a part of the new cult of civilization, an idol of life, and "increases of the distinction between Him who in the Gospel said that He was 'the life,' and the abstract fragment which today in biomanagement is called a life." (Illich 2017 288) Illich recognizes in the united efforts of the Church and medicine to transform each human being into "a life" "an implicit denial of the one who said to Martha, 'I am the life.'" (Duden 2002 227)

Along with distancing himself from the official position of the Church, Illich consistently differentiates his line of arguments from the transformative mode of negation. He calls for accepting all diseases as a socially created reality and recognizes the inconsistency in the partial destruction of medical norms in the anti-psychiatry movement and the other implications of Canguilhem's approach (Illich 1974 117). Illich rejects the option to transform the existing order of medical care through refining the notion of disease and health: "[T]his is a true crisis because it admits of two opposing solutions, both of which make present hospitals obsolete. The first solution is a further sickening medicalization of health care, expanding still further the control of the medical profession over healthy people. The second is a critical, scientifically sound, de-medicalization of the concept of disease." (Illich 1974 116)

Criticizing medicine for the separation of the mental faculties from actions and decisions, another feature of the socialist-Christian medicine presented in Illich's theorizing, and the call for medical knowledge that would be "engaged, relational, and pragmatic," (Williams 1994 220) gains new motives. Illich chooses genetic counseling to illustrate his thesis concerning the ambiguous power of medicine which had achieved incredible control but remains unable to help:

> In these meetings [with genetic counselors], you shift from information on fertilization and a summary of Mendel's laws to the drawing up of genetic family trees, up to inventory of risks and a walk through the garden of "monstrosities." Every time a woman asks if something could happen to her, the doctor replies: "Madam, we cannot exclude it with certainty." But, for sure, such an answer leaves its mark. For this ceremony possess an inevitable symbolic effect: it forces the pregnant woman to take a "decision" by identifying herself and her unborn child with a probability configuration. I am not talking about the decision for or against the continuation of her pregnancy, but the obligation of the woman to identify [that] she is forced to take an oxymoronic decision, a choice that one pretends to be humane but which, on the contrary, reduces her to a mere, inhuman number. (Illich 2017 289)

Like Canguilhem, Illich calls for studying the history of medical ideologies, but limits the timeline of such historicizing to current ideologies in order to understand "the degree to which we are prisoners of the medical ideology in which we were brought up." (Illich 1974 117) Defining this prison as "a suffocating void" Illich urges people to fill it with new meaning through historicizing health "as the reverse of salvation, as a societal liturgy serving an idol of life that extinguishes the subject." (Illich 2017 288) This argument is deepened by Duden, who sees the history of the body as aimed at understanding "how each historical moment is incarnated in an epoch-specific body and how we can decipher the body of subjective experience as a unique 'enfleshment' of an age's ethos." (Duden 2002 223)

6 Conclusion

The reproduction of vitalist arguments within the debates between theologians and biologists since the 1960s has added new forms of coherence and underscores the eclectic forms acquired by vitalism between the eighteenth and nineteenth centuries. Historicizing vitalist arguments in the debate between biologists and theologians introduces new contexts for interpreting the progress of human genetics as a science and the source of influential expertise for population policy. The attribution of negative value to primal instinct as an obstacle to the progress of human civilization not only united different camps of geneticists but also forged bonds between them and many of those affiliated with Christianity as main agents of developmental idealism. The ongoing expansion of boundaries between scientific and theological vitalist arguments in favor of a strong opposition between health and disease predisposes a kind of public consensus regarding the necessity to practice surveillance over reproduction.

While this societal context has increased the public legitimacy of medicine, and especially genetics, it has generated multiple contradictions in the epistemological grounds of medicine and genetics in particular, which has led to their critical revision.

Two main movements, the so-called 'normative vitalism' mainly associated with French philosophers Georges Canguilhem, Michel Foucault, and Gilles Deleuze, and the movement for the deinstitutionalization of health care within negative theology presented by Ivan Illich, have launched two different epistemologies of medical knowledge to address the question of transforming or deconstructing the opposition between health and disease in favor of sustainable individual autonomy. Like developmental idealism, normative vitalism and the deinstitutionalization of medicine rely on the revision of vitalist arguments. Normative vitalism replaces the opposition between health and disease with a refined scale of anomalies and pathologies that would provide the grounds for thoughtful separation between objective description and subjective judgement regarding health. Deinstitutionalization attacks various practices of modern medicine aimed at submitting vital forces to the aims of adaptation.

Both approaches have introduced the historicization of medicine as a prerequisite for its emancipation from medical ideologies, including developmental idealism. If normative vitalism focuses on the emancipation of medicine from medical ideologies as an ongoing process of creating the internalist history of medicine and its progress, deinstitutionalization focuses on the various practices of alienating people from their corporeality and expropriating individual experience through medicine. Both strategies of historicization underscore the unique role of vitalist arguments in connecting internalist and externalist narratives of medicine in to order to prevent a misreading of the history of medicine.

Further development of the type of historical reconstruction undertaken in this chapter is relevant to exploring issues that address the task of remapping the composition of driving forces and actors in at least three ways: (1) the transfer from eugenics to genetics in different regions and the exploration of intercountry cooperation through the lenses of adapting various modes of ascribing negative value and the strategies of boundary work to vital forces; (2) the history of deconstructing Nazi medicine (more generally, the history of professions and institutions in structural violence) and embedding this historicization into the debates regarding race, the limits of bioethics, and the role of medical science in historical justice; and (3) the role of theological arguments and their agencies in developing medical science and the dissemination of its implications.

Acknowledgments The author is grateful to Christopher Donohue and the members of the Philosophy of Biology Circle for the valuable comments and general support in developing this text. The research for this chapter was sponsored by the FWF Austrian Science Fund (2674-G28) as part of the project 'Die Rassenkunde: Unentdeckte Macht des Aufbaus der Nationen '.

References

Cardinal Barragán and Javier Lozano. 2005. *Intervention on the Occasion of the Presentation of the 20th International Conference on "The Human Genome" (November 15, 2005)*. Available online: http://www.vatican.va/roman_curia/pontifical_councils/hlthwork/documents/rc_pc_hlthwork_doc_20051115_barragan-genoma_it.html.

Benton, Edward. 1974. Vitalism in Nineteenth-Century Scientific Thought: A Typology and Reassessment. *Studies in History and Philosophy of Science* 4 (5): 17–48.

Bhaskar, Roy. 1993. *Dialectic: The Pulse of Freedom*. London: Verso.

Bianco, Giuseppe. 2013. The Origins of Georges Canguilhem's 'Vitalism': Against the Anthropology of Irritation. In *Vitalism and the Scientific Image in Post-Enlightenment Life Science, 1800 – 2010*, ed. S. Normandin and C.T. Wolfe, 243–270. Dordrecht: Springer.

Canguilhem, Georges. 1988. *Ideology and Rationality in the History of the Life Sciences*. Cambridge MA: MIT Press.

———. 1991. *The Normal and the Pathological*. Princeton: Zone Books.

———. 2009. *Knowledge of Life*. New York: Fordham.

Castel, J.-G. 1973. Legal Implications of Biomedical Science and Technology in the Twenty-First Century. *Canadian Society of Forensic Science Journal* 6 (4): 184–201.

Clarke, Agnus. 1997. Outcomes and Process in Genetic Counselling. In *Genetics Society and Clinical Practice*, ed. P. Harper and A. Clarke, 165–178. New York: Garland Science.

Contact Practical Theology. 1968. *Introduction*. 22(1)

Creaven, Sean. 2002. The Pulse of Freedom? Bhaskar's Dialectic and Marxism. *Historical materialism* 10 (2): 77–141.

Deane-Drummond, Celia. 2012. *Genetics and Christian Ethics*. Cambridge: Cambridge University Press.

Debru, Clod. 2011. The Concept of Normativity from Philosophy to Medicine: An Overview. *Medicine Studies* 3: 1–7.

Dietrich, Donald J. 1992. Catholic Eugenics in Germany, 1920–1945: Hermann Muckermann, S.J. and Joseph. *Mayer Journal of Church and State* 34 (3): 575–600.

Duden, Barbara. 2002. The Quest for Past Somatics. In *Challenges of Ivan Illich*, ed. L. Hoinacki and C. Mitcham. Albany: Suny Press.

Elden, Stuart. 2019. *Canguilhem*. Cambridge: Polity Press.

Gieryn, Thomas F. 1999. *Cultural Boundaries of Science: Credibility on the Line*. Chicago: The University of Chicago Press.

Gillette, Aaron. 2014. Agostino Gemelli and the Latin Eugenics Movement. *Römische Quartalschrift*, 109 (1–2): 92–102.

Hefner, Philip. 1993. *The Human Factor: Evolution, Culture, and Religion*. Fortress Press.

Huneman, Philippe. 2008. Montpellier Vitalism and the Emergence of Alienism in France (1750–1800): the Case of Passions. *Science in Context* 21 (4): 615–647.

Illich, Ivan. 1970. *Deschooling Society*. New York: Harper & Row.

———. 1974. *Medical Nemesis. The Expropriation of Health*. New York: Pantheon Books.

———. 2017. The Obsession with Perfect Health. *Journal of Cultural Research* 21 (3): 286–291.

Jacobs, Patricia. 1968. Genes and Society. *Contact* 22 (1): 20–24.

Knorr-Cetina, Karin. 1999. *Epistemic Cultures: How the Sciences Make Knowledge*. Cambridge, MA: Harvard University Press.

Koch, Lene. 2004. The Meaning of Eugenics: Reflections on the Government of Genetic Knowledge in the Past and the Present. *Science in Context 17* (3): 315–331.

Koonin, Eugene V., and Yuri I. Wolf. 2009. Is Evolution Darwinian or/and Lamarckian? *Biology direct* 4 (42).

Lakatos, Imre. 1970. History of Science and Its Rational Reconstruction. *Proceedings of the Biennial Meeting of the Philosophy of Science Association* 1: 91–136.

Lederberg, Joshua. 1966. Experimental Genetics and Human Evolution. *The American Naturalist* 100 (915): 519–531.

McLaren, Anne. 1968. The Control of Reproduction. *Contact* 22 (1): 13–19.

Muckermann, Hermann. 1934. *Eugenik*. Berlin-Bonn: Ferdinand Dümmler.

Nickolson, Peter P. 1990. *The Political Philosophy of the British Idealists: Selected Studies*. Cambridge: Cambridge University Press.

Okasha, Samir. 2018. *Agents and goals of evolution*. Oxford: Oxford University Press.

Ornstein, Leonard & Lederberg, Joshua (1967) The Population Explosion, "Conservative Eugenics," and Human Evolution. *Bulletin of the Atomic Scientists*, 23 (6): 57–61.

Osborne, Thomas. 2016. *Vitalism as Pathos*. Biosemiotcs 9: 85–205.

Pasquinelli, Matteo. 2015. What an Apparatus is Not: On the Archaeology of the Norm in Foucault, Canguilhem, and Goldstein. *Parrhesia* 22: 79–89.

Patton, Paul. 2018. Philosophy and Control. In *Control Culture: Foucault and Deleuze after Discipline*, ed. F. Beckman, 193–210. Edinburg: Edinburg University Press.

Rabinow, Paul, and Nikolas Rose. 2006. Biopower today. *BioSocieties* 1: 195–217.

Ramsden, Edmund. 2006. Confronting the Stigma of Perfection: Genetic Demography, Diversity and the Quest for a Democratic Eugenics in the Post-war United States. *Working Papers on the Nature of Evidence: How Well Do "Facts" Travel?*, 12.

Ramsey, Paul. 1970. *Fabricated Man The Ethics of Genetic Control*. New Haven: Yale University Press.

Shan, Yafeng. 2020. *Doing Integrated History and Philosophy of Science: A Case Study of the Origin of Genetics*. Springer.

Sjørup, Lene. 1999. Religion and Reproduction: The Vatican as an Actor in the Global Field. *Gender, Technology and Development* 3 (3): 379–410.

Thornton, Arland, Georgina Binstock, Kathryn M. Yount, Mohammad Jalal Abbasi-Shavazi, Dirgha Ghimire, and Yu Xie. 2012. International Fertility Change: New Data and Insights From the Developmental Idealism Framework. *Demography* 49 (2): 677–698.

Turda, Marius, and Aaron Gillette. 2014. *Latin Eugenics in Comparative Perspective*. London\ New York: Bloomsbury Academic.

Weir, Lorna. 1998. Cultural Intertexts and Scientific Rationality: The Case of Pregnancy Ultrasound. *Economy and Society* 27 (2–3): 249–258.

Williams, Elizabeth A. 1994. *The Physical and the Moral. Anthropology, Physiology, and Philosophical Medicine in France, 1750–1850.* Cambridge University Press.

Wolfe, Charles T. 2008. Introduction: Vitalism Without Metaphysics? Medical Vitalism in the Enlightenment. *Science in Context* 21 (4): 461–463.

Wolfe, Charles T., and Motoichi Terada. 2008. The Animal Economy as Object and Program in Montpellier Vitalism. *Science in Context* 21 (4): 537–579.

What Is Living and What Is Dead in Political Vitalism?

Cat Moir

Abstract Does vitalism inherently imply a specific politics, and if so, what is it? In this chapter, I aim to offer at least some possible answers to this question by examining historical and contemporary discussions around the politics of vitalism. In so doing, I offer an account of what vitalism is as a set of scientific and philosophical ideas about the nature of life and its status as an object of study. It is precisely because vitalism is concerned with the question of life that it implies political considerations from the get-go. However, some of the more problematic political consequences of what has often been referred to (sometimes erroneously or confusedly) as vitalism stem, I argue, from the attribution of vital powers to the non-living. This infusion of vitality into everything may seem egalitarian in its apparent levelling out of differences between forests, objects, spirits, the dead, and whole societies. Yet if everything is living, then the specificity of the living, the living itself, disappears. Whatever equality may or may not be purchased from this perspective, then, I argue that it can no longer properly be called vitalist.

On 1 January 2019, US magazine *Commune* published the manifesto of a group calling itself the Vitalist International. Entitled 'Life Finds a Way', the manifesto—referencing the slogan of the indigenous-led protests against the Dakota pipeline project, *mni winconi*, 'water is life'—called for a revolution in the name of life itself. The vitalism invoked by the International emphasises unity between all forms of life against the 'great structures of power—race, gender, private property, the state', which are viewed as 'the guarantors of separation'. In naming the 'resonance between the rivers, the trees, and the forests', the vitalism invoked has predictable ecological overtones. However, in its identification of resonances between 'objects, spirits, the dead and those banished to social death', with its demand for 'the coordination of human bodies with bodies of thought, bodies of water, with bodies of buffalo charging at police, of life forms with art forms', the political force of vitalism envisaged here is also seen as extending far beyond the unity of the living (Vitalist International 2019).

C. Moir (✉)
University of Sydney, Sydney, Australia

C. Donohue, C. T. Wolfe (eds.), *Vitalism and Its Legacy in Twentieth Century Life Sciences and Philosophy*, History, Philosophy and Theory of the Life Sciences 29, https://doi.org/10.1007/978-3-031-12604-8_13

Commune is a magazine of the political left that emphasises aesthetic experience and popular participation as the levers of a necessary revolution.[1] Its publication of the vitalist manifesto nevertheless aroused suspicion on the part of other leftists, who view vitalism as an ideological tributary of fascism. Josie Sparrow, writing in *New Socialist* in March 2019, took such a critical line in her piece 'Against the New Vitalism,' where she identified the manifesto as part of a growing trend on the left to rehabilitate a vitalism seen as not only *historically* tainted by 'political genealogies' connecting it to fascism, but also compromised by intrinsic 'ideological continuities' with the latter. Sparrow argues in this respect that thinkers such as Nietzsche, Klages and Heidegger are connected to historical and contemporary forms of eco-fascism by 'the positing of some deep, trans-temporal/ahistorical force or power' that supposedly underpins and supersedes the reality of social structures (Sparrow 2019).

The manifesto and Sparrow's response raise important questions for considering the political reception of vitalism. For although forms of vitalist philosophy have undoubtedly been historically allied with right-wing agendas, they have also, as Sparrow notes, been invoked in the service of emancipatory forms of politics, whether by Gilles Deleuze, or more recently feminist new materialist thinkers such as Jane Bennett and Rosi Braidotti. The question of whether and to what extent these self-declared progressive or left-leaning invocations of vitalism are *successfully* progressive or emancipatory bears on the deeper issue of whether vitalism does, as Sparrow suggests, display elective affinities with reactionary political ideologies. Does vitalism inherently imply a specific politics, and if so, what is it?

Here, I aim to offer at least some possible answers to this question by examining historical and contemporary discussions around the politics of vitalism. In so doing, I offer an account of what vitalism is as a set of scientific and philosophical ideas about the nature of life and its status as an object of study. It is precisely because vitalism is concerned with the question of life that it implies political considerations from the get-go. However, some of the more problematic political consequences of what has often been referred to (sometimes erroneously or confusedly) as vitalism stem, I argue, from the attribution of vital powers to the non-living. This infusion of vitality into everything may seem egalitarian in its apparent levelling out of differences between forests, objects, spirits, the dead, and whole societies. Yet if everything is living, then the specificity of the living, the living itself, disappears. Whatever equality may or may not be purchased from this perspective, then, it can no longer properly be called vitalist.

---

[1] Cf. https://communemag.com/about/. In 2020, the magazine was involved in a controversy over rape allegations against a member of its editorial collective, with statements from other leftist groups and a petition calling for its dissolution on account of a supposed failure of processes within the organization to address the allegations.

1   Vitalism and Its Avatars

The matter of definition is one of the basic problems of engaging with the politics of vitalism. Vitalism is all too often rather loosely defined, so that it becomes indistinguishable from related but not co-extensive concepts, particularly holism (the privileging of wholes over parts), hylozoism (the idea that all matter is alive), *Lebensphilosophie* (the philosophical posture according to which life and its affirmation should be the primary object of philosophy), and biopolitics (the underpinning claim of which is that biological life is the ultimate object of politics).[2] As I aim to show here, some of the objections made against theories that describe themselves or are described as forms of political vitalism are actually criticisms of one or more of these related positions.

I agree with Charles T. Wolfe when he suggests that '"vitalism" be restricted to theories in which the difference between "life" and "nonlife" (living matter and nonliving matter, living bodies and dead bodies, bodies and machines, biology and physics, etc.) is crucial, however this difference is detailed and laid out' (Wolfe 2021: 2). When the discourse of vitalism emerged in the late eighteenth century, it was predicated on the idea that life could not be explained purely in terms of mechanism or mathematics; that the laws of life were not reducible to the laws of the existing sciences of physics and chemistry. Thus, although it would later be sharply distinguished from biology as an unscientific, even supernaturalist, approach to the study of life, in fact the discourse of vitalism emerged in the context of the institution of the life sciences with figures such as Kant, Blumenbach, and Buffon (Steigerwald 2013: 51–76; Wolfe, Forthcoming). The specificity of life as compared with non-life, then, is the fundamental tenet of vitalism.

In many respects, vitalism's meaning thus defined covers both what might be considered more modest claims for the necessity of a special science—biology—to describe the laws of life, and the more ambitious, metaphysical forms of vitalism that seek to explain life's specificity by positing the existence of some kind of *élan vital* or vital principle. Though these two claims are connected, however, they are not the same. The first can be seen simply as a version of the claim to what Lovejoy (1911) calls 'scientific autonomism', which asserts or implies the logical discontinuity of scientific laws. Lovejoy classically made this point by arguing that if the basic premise of mechanism is that 'the explanations of organic processes can eventually be found in the laws of some more "fundamental" science whose generalizations are more comprehensive than those of biology'—in other words if biological laws could be shown to be special cases of chemical or physical laws—then the vitalist counter-claim would be that, even given complete empirical data concerning the inorganic processes involved in organic processes, the latter cannot be reduced

---

[2] In Sparrow's critique, *Lebensphilosophie* and vitalism are taken to be literally synonymous, a trend that can also be observed in the English translation of Georg Lukács' *Die Zerstörung der Vernunft*, where the German term *Lebensphilosophie* is given simply as vitalism. For more on the conceptual history of *Lebensphilosophie* and its relation to vitalism, see Bianco 2019.

to or deduced from the laws of the inorganic. Here, the specificity of life is asserted without making any grand claim for a vital principle. An extrapolation from this position occurs when it is asserted that living organisms not only adhere to unique laws of their own, but that these laws cannot even be stated without positing the existence of special forces or agents—Bergson's élan vital, Driesch's entelechy— that animate living beings. Wolfe (2011) has distinguished between these forms of vitalism as 'functional' and 'substantival' respectively.[3]

To some extent, vitalism's political valency has always hinged on this assertion of an invisible power animating life. Peter-Hans Reill (1989) has argued that the shift in the eighteenth century from mechanistic theories of force to vitalistic theories of active power lent scientific credence to a new worldview that broke with traditional theological conceptions according to which power originated from an omniscient ruler dominating an inert, material nature. The dynamic conception of nature that accompanied what Wolf Lepenies (1976) has called the 'end of natural history' became associated, as Reill explains, with the idea that power 'is fluid [...] and hence cannot be associated with any solid, static body'. When translated into political language, this clearly challenged the idea that power was the preserve of an established elite. Thus, eighteenth century vitalism became the political correlate of 'action and freedom of individual choice' (Reill 1989: 212). Reill thus associates the political connotations of early vitalism with the emergence of what might be called liberal individualism, over against the conservative associations of both an earlier, mechanistic view of nature, and the quasi-mystical biologistic holism with which vitalism became associated in the latter half of the nineteenth century (Reill 1989: 208).

2   Ernst Haeckel's Monism Between Mechanism and Vitalism

The association of a loosely defined vitalism with the reactionary politics of the nineteenth century can be found in Daniel Gasman's classic study *The Scientific Origins of National Socialism* (Gasman 2004 [1971]: 64; 78). There, Gasman argued that the zoologist and populariser of Darwin in Germany, Ernst Haeckel, helped to lay the groundwork for the emergence of Nazi ideology by popularising a crudely biologistic worldview called monism, which combined elements of social Darwinism with mystical ideas of an invisible force animating matter and world history. For Gasman, Haeckel's tacit embrace of vitalism was a crucial element of his worldview, and helped to pave the way ideologically for German fascism by valorising nationalistically inflected concepts of natural wholeness and vitality. Yet apart from being the teacher of Hans Driesch, whose contribution to the history and

---

[3] Wolfe also argues for a third form of vitalism not dealt with here, which he calls 'attitudinal', and associates primarily with Georges Canguilhem.

theory of vitalism is undisputed, Haeckel is much more commonly associated with mechanism or materialism than with vitalism (Freyhofer 1982; Normandin & Wolfe 2010: 8; Gregory 1971: x).[4] Questions thus remain as to what extent and in what way Haeckel can be considered a vitalist, and what role vitalism plays, if any, in the direct or indirect political uses of his ideas.

Certainly, rhetorically speaking, Haeckel was no vitalist. Throughout his career, he dismissed vitalism as an outdated, mystical and unscientific approach to the study of life and evolution. He explicitly rejected the idea of pluralism in the sciences, arguing that all of nature could be understood according to the laws of mechanical causation that govern the disciplines of physics and chemistry. In his early work *Generelle Morphologie der Organismen* (1866: XIV-XV), Haeckel aimed to bring biology, 'the whole science of developed and emerging forms of organisms to the same level of monism in which all other natural sciences have sooner or later found their steadfast foundations', namely on the basis of 'mechanical-causal justification' (*ibid.*). He explicitly rejected any recourse to notions of a teleological 'vital principle' or a 'Bauplan' governing the functioning of organisms, as well as the idea that the 'force' supposedly responsible for life is substantially distinct from chemical and physical forces (Haeckel 1866: 97).

Even in this early work, however, and despite Haeckel's own avowed intentions, aspects of a metaphysical vitalism can be perceived. For though he shunned the idea of a special 'Lebenskraft', Haeckel *did* claim that all biological life was deeply connected by an underlying substance. Though Haeckel was a self-proclaimed Darwinist, his orthogenetic vision in fact drew heavily on pre-adaptive, progressive theories of evolution—that is, on paradigms that appear more teleological-vitalistic than causal-mechanistic. Lamarck and, above all, Goethe—the originator of the idea of a *Bauplan*, whom Haeckel frequently quotes—feature as centrally in Haeckel's theoretical framework as Darwin. The biogenetic law Haeckel puts forward in the *Generelle Morphologie*, according to which each stage of an embryo's development represents the adult form of an evolutionary ancestor, implies a form of empirically imperceptible organic memory—similar to Richard Semon's later theory of engrams—that is difficult to reconcile with his formal rejection of teleology and metaphysical vitalism.

In Haeckel's later work, the influence of vitalism is more pronounced. Insisting in *Kristallseelen: Studien über das anorganische Leben* (1917: VIII) on 'the fundamental unity of all natural phenomena', Haeckel attributes features usually only associated with living beings to inorganic nature. Specifically, he argued that crystals are a form of 'inorganic life' possessing inner feeling states and even sexuality. With reference to crystals, Haeckel claimed that his monism broke down 'the artificial boundaries between inorganic and organic nature, between life and death [...]. All substance possesses life', he argued, 'inorganic and organic; all things have a soul, crystals just as organisms'. Haeckel's claim here that 'all substance possesses

---

[4] Frederick Gregory does not include Ernst Haeckel in his study of nineteenth-century German materialism, partly on political grounds: he argues that the mechanical materialists were more sympathetic to the German liberal tradition and could not stomach Haeckel's anti-Semitism.

life' is hylozoist rather than strictly speaking vitalist: he attributes life to matter in general, rather than distinguishing between living and non-living beings. Needless to say, hylozoism is far from functional vitalism, and it is not identical to, though it comes close to, substantival or metaphysical vitalism, which, like its functional counterpart, maintains the specificity of the laws of life and/or the life sciences. In seeking to explain this specificity, however, substantival vitalism has recourse to a speculatively posited life force, in which it is distinguished from hylozoism only in the fact that the hylozoist sees the vital force operative in all matter.

In Haeckel's late work, therefore, his monism does come close to a form of vitalism in this specific respect. For if functional vitalism is associated with scientific autonomism and a pluralism among the laws governing the objects of the distinct sciences, whereas mechanism is conceived as monistic, proposing the possibility of an explanatory 'reduction' of biological and other phenomena to causal-mechanistic laws, it is surely too simple to align vitalism solely with pluralism and mechanism with monism. Although Haeckel continued to insist that life was ultimately explicable in terms of the same laws as the rest of non-living nature, rather than seeking to explain life in physico-chemical terms—in 'reductive' mechanistic fashion, as it were—Haeckel increasingly projected biological and psychological characteristics onto pre-organic nature. As such, his monism arguably comes much closer to a form of vitalist holism than to mechanist materialism. In fact, Haeckel's monism is symptomatic of a *decline* in the currency of materialistic and mechanistic paradigms in Europe in the late nineteenth and early twentieth centuries, which was correlated with the gradual triumph of evolutionary theory and the dynamic view of the natural world that it implied (Holt 1971: 267).

The political implications of Haeckel's monist worldview have been the subject of vigorous debate. There is little doubt that Haeckel himself saw monism as entailing such implications. His first target in this respect was traditional religion; little wonder, perhaps, given that Darwin's theory of evolution, which Haeckel broadly and at least rhetorically supported even if he did not follow Darwin on all points, refuted traditional religious teaching concerning the origin of human beings and their place in the cosmos. The organisation Haeckel founded, the Monist League, aimed to combat vitalism precisely because it saw it as a form of 'intellectual and political reaction' linked to outdated views of a natural world infused with divine spirit (cited in Holt 1971: 269). Far from rejecting religion, however, Haeckel saw monism as capable of connecting religion and science. He argued in this vein that

> 'A broad historical and critical comparison of religious and philosophical systems…leads as a main result to the conclusion that every great advance in the direction of profounder knowledge has meant a breaking away from the traditional dualism (or pluralism) and an approach to monism' (Haeckel 1894: 15).

Haeckel's claim that profounder knowledge in both religion and philosophy—and by extension, science—moves towards monism is consistent with his belief in the unity of the sciences, which as we saw above is often associated with mechanism rather than vitalism. However, when Haeckel concludes from this insight that 'God is not to be placed over against the material world as an external being, but must be placed as a "divine power" or "moving spirit" within the cosmos itself', we can see once again that his own position appears rather as a more extreme, expansive version of vitalism than as a rejection of it (*ibid.*). Whereas the vitalist in Haeckel's view merely sees the living world as moved by some invisible, divine force, for Haeckel the whole of reality is thus moved—by something that in the last instance appears as a kind of quasi-Spinozist God-as-matter. As he argues further in his writings on monism as a bridge between science and religion, 'all the wonderful phenomena of nature around us, organic as well as inorganic, are only the products of one and the same original force, various combinations of one and the same primitive matter' (Haeckel 1894: 16). That Haeckel identifies a mysterious evolutionary vital force in all matter rather than only in living matter makes it no less a vital force for the fact.

As we have seen, Haeckel extended the idea of an orthogenetic evolution also to non-organic nature, such that for him there was no mystery concerning the origin of life, which he claimed was a recurrent argument of the vitalist (Haeckel 1894: 93). He also extended these laws into the social and political realm, however. As Todd H. Weir (2012: 2) has pointed out, for Haeckel, Darwinian evolution was a 'master theory that explained not only the multiplicity of biological life, but also developments in human consciousness and civilization, and linked them all together into a single meaningful totality'. Haeckel (1916: 116) articulates this idea explicitly in his essay 'Eternity: World-war Thoughts on Life and Death, Religion, and the Theory of Evolution', where he claims that 'Civilization and the life of nations are governed by the same laws as prevail throughout nature and organic life'. Elsewhere, he makes it plain precisely what political flavour he believed Darwinism has when applied to the social world. In a piece responding to his contemporary the scientist and politician Rudolf Virchow, who had denounced Darwin's theory of natural selection as dangerously socialist in orientation, Haeckel argues that 'Darwinism is anything but socialist! If one wants to attribute a particular political tendency to this English theory—which is entirely possible—this tendency can only be an aristocratic one, not at all democratic and least of all socialist!' (Haeckel 1908 [1878]: 68).

To the extent that Haeckel saw evolutionary forces at work in the social world, it is fair to call him a social Darwinist. In his *Natürliche Schöpfungsgeschichte* (1889), for instance, Haeckel represented the human species according to a racial hierarchy that has caused many commentators to present him as a forerunner to National Socialist ideology (Gasman 2004; Weikart 2004). Indeed, Haeckel argues that the group he calls the 'Midlanders', which he claims includes the 'caucasian', 'basque', 'hemosemitic' and 'indogermanic' sub-groups, of which he sees the indogermanic as the most advanced, 'surpassed all other [human] races and species in the struggle for existence thanks to its more advanced brain development, and now its web of dominance extends across the entire planet' (Haeckel 1889: 739).[5] However, as we have seen, Haeckel's evolutionism was far from straightforwardly Darwinist, and indeed some of the non-Darwinist elements of his evolutionism—his belief in

---

[5] See 726–727 for the visualisation of Haeckel's racial hierarchy.

progressive development, and morphological hierarchy—were most problematic when applied to the human world. If he invokes the classic social Darwinist refrain of the 'struggle for existence' in an attempt to justify indogermanic supremacy, his position on race was at least as much the result of the specific combination of holism and progressivism at the heart of his hylozoistic monist philosophy as it was of any Darwinian concept of adaptation. It was by erasing the distinction between life and non-life that Haeckel could argue that everything from crystals to human societies operated in accordance with 'natural', transhistorical evolutionary laws.[6]

The objections against the political implications of Haeckel's monism thus cannot be levelled at him as a vitalist, strictly speaking, if vitalism is defined by the distinction between the living and non-living realms. In fact, Haeckel's projection of the laws of life *beyond* their biological domain of application can largely be seen to be responsible for the political consequences on account of which his monism has been criticised. Haeckel saw all material things as imbued with life, though not equally complex; he saw evolution as progressive, though not straightforwardly continuous. It was on this basis that he claimed some human races were less intelligent than many higher animals, just as he argued that a crystal could be more complex than some simple organic life forms (Richards 2008: 157–158).

Though there are undoubtedly genealogical connections between Haeckel's work and fascist ideology, it is important to remember that reception does not necessarily imply retroactive culpability. Moreover, even if some monists adopted a nationalist stance before the First World War (despite his avowed pacifism, both Haeckel and his contemporary, the chemist and fellow monist Wilhelm Ostwald, signed the statement of ninety-three intellectuals defending militarism as a means of protecting German culture), there were also strong links between monism and pacifism in this period (cf. Holt 1975). Although many monists espoused eugenic theories and some were influenced by the chauvinist right in the early decades of the twentieth century, a majority supported pacifism and the ideas of the moderate and radical left (Dickinson 2001: 225). Many monists, such as Helene Stöcker, were also feminists and advocates for women's rights. Even if pacifism and eugenicism, including in its explicitly racist forms, were not mutually exclusive among monists and others, the position of the Monist League was strongly anti-Nazi: throughout the 1920s, the monists repeatedly insisted that Darwin was a liberal, pacifist, and internationalist, warned of the rise of racism and superstition, and denounced anti-Semitism as unscientific.[7] In 1933, when many German organizations strategically aligned themselves with the Nazi regime, the Monist League disbanded rather than submit to *Gleichschaltung* (Holt 1975:37).

---

[6] See Rupke 2019 for more on the ways in which Darwin's and Haeckel's theories became gradually nationalized largely as a result of political enmities arising from the First World War, and then in the context of National Socialism when Haeckel's evolutionary heroes such as Goethe and Kant, and also Haeckel himself, received an enthusiastic reception from Nazis.

[7] For more on the relation between pacifism and eugenics, see Weikart 2003: 273–294; for examples of monist denunciation of racism and anti-Semitism, see the appeals in *Die Stimme der Vernunft* 17 (1932): 1, 54, 93, 169 and 18 (1933): 57–58.

3 Hans Driesch's Vitalism and the Politics of Holism and Occultism

If Haeckel represents a borderline case for the problems of vitalism and its possible attendant politics, that of his student Hans Driesch is surely clearer cut. Driesch can be considered a vitalist in both the functional and substantival senses. Driesch's early experimental work on embryology challenged the prevailing mechanistic theories of August Weismann and Wilhelm Roux, supporting the argument that biological processes were subject to specific laws of their own. The non-mechanical causal force that he believed underpinned life processes, which he called entelechy, was, however, a highly speculative construct that resembled in important respects the psychoid force that Haeckel saw animating matter. Driesch's concept of entelechy was, however, more obviously vitalist in the sense that he restricted it to the realm of organic life.

Driesch first became acquainted with Haeckel's evolutionary ideas as an adolescent attending lectures at the newly opened *Kosmos Institut* in Hamburg where he grew up. At that time, he was convinced by Haeckel's account of evolution and made up his mind to study zoology with Haeckel at university (Innes 1973: 29). However, thanks to his thorough Gymnasium grounding in mathematics and physics, Driesch soon began to question Haeckel's approach to biology. For instance, he later claimed that he failed to follow Haeckel's proofs for phylogenetic trees, because they did not resemble the deductive proofs he had learned in mathematics (Innes 1973: 30). Far from concluding, however, that his failure to follow Haeckel's proofs was evidence of their inaccuracy, Driesch believed that he must not yet be advanced enough to follow Haeckel's reasoning, a decision that led him to study in the first instance not with Haeckel in Jena, but with August Weisman in Freiburg. His education with Weisberg was significant, not least because his theory of entelechy as he articulated it in relation to embryological cell division was a direct response to the mechanistic Roux-Weisman theory of cell division. As Innes notes, however, Driesch's decision is also revealing because it indicates that even if he doubted the accuracy of Haeckel's more speculative claims about evolutionary forces, he was still open to the possibility that such explanations could be plausible (*ibid.*).

The aim of Driesch's dissertation, which Haeckel supervised, was to analyse the different methods of colony formation in two species of freshwater hydroids. Driesch's concern with what constitutes an individual organism, which he would articulate in more detail later in his lectures on *The Science and Philosophy of the Organism*, is clear already here, and reveals the influence of Kant's conception of organism on his thinking. One of Driesch's main concerns in the dissertation was how the morphology of individual hydroids functioned relative to the interests of the colony, and how the definition of the boundaries of the individual could affect the perception of that relationship. Driesch argued that since different individuals have different functions in the colony, such as feeding or reproduction, the development of form in the individual could only be understood with reference to its function in the colony. In other words, structural features could only be understood as a function of the whole (Innes 1973: 32–33). Thus, one can find traces even in Driesch's early work of the holist dynamics that underpinned Haeckel's monism, even if Driesch was clearly less afraid of teleological explanation than Haeckel. Indeed, one of the central principles of Driesch's vitalism was that there are also laws of wholeness and finality in nature, i.e. not only purely causal connections. Unlike Haeckel, Driesch placed finality on an equal footing with causality (Driesch 1939).

Although Driesch's vitalism is sometimes seen as representing a break with his early experimental work, one can find evidence of both functional and substantival forms of vitalism even in his earliest theoretical writings. In 1890, just a year after Driesch published his dissertation, he wrote a short paper called 'The System of Biology' in which he criticised the separation of biology into two separate faculties in German universities, each of which emphasised different approaches to the study of life. Driesch's argument is that medical faculties emphasised physiological questions as they relate to the practice of medicine, and was based on an experimental method, while the more philosophical faculties teaching botany and zoology focused on morphological questions, which were explained with historical-comparative methods. Yet Driesch believed that organisms must be studied from both physiological and morphological perspectives, i.e. as a totality. Driesch's argument that institutional isolation prevents the formulation of heuristic scientific questions may seem 'monist' in its implications, but in fact the unity Driesch aimed at here was a unity of biology as a discipline that operates according to distinct, internally coherent principles.

According to Driesch, the only researchers in Germany currently studying the organism as a totality in this way were those associated with the programme of developmental mechanics (*Entwicklungsmechanik*) advanced by Wilhelm Roux at one of the state-sponsored research institutes that operated outside the traditional university faculties and departments (Innes 1973: 33–34). Developmental mechanics was concerned with functional differentiation in organisms. At the embryonic stage, this concerned the question of how specific cells came to form specific organs. Roux supported a mechanistic view known as the Roux–Weismann mosaic theory, because the same idea had also been proposed by the cytologist and evolutionary theorist—and Driesch's former teacher—August Weismann. The mosaic model claimed with each cell division hereditary units were apportioned in such a way that each cell generation received increasingly specialized particles. By the time differentiation was complete, each cell type (muscle, nerve, skin) contained only the particles determining that cell's specific characteristics (Allen 2005: 270). In order to test this hypothesis, the developmental mechanists attempted to influence the sequence of normal embryonic developmental stages using mechanical interventions.

In his account of this research programme in 'The System of Biology', Driesch insisted, contra the developmental mechanists, that a teleological element is built into the theory of the development of form from the outset. 'The essential feature of life', he argues there, is the typical order of specific differentiated forms; there is absolutely no sense in which one could talk about such and such an amount of "eagle" or "lion" or "earthworm"'. However, if this observation led him to reject the concept of a 'life substance', he replaced the latter with that of 'entelechy', which was supposed to explain the 'mechanism' by which an organism becomes what it is as a whole (cited in Innes 1973: 35). Unlike the force that Haeckel saw animating all matter from the inside, as it were, Driesch imagined entelechy as in itself an immaterial factor primarily acting on matter from *outside*.[8]

Driesch further developed his idea of entelechy in dialogue with some of the developmental mechanists' experimental observations. One prediction of the mosaic hypothesis was that if one of the first two blastomeres—the first two cells formed by the cleavage of the fertilized ovum—was killed or removed, it would result in half-larvae, since half the determiners for the cell's specific function had already been parcelled out into each daughter cell. Roux tested this prediction by puncturing one of the first two blastomeres of the frog egg with a hot needle and raising the subsequent developing larvae as far as they could develop. The results fitted Roux's predictions, with some half-embryos developing. This supported the thesis that differentiation could be explained as a mechanical process resulting in an adult mosaic organism, with each cell type containing only the active determiners for its special characteristics (Allen 2005 270–271).

In 1891, Driesch, working at the Stazione Zoologica in Naples, performed a series of experiments similar to Roux's using sea urchins. Instead of killing one of the first two blastomeres of the sea urchin embryo, Driesch separated them by

---

[8] This is an important and insufficiently recognized point in relation to Driesch's theory of entelechy, which is often conceptualized only as an internal forming force. Driesch classically defined entelechy as a non-spatial, intensive, and qualitative force that governs morphogenesis (Driesch 1908: 144: 226). He described it as an 'intensive manifoldness' that produces manifoldness in space. As such, it is perhaps unsurprising that many commentators on Driesch have interpreted entelechy as a force that drives morphogenesis from within the organism. When Freyhofer argues that 'all organisms seem to follow a master plan, a shared entelechy' (1982: 35), he implies that entelechy is something that works from within organisms, a claim that is reiterated later when he argues that Driesch's entelechy 'regulates, causes, manifests, acts…in material nature' (48). Similarly, Innes argues that entelechy is 'a specifically teleological construct which refers to the principle by which the parts are constructed in reference to the whole' (1973: 36), implying, since it is the construction of organisms that is the issue here, that this principle forms organisms from the inside out. Moving Driesch's concept of entelechy closer to Aristotle, as a form of 'being-at-work', Armando Aranda-Anzaldo identifies entelechy with the specific way in which organisms are at work in the act of morphogenesis (2011: 337). Recent appropriations of Driesch also posit entelechy as an internal forming force. Jane Bennett (2010: 71), for instance, sees entelechy 'animating matter', working as a 'driving force' from within on the bodies of organisms. Normandin and Wolfe are quite explicit that entelechy is 'a teleological nature *in* living things' (2013: 8, emphasis mine). Certainly, Driesch discusses entelechy in such terms; however, entelechy also has another function, working on matter from outside. In *The History and Theory of Vitalism*, for instance, Driesch speaks of entelechy as 'suspending possible change and relaxing suspension', and states that entelechy 'allows that to become real which it has itself held in a state of mere possibility' (Driesch 1914: 204–205). This perspective sees entelechy as working rather *on* living things from *outside*—not that these two functions are mutually exclusive. For more on this, see (Moir 2020: 143–145).

vigorously shaking. If Roux's mosaic principle were true, Driesch predicted that he should also see half-embryos develop. In fact, however, he found that the separated sea urchin blastomeres each developed into a complete, though smaller-than-average sized embryo. Driesch interpreted the results as contradicting Roux's mechanical model. The sea urchin embryo acted as what Driesch called a 'harmonious equipotential system', and for the next 7 years he carried out experiments in an attempt to understand the causes of this phenomenon. When, by the early 1900s, he had failed to find a mechanistic, physico-chemical explanation, he concluded that embryonic development was guided by what he had already designated as entelechy, 'an organizing, directive force that consumed no energy, was immaterial, but was the factor that distinguished living from non-living matter' (Allen 271). Even if Driesch's refutation of the Roux-Weisman mosaic theory would later be proven more rigorously by Hans Spemann (whose concept of the 'organizer' is another 'vitalistic' concept in modern biology, cf. Peterson 2016), his commitment to vitalism was always highly controversial among his colleagues in academic biology.[9] Even erstwhile collaborators like Jacques Loeb, who had esteemed Driesch's work despite being the arch-reductionist of his time, saw Driesch's vitalism as a form of metaphysics unacceptable in natural science (*ibid.*).

Indeed, Driesch eventually abandoned experimental natural science altogether to work on the philosophy of vitalism and on parapsychology, the study of occult phenomena such as telekinesis, clairvoyance, and telepathy. If Driesch conceptualised entelechy as an immaterial force in contrast with Haeckel's self-animating matter, he nevertheless followed Haeckel in attributing 'psychoid' characteristics to this force. Driesch conceived of entelechy as having two components: while the forming entelechial force worked on matter from the outside, Driesch also saw the reaction-determining factor in actions as a type of entelechy, which he called 'psychoid', or even 'soul', and located inside the organism. For Driesch, 'psychoid' or 'soul' was 'the name given to the entelechy of an organism, insofar as its actions are subject to it' (Driesch 1939: 269). Organisms, according to Driesch, therefore had a dual character: on the one hand, their form was the result of the action of entelechy in its externally shaping mode, while their action was the result of psychoid side of entelechy, which occupies something like the place of a psychological unconscious.

The broader concept of the 'soul' encompassed for Driesch both the unconscious, entelechial moment of organismic action, and the conscious moment in which the soul 'illuminates itself' (Driesch 1939: 271). Driesch saw this conception of the organism's dual character resulting in a form of 'psycho-physical parallelism, which is admittedly something quite different from the old psycho-mechanical parallelism, which wanted conscious experience to be the "other side" of brain mechanics'. He perceived a 'causality between entelechy and ego-being' in the way that the psychoid dimension of entelechy interacted with the conscious part of the soul. And if he denied that there could be a direct causality between mind and matter ('entelechy is always the mediator here'), he nevertheless believed that 'parapsychological

---

[9] Thanks to Charles T. Wolfe for the Peterson reference.

facts' such as 'telepathy and thought transmission, speak of a direct causal relationship between different ego-beings' (*ibid.*). Driesch pushed these arguments further in his 1925 work *The Crisis in Psychology*, where he claimed that psychology and biology were—and should be—moving away from the mechanistic paradigm and towards 'totality concepts' such as soul and entelechy, and called on psychology to take seriously the investigation of parapsychological phenomena.[10] Driesch's interest in the occult and in parapsychology not only saw his vitalism discredited scientifically—in the 1930s, the Gestalt psychologist Max Wertheimer, claimed that his adversary Driesch had 'gone over to the spiritualists'—it has also led to associations of his vitalism with fascist politics (cf. Harrington 1996: 124).[11]

There is no doubt that Driesch personally opposed fascism. As a pacifist and a member of the League for Human Rights, Driesch opposed both world wars and despised racism and anti-Semitism: ultimately, Driesch's pacifism and support for Jewish people saw him forced from his position in 1933 when the Nazis came to power. The question here, however, is not so much whether Driesch personally opposed Nazism as whether his vitalism can in some sense be seen to lend itself to Nazi ideology. On closer examination, it is clear that it is not strictly speaking Driesch's vitalism that is aligned with reactionary, totalitarian, and ultimately fascist politics—certainly not if we define vitalism functionally as the insistence that life and the life sciences are subject to distinctive laws that cannot be reduced to those of physics and chemistry. The elements of Driesch's project that *are* supposedly politically problematic are rather occultism and holism, neither of which are coextensive with vitalism.

Concerning the occultist point, Horst Freyhofer (1982: 158) has observed that the decline in material living conditions in Germany during the 1920s and 1930s (whether this was construed nationalistically, in terms of Germany's defeat in the First World War, or socially, as a failure of revolution and/or welfarism), 'seemed to demonstrate that men were subject to forces either unknown or less known than claimed by many, or that they were subject to these forces to a greater degree than thought so far'. Fascists, in particular, argued that these forces were not the forces of reason—not Kantian categories or Marxian laws of history—but those of will, the unconscious, or the historical teleology of destiny.

Certainly, the substantival vitalist claim that life is driven by invisible forces can be seen to lend itself to extrapolation in an occultist direction. As Freyhofer writes, '[v]italists and fascists both affirm the existence of a transcendental, autonomous, and capricious force that controls all life, most apparently the life of men. And both think that some people can know and handle these forces better than most people' (Freyhofer 1982: 161). For Driesch the vitalist parapsychologist, these 'elite' initiates were the experimenters, mediums, clairvoyants, and hypnotists who mediated

---

[10] For a succinct account of Driesch's *Crisis in Psychology*, see Allesch 2012.

[11] In a critical review of Harrington's book, Jüri Allik and Wolfgang Drechsler claim that Harrington's association of holism with fascism relies on a methodologically clumsy whig-historical account of the history of ideas, similar to the problems that beset the Sonderweg thesis. Cf. Allik and Drechsler 1999.

between individual souls and the collective dimension of the psychoid; for fascists, the elite intermediaries were the *Führer* and his hierarchy. The political criticism of vitalism can thus be seen to be valid to the extent that, in its metaphysical form, vitalism does indeed claim that life is driven by unconscious, largely or wholly empirically unavailable forces that transcend history and are therefore outside of human control.

Yet some of Driesch's own claims contradict Freyhofer's assertion that, for fascists, 'elites constitute the ultimate source of authority in all matters of human existence' whereas for vitalists 'the ultimate authority remains with the transcendental forces themselves' (*ibid.*). In his autobiography, written between 1938 and 1940 but only published posthumously, Driesch denies the Nazi claim that what he calls the 'despicable pogroms of winter 1938' were a case of the 'popular soul' 'boiling over' against the Jews. Rather, Driesch argues that the pogrom was 'deliberately ordered' by the Nazis, an accusation that he recognises could land one in a concentration camp if uttered publicly (Driesch 1951; cf. Krall 2015: 111–112). Driesch's attribution of responsibility ('authority') for the pogrom to the National Socialists themselves does not make him a fascist; rather, it demonstrates a deflationary attitude on his part vis-à-vis the role of spiritual forces in human history. Clearly, for Driesch, the vitalist claim that life is shaped and driven by unconscious, immaterial, only indirectly observable entelechial and psychoid factors did not translate into an apology for political atrocity as the effect of some vaguely defined transcendent power.

That National Socialists *did* invoke occultism in support of their programme (albeit with an intensity and enthusiasm that varied according to the moment and the personality: Hitler was well-known to be less committed to the incorporation of occultism into Nazi ideology than Rosenberg, for instance) makes it no less true that the almost causal association between occultism and Nazism is largely a post-war historiographical narrative. As Heather Wolffram (2003) has argued, the focus on irrationalism as an explanatory factor in the rise of National Socialism was popularised as part of the German *Sonderweg* thesis by writers such as George L. Mosse and Nicholas Goodrick-Clarke.[12] Reassessing this history, Wolffram claims that 'far from being an irrational and politically conservative backlash against modernity, occultism was often the site of complex negotiations over religion, politics, and scientific knowledge, which aimed not only at critique, but also at reform' (2003: 150). In this context she emphasises that 'for Driesch, as for a number of European scholars, parapsychology provided scientific validation of a belief in the naturalness of liberalism and pacifism' (*ibid.*). With this in mind, any straightforward association between Drieschian vitalism and fascist politics appears highly tenuous.

The holistic dimension of vitalism, too, has been claimed to harbour politically questionable implications. There is certainly reception-historical evidence for this point, as Anne Harrington has shown. She points to a number of Nazi figures and

---

[12] For recent accounts of the genesis and course of the German Sonderweg debate, see Everett (2015), Kocka (2018), and Walser Smith (2008).

sympathisers who explicitly hailed Driesch as laying the groundwork for a holistic view of society and cosmos that they believed accorded with their own ideology. The philosopher of biology, Adolf Meyer-Abich, who played an active role in defining the significance of holistic biology for National Socialism, declared in 1935 that '"Holism stands . . . on the shoulders of [Driesch's] vitalism"'. In *Das Weltbild der Biologie* (1941), Arthur Neuberg wrote that Driesch instilled in the era the conviction that '[n]o part exists preformed . . . everything can become anything and the whole determines the function that the part must undertake'. Meanwhile, in a 1936 letter to Driesch, the philosopher Paul Gast told his friend that his (Driesch's) terminology 'had been adopted in some National Socialist discussions of the Volk as a vital, biological whole'. Some were arguing, Gast reported, that '"Driesch's vitalistic, holistic-philosophy would be a splendidly appropriate scientific underpinning for National Socialist terminology"' (Harrington 1996: 190).

Harrington eschews the idea of some intrinsic connection between holism and fascism, arguing that 'the Nazification of German holism' occurred 'not through some process of intellectual determinism, but because people—both opportunistic and out of conviction—came to "see" and argue that certain political conclusions must be drawn from the antimechanistic impulses of the interwar years' (Harrington 1996: 189). As she acknowledges, Driesch sought to make 'the language of wholeness and vitalism serve, not a fascist ideology, but a pacifist, democratic, humanist politics' (Harrington 1996: 190). Significantly, the way in which Driesch did so reveals a distinction between his vitalist holism and a more expansive form of holism that sees vital principles applied simplistically to the social and political spheres.

The Nazi application of biological laws to human society is one ground for associating vitalism and/or holism with fascism. Yet if the application of biological laws beyond biology is certainly holist, it is not vitalist, as the case of Driesch's criticism of his friend and collaborator Jacob von Uexküll's 1920 book *Staatsbiologie* [*Biology of the State*] makes clear. In *Staatsbiologie*, Uexküll conceptualised the state as a living organism and aimed to diagnose certain pathologies that threatened the functioning of a healthy state. Much like Haeckel's conclusion that the political orientation of Darwinism was aristocratic, Uexküll advanced the anti-democratic argument that the healthiest form of organization of the state was monarchy. In a quietly critical review of Uexküll's book written in 1921, Driesch declared that Uexküll's central claim that the state is properly understood as an organism was untenable. For Driesch, any particular empirical state could not, by definition, be considered a true biological whole because states possessed no independent, creative entelechy. Indeed, the only 'super-personal' organismic entity that Driesch was prepared to recognize—and, even here, only cautiously—was a concept of mankind that recognized no national or völkisch boundaries (Harrington 1996: 61). Driesch reiterated this idea of human unity in his autobiography, where he argued, contra the Nazis, that 'racially determined differences, both physical and mental, are of almost no significance in comparison with what all human beings share in common' (cited in Krall 2015: 112). Driesch's assertion here of what Marx called the species-being of humanity is certainly vitalist in that it emphasises the unity and totality of the life form. However, by rejecting the validity of projecting the laws of life into the realm of the social, Driesch opposes a vitalistic form of holism to Uexküll's more expansive and—on Driesch's account—*non-vitalist* holism.

If Driesch rejected Uexküll's extension of the laws of life itself to the social and political realm, he nevertheless did see the ultimate unity of all humankind beyond differences of race and culture as entailing political consequences. In *Die sittliche Tat* (1927), Driesch reflects on the political form most appropriate to this conception of human unity, and it is decidedly anti-nationalist. Influenced by the experience of the First World War, he writes that

> 'There should be one state only, for there is only one mankind. But since historically states did not develop as they should have, but through violent acts of a few individuals or groups, we now have many empirical states. Every person should see to it that the many states become one state....But the best that could happen to the one state, and therewith to each particular empirical state, is that it would dissolve itself...that anarchy would become its "constitution," because it needs no constitution. But this presupposes that "by nature" all men always act in accordance with the good which unfortunately is not the case. Still, while working toward the one state everyone must work toward obsolescence of the "state" as legal institution: the ultimate "goal" of the state is its own dissolution' (Driesch 1927: 98).

If Driesch's claim that states 'did not develop as they should have' might at first sight be seen to echo the Haeckelian idea of an orthogenetic law governing not only (in)organic nature, but also 'civilization and the life of nations', this impression is quickly dispelled by Driesch's insistence on the role of human agency in constructing the socio-political world. Not only does Driesch see his vitalist conviction in the unity of human life as ultimately necessitating some form of single-state globalism on the political plane; unlike the National Socialists, whose application of the laws of life to the social world was associated with the idea of the rule of the strongest and with war between nations, Driesch's political vision involves an explicitly anarchic component.

One can challenge the fundamental association of holism and fascism on a number of grounds. Among other things, there are at least three distinct, perhaps related, but certainly more and less maximalist versions of holism. From the perspective of the political instrumentalization of holism, it presumably makes a difference whether one departs from the idea, which Lovejoy identifies with mechanism, that all phenomena can ultimately be explained in terms of causal physico-chemical laws, or from the attribution of the laws of life to the whole of natural and social reality as in the case of Haeckel's monism, or from the more restrictive vitalist claim that living beings can only be understood as self-contained wholes. If all vitalism is holist in the latter sense, not all holisms are vitalist, such that even if holism can be said to display an elective affinity with fascism (or perhaps better, totalitarianism, which in the Soviet context is usually associated with mechanism rather than organicism/vitalism), an intrinsic connection between fascism and vitalist holism remains to be shown (see Ash 1995 and, especially, Joravsky 1961 for more on mechanistic holism in the Soviet context).

4 Afterlives of Political Vitalism

Let us return in conclusion to Sparrow's critique, which does indeed posit something like an intrinsic connection between vitalism and fascism. There, historical figures including Driesch, Heidegger, Klages, Nietzsche and others sit beside contemporary thinkers and activists: the Vitalist International, but also new materialists such as the political theorist Jane Bennett. In her book *Vibrant Matter* (Bennett 2010), Bennett develops a theory of matter itself as inherently vibrant or lively, imbued with power and agency in ways that blur the boundary between life and non-life. She encourages us to see both non-human and non-living material things as imbued with life and agency, capable of making things happen. For Bennett, political theory has for too long been focused on human agency, forgetting that we too are matter, and that matter itself is in some sense alive or at least possess creative capacities that are not entirely within our ken or control.

The stakes of Bennett's project are thus highly political. Questions of ecology and racial justice, in particular, depend in Bennett's view on an appreciation of thing-power.

She sees the 'figure of an intrinsically inanimate matter' as 'one of the impediments to the emergence of more ecological and more materially sustainable modes of production and consumption' (2010: ix). Encouraging us to see the natural world, including in its inorganic facets, as alive may, she suggests, foster a more sustainable environmental politics.

In relation to matters of social justice, Bennett makes the following argument (2010: 13):

> 'If matter itself is lively, then not only is the difference between subjects and objects minimized, but the status of the shared materiality of all things is elevated. All bodies become more than mere objects, as the thing-powers of resistance and protean agency are brought into sharper relief. Vital materialism would thus set up a kind of safety net for those humans who are now, in a world where Kantian morality is the standard, routinely made to suffer because they do not conform to a particular (Euro-American, bourgeois, theocentric, or other) model of personhood. The ethical aim becomes to distribute value more generously, to bodies as such.'

Yet despite its self-declared investment in the project of an emancipatory politics, Bennett's vitalism—or vital materialism—comes in for criticism on account of what Sparrow, following feminist theorist Nikki Sullivan, calls 'white optics'. Bennett's explicit distinction of thing-power from 'Flower Power, Black Power [and] Girl Power' serves, Sparrow claims, to 'belittle and diminish both feminist and Black liberatory movements'. She cites Sullivan's (2019: 310) point here that new materialism relies on the 'appropriation of both the (bad) feminist other [or Black, or any other so-called "cultural" identity that resists the flattening binary of human/nonhuman], and the (good) non- human other' as a means of ordering and classifying the world towards certain ends, a process that paradoxically reasserts colonial power.

The white optics of Bennett's vital materialism according to Sparrow thus stem from the fact that it is what has often been called a flat ontology: it 'creates a flattened image of the monolithic "human", and then works to performatively disavow it.' She continues:

'*Humanity*, in these discourses, is always the white settler-subject: the consumer, the polluter, the extractor; the creator of an "inauthentic" culture that is both reviled and mobilised by those doing the reviling. Attempting to escape this contradiction, vitalists and new materialists alike must posit a more authentic realm of individual entities that, in their withdrawnness, resist the relational assimilation of culture. It is not in their interest to consider differences *within* the category of "human".'

It is on this basis that Sparrow argues Bennett's revival of vitalism harbours an implicit connection to ecofascism understood as the positing of a romanticized and/or idealized natural state of life to which we must return or perhaps which we must still attain in order to achieve social and ecological equality qua harmony.[13]

To take another example of a form of contemporary new materialism inspired by vitalism, we might turn to Rosi Braidotti's theorization of the posthuman. Braidotti argues for what she calls a 'Zoe-centred egalitarianism' as 'a materialist, secular, grounded and unsentimental response to the opportunistic trans-species commodification of Life that is the logic of advanced capitalism.' The key notion, she continues, 'is embodiment on the basis of neo-materialist understandings of the body, drawn from the neo-Spinozist philosophy of Gilles Deleuze and Felix Guattari, but re-worked with feminist and postcolonial theories. Embracing their version of vital bodily materialism, while rejecting the dialectical idea of negative difference, this theoretical approach changes the frame of reference' (2013: 22). Braidotti's reference to 'zoe-centred egalitarianism' calls to mind Giorgio Agamben's biopolitical distinction between bios as a particular mode of life, i.e. the life of a citizen in a given society with certain rights etc., and 'bare life' (zoe), or life in its brute physicality, without (whether prior to or deprived of) any sense of social belonging or rights (cf. Agamben 1998: 8). Braidotti distinguishes her 'positive' account of zoe from Agamben's formulation (2013: 121), her use of the term signalling the politicization of what from an Agambenian perspective might be seen as the pre- or extrapolitical: living beings in their very existence as such are considered equal in this vision, outside the realm of 'anthropos, that is to say bios': outside the constructed realm of society and culture.

Yet Braidotti's posthuman perspective looks not only beyond the human to animal and plant life, but also to inorganic nature, which is seen here too as inextricably connected to life. 'The posthuman predicament', she argues, forces 'a displacement of the lines of demarcation between structural differences, or ontological categories, for instance between the organic and the inorganic, the born and the manufactured, flesh and metal, electronic circuits and organic nervous systems' (2013: 89). Here the boundaries between the living and non-living become blurred, if they do not

---

[13] Bennett's attribution of agency to all matter does tend to work against her rhetorical commitment to the Deleuzian principle of ontological one-ness but formal diversity (cf. p. xi).

disappear, such that the power, certainly of non-human creatures but also of things, is considered equal to that of living (human) beings. Here the parallel with Bennett's vital materialism is clear. For Bennett (2010: x), not only are 'stem cells, electricity, food, trash, and metals…crucial to political life (and human life per se)', but 'dead rats, bottle caps, gadgets, fire, electricity, berries, [and] metal' are also living agents (2010: 107).

The naturalisation of bodily—and in relation to inorganic non-living phenomena, perhaps also evental or existential—equality opens up an interesting philosophical problem: if the premise of vitalist materialism is that all things are ontologically equal (positive difference), and that it is (negative) difference and inequality that are historically/socially constructed, then social *equality*, where and to the extent that it exists, must also be seen as historically constructed, i.e. fought and won. What, then, is the status of the relation between ontological equality/positive difference and historical inequality/negative difference? That is a question that new materialism is yet to answer.

The same problem, however, can be stated in more practical political terms, as Sullivan and Sparrow do: if equality is simply naturalised—which is to say that all things are considered naturally equal, thus equality is considered to exist already, at least at an ontological level—this threatens to erase the reality of very real historical struggles for social equality, and perhaps even to imply that no such further struggles are necessary. The question as to the implied political consequences of Bennett's and Braidotti's vital materialisms—as opposed to the avowed political positions of the authors themselves—is arguably more legitimate than the same question is when put to the philosophy of someone like Driesch, whose primary object was not the political per se.

One can take these criticisms of Bennett's vital materialism as valid and still pose the question as to whether they are in fact criticisms of vitalism, as they claim to be, or rather of some other philosophical position. Certainly, Bennett and Braidotti both invoke vitalism in defence of their positions, and describe those positions as in important respects vitalist. Yet this is not a vitalism concerned with 'the difference between "life" and "nonlife" (living matter and nonliving matter, living bodies and dead bodies, bodies and machines, biology and physics, etc.) […] however this difference is detailed and laid out' (Wolfe 2021: 2). For Braidotti (2013: 60), a 'vitalist approach to living matter displaces the boundary' between bios and zoe. Vitalism in this sense stands for a 'generative vitality' (*ibid.*) that bears more resemblance to Haeckel's monistic extension of living power to the realm of non-living matter than to Driesch's life-specific—if immaterial—entelechy. In other words, applied to Braidotti, the objection the equalization of power making human action equivalent to that of rivers, trees, and forests—to return to the formulation in the *Vitalist Manifesto*—is a criticism of holism and/or hylozoism in Braidotti, not of vitalism. The sale can be said for Bennett. Like vitalists such as Driesch, Bennett posits 'the presence of some kind of energetic, free agency whose spontaneity cannot be captured by the figure of bodies or by a mechanistic model of nature' (Bennett 2010: 61). Yet if, from the perspective of this form of vitalism, 'matter seemed to require a not-quite-material supplement, an elan vital or entelechy, to become animate and mobile', Bennett—claiming to follow Deleuze and Guattari in this respect—insists that 'materiality needs no animating accessory. It is figured as itself the 'active principle' (*ibid.*). Here again, if matter itself is alive, then the characteristic vitalist distinction between life and non-life falls away. Thus again, the critique of the levelling out of power relations between living and non-living things is no longer a critique of vitalism, but of a monistic hylozoism or perhaps better in this case, animism.

Braidotti and Bennett's common reference to Deleuze and Guattari is telling. Certainly, the sense in which Deleuze and Guattari were vitalists is not itself straightforward (cf. Ansell-Pearson 1999: 4). To be sure, Deleuze refers to his own work as vitalist and vitalistic at isolated points in his oeuvre, perhaps most notably in a 1988 interview often quoted in this context, where he states: 'Everything I've written is vitalist, at least I hope it is' (1990: 196). The context in which Deleuze makes that assertion, though, has little to do with theorizing the specificity of living beings, whether in terms of Drieschian entelechy or other (cf. Protevi 2012: 247). Rather, Deleuze is discussing the temporality of life versus art in terms that arguably bring his project close to a philosophy of life (*Lebensphilosophie*) that seeks to defend what Braidotti calls 'generative vitality' as a subject of philosophical reflection and to theorise it in terms that seek to escape from the strictures of sterile, commodified academic discourse. But as we have seen, *Lebensphilosophie* is not vitalism, despite frequent confusion. *Lebensphilosophie* seeks to affirm life philosophically, not to define it scientifically. Though the status of vitalism for Deleuze and Guattari is not my main concern, it is perhaps helpful to note, as John Protevi (2012: 247) has, that what is central to 'Deleuze's notion of vitalism is the "life" that encompasses both organisms and "non-organic life."' It should, then, be clear that Deleuze—nor Braidotti nor Bennett, to the extent that they follow him—is no vitalist in the sense intended here.

Theories and philosophies of vitalism are concerned with the distinction between life and non-life in some form. Given this emphasis, vitalism always has political stakes: where the boundary between life and non-life is drawn is, as theorists of biopolitics have shown, a supremely political matter; as perhaps is the will to draw it, though in scientific terms there are very good reasons for seeking to do so. Historically, vitalism has been associated with conservatism and fascism in ways that are sometimes genealogically defensible, sometimes only with difficulty or not at all. Conceptual confusion also complicates the task of onto-political critique. The political objections that have been made against new materialism (cf. Moir 2020)— that by levelling agency it denies the specificity of the historical—may be valid, but they are not critiques of vitalism. If anything, it is the extension of the attributes of life to the realm of the non-living that raise the political problems so often associated with vitalism.

References

Agamben, Giorgio. 1998. *Homo Sacer. Sovereign Power and Bare Life*. Stanford: Stanford University Press.

Allen, Garland E. 2005. Mechanism, Vitalism and Organicism in Late Nineteenth and Twentieth-Century Biology: The Importance of Historical Context. *Studies in History and Philosophy of Biological and Biomedical Sciences* 36: 261–283.

The Vitalist International (Atlanta Faction). 2019. *'Life Finds a Way. A first Manifesto from the Vitalist International*. Commune, 01/01/2019, accessed at: https://communemag.com/life-finds-a-way/ on 15 June 2021.

Allesch, Christian G. 2012. Hans Driesch and the Problems of "Normal Psychology". Rereading his *Crisis in Psychology* (1925). *Studies in History and Philosophy of Biological and Biomedical Sciences* 43: 455–461.

Allik, Jüri, and Wolfgang Drechsler. 1999. German Holism Revisited: Really? *Culture & Psychology* 5 (2): 239–247.

Ansell-Pearson, Keith. 1999. *Germinal Life: The Difference and Repetition of Deleuze*. London: Routledge.

Aranda-Anzaldo, Armando. 2011. Assuming in Biology the Reality of Real Virtuality (A Come Back for Entelechy?). *Ludus Vitalis* XIX:36: 333–342.

Ash, Mitchell G. 1995. *Gestalt Psychology in German Culture, 1890–1967: Holism and the Quest for Objectivity*. Cambridge: Cambridge University Press.

Bennett, Jane. 2010. *Vibrant Matter: A Political Ecology of Things*. Durham: Duke University Press.

Bianco, Giuseppe. 2019. Philosophies of Life. In *The Cambridge History of Modern European Thought*, ed. Peter E. Gordon, 153–175. Cambridge: Cambridge University Press.

Braidotti, Rosi. 2013. *The Posthuman*. Cambridge: Polity Press.

Deleuze, Gilles. 1990. *Pourparlers (1972–1990)*. Paris: Éditions de Minuit.

Dickinson, Edward Ross. 2001. Reflections on Feminism and Monism in the Kaiserreich, 1900–1913. *Central European History* 34 (2): 191–230.

Driesch, Hans. 1908. *The Science and Philosophy of the Organism*. London: Adam and Charles Black.

———. 1914. *The History and Theory of Vitalism*. trans. C.K. Ogden. London. Macmillan & Co.

———. 1925. *The Crisis in Psychology*. Princeton, NJ: Princeton University Press.

———. 1927. *Die Sittliche Tat*. Reinicke: Leipzig. E.

———. 1939. Entelechie und Seele. *Synthese* 4 (6): 266–279.

———. 1951. Lebenserinnerungen. *Aufzeichnungen eines Forschers und Denkers in entscheidender Zeit*. Munich/Basel. Ernst Reinhardt.

Everett, Annie. 2015. The Genesis of the Sonderweg. *International Social Science Review* 91 (2): 1–42.

Freyhofer, Horst. 1982. *The Vitalism of Hans Driesch: The Success and Decline of a Scientific Theory*. Peter Lang.

Gasman, Daniel. 2004. [1971]. *The Scientific Origins of National Socialism*. New Brunswick, NJ: Transaction.

Gregory, Frederick. 1971. *Scientific Materialism in Nineteenth-Century Germany*. Dordrecht: Reidel.

Haeckel, Ernst. 1866. *Generelle Morphologie der Organismen: Allgemeine Grundzüge der organischen Formen-Wissenschaft, mechanisch begründet durch die von Charles Darwin reformirte Descendenz-Theorie*. Vol. 1. Berlin: Georg Reimer.

———. 1889. Natürliche Schöpfungsgeschichte: Gemeinverständliche wissenschaftliche vorträge über die entwickelungslehre im allgemeinen und diejenige von Darwin, *Goethe und Lamarck im besonderen*. Berlin. Georg Reimer.

———. 1894. *Monism as Connecting Religion and Science*. trans. J. Gilchrist. London. Charles and Adam Black.

─────. 1908 [1878]. *Freie Wissenschaft und freie Lehre: Eine Entgegnung auf Rudolf Virchow's Münchener Rede über die "Freiheit der Wissenschaft im modernen Staat"*. Leipzig: Alfred Kröner Verlag.

─────. 1916. *Eternity: World-war Thoughts on Life and Death, Religion, and the Theory of Evolution*. New York: Truth Seeker.

─────. 1917. Kristallseelen. *Studien über das anorganische Leben*. Leipzig: Alfred Kroner.

Harrington, Anne. 1996. *Reenchanted Science: Holism in German Culture from Wilhelm II to Hitler*. Princeton, NJ: Princeton University Press.

Holt, Niles R. 1971. Ernst Haeckel's Monistic Religion. *Journal of the History of Ideas* 32 (2): 265–280.

─────. 1975. Monists & Nazis: A Question of Scientific Responsibility. *The Hastings Center Report* 5 (2): 37–43.

Innes, Shelley Anne. 1973. *Hans Driesch and Vitalism: A Reinterpretation*. MA Thesis. Simon Fraser University.

Joravsky, David. 1961. *Soviet Marxism and Natural Science, 1917–1932*. New York: Columbia University Press.

Kocka, Jürgen. 2018. Looking Back on the Sonderweg. *Central European History* 51: 137–142.

Krall, Stephen. 2015. Hans Driesch, der Vitalist: Zwischen Biologie, Philosophie und Parapsychologie. *Zeitschrift für Anomalistik, Band* 15: 110–129.

Lepenies, Wolf. 1976. D*as Ende der Naturgeschichte: Wandel kultureller Selbstverständlichkeiten in den Wissenschaften des 18. und 19. Jahrhunderts*. Munich/Vienna. Hanser Verlag.

Lovejoy, Arthur O. 1911. The Meaning of Vitalism. *Science* 33 (851): 610–614.

Moir, Cat. 2020. *Ernst Bloch's Speculative Materialism: Ontology, Epistemology, Politics*. Leiden/Boston: Brill.

Normandin, Sebastian, and Charles T. Wolfe. 2010. Vitalism and the Scientific Image: An Introduction. In *Vitalism and the Scientific Image in Post-Enlightenment Life Science*, ed. Normandin and Wolfe, 1800–2010. Springer.

Peterson, Erik L. 2016. *The Life Organic: The Theoretical Biology Club and the Roots of Epigenetics*. Pittsburgh: University of Pittsburgh Press.

Protevi, John. 2012. Deleuze and life. In *The Cambridge Companion to Deleuze*, ed. D. Smith and H. Somers-Hall, 239–264. Cambridge: Cambridge University Press.

Reill, Peter-Hans. 1989. Anti-Mechanism, Vitalism, and Their Political Implications in Late Enlightened Scientific Thought. *Francia. Forschungen zur westeuropäischen Geschichte* 16: 2.

Richards, Robert J. 2008. The Tragic Sense of Life. In *Ernst Haeckel and the Struggle over Evolutionary Thought*. Chicago/London: University of Chicago Press.

Rupke, Nicolaas. 2019. The Break-Up Between Darwin and Haeckel. *Theory in Biosciences* 138: 113–117.

Sparrow, Josie. 2019. Against the New Vitalism. *New Socialist*. Accessed at https://newsocialist.org.uk/against-the-new-vitalism/ on 15 April 2021.

Steigerwald, Joan. 2013. Rethinking Organic Vitality in Germany at the Turn of the Nineteenth Century. In *Vitalism and the Scientific Image in Post-Enlightenment Life Science, 1800-2010*, ed. Normandin and Wolfe, 51–76. Springer.

Sullivan, Nikki. 2012. The Somatechnics of Perception and the Matter of the Non/Human: A Critical Response to the New Materialism. *European Journal of Women's Studies* 19 (3): 299–313.

von Uexküll, Jakob. 1920. *Staatsbiologie (Anatomie-Physiologie-Pathologie des Staates)*. Berlin: Verlag von Gebrüder Paetel.

Walser Smith, Helmut. 2008. When the Sonderweg Debate Left Us. *German Studies Review* 31 (2): 225–240.

Weikart, Richard. 2003. Progress Through Racial Extermination: Social Darwinism, Eugenics and Pacifism in Germany, 1860-1918. *German Studies Review* 26 (2): 273–294.

─────. 2004. *From Darwin to Hitler: Evolutionary Ethics, Eugenics, and Racism in Germany*. New York: Palgrave Macmillan.

Weir, Todd H., ed. 2012. *Monism: Science, Philosophy, Religion, and the History of a Worldview.* New York: Palgrave Macmillan.

Wolfe, Charles T. 2011. From Substantival to Functional Vitalism and Beyond: Animas, Organisms, and Attitudes. *Eidos* 14: 212–235.

———. 2021. Vitalism in Early Modern Medical and Philosophical Thought, in *Encyclopedia of Early Modern Philosophy and the Sciences,* eds. D. Jalobeanu and C.T. Wolfe.

———. Forthcoming. Vitalism and the Construction of Biology: A Historico-Epistemological Reflection. In *Philosophy, History, and Biology: Essays in Honor of Jean Gayon,* ed. P.-O. Méthot. Springer.

Wolffram, Heather. 2003. Supernormal Biology: Vitalism, Parapsychology and the German Crisis of Modernity, c. 1890-1933. *The European Legacy* 8:2: 149–163.

Index